## SECOND EDITION

# EMBEDDED MULTIPROCESSORS

## Scheduling and Synchronization

# Signal Processing and Communications

# EMBEDDED MULTIPROCESSORS
## Scheduling and Synchronization

Sundararajan Sriram
Shuvra S. Bhattacharyya

**CRC Press**
Taylor & Francis Group
Boca Raton   London   New York

CRC Press is an imprint of the
Taylor & Francis Group, an **informa** business

CRC Press
Taylor & Francis Group
6000 Broken Sound Parkway NW, Suite 300
Boca Raton, FL 33487-2742

First issued in paperback 2017

© 2009 by Taylor & Francis Group, LLC
CRC Press is an imprint of Taylor & Francis Group, an Informa business

No claim to original U.S. Government works

ISBN 13: 978-1-138-11417-3 (pbk)
ISBN 13: 978-1-4200-4801-8 (hbk)

### Library of Congress Cataloging-in-Publication Data

Sriram, Sundararajan, 1968-
 Embedded multiprocessors : scheduling and synchronization / authors, Sundararajan Sriram, Shuvra S. Bhattacharyya. -- 2nd ed.
  p. cm.
 "A CRC title."
 Includes bibliographical references and index.
 ISBN 978-1-4200-4801-8 (hardcover : alk. paper)
  1. Embedded computer systems. 2. Multiprocessors. 3. Multimedia systems. 4. Scheduling. 5. Memory management (Computer science) I. Bhattacharyya, Shuvra S., 1968- II. Title.

TK7895.E42S65 2009
004.16--dc22                                                      2008050949

**Visit the Taylor & Francis Web site at**
**http://www.taylorandfrancis.com**

**and the CRC Press Web site at**
**http://www.crcpress.com**

*To my parents, and Uma*

Sundararajan Sriram


*To Arundhati*

Shuvra S. Bhattacharyya

# FOREWORD

Embedded systems are computers that are not first and foremost computers. They are pervasive, appearing in automobiles, telephones, pagers, consumer electronics, toys, aircraft, trains, security systems, weapons systems, printers, modems, copiers, thermostats, manufacturing systems, appliances, etc. A technically active person today probably interacts regularly with more embedded systems than conventional computers. This is a relatively recent phenomenon. Not so long ago automobiles depended on finely tuned mechanical systems for the timing of ignition and its synchronization with other actions. It was not so long ago that modems were finely tuned analog circuits.

Embedded systems usually encapsulate domain expertise. Even small software programs may be very sophisticated, requiring deep understanding of the domain and of supporting technologies such as signal processing. Because of this, such systems are often designed by engineers who are classically trained in the domain, for example, in internal combustion engines or in communication theory. They have little background in the theory of computation, parallel computing, and concurrency theory. Yet they face one of the most difficult problems addressed by these disciplines, that of coordinating multiple concurrent activities in real time, often in a safety-critical environment. Moreover, they face these problems in a context that is often extremely cost-sensitive, mandating optimal designs, and time-critical, mandating rapid designs.

Embedded software is unique in that parallelism is routine. Most modems and cellular telephones, for example, incorporate multiple programmable processors. Moreover, embedded systems typically include custom digital and analog hardware that must interact with the software, usually in real time. That hardware operates in parallel with the processor that runs the software, and the software must interact with it much as it would interact with another software process running in parallel. Thus, in having to deal with real-time issues and parallelism, the designers of embedded software face on a daily basis problems that occur only in esoteric research in the broader field of computer science.

Computer scientists refer to use of physically distinct computational resources (processors) as "parallelism," and to the logical property that multiple activities occur at the same time as "concurrency." Parallelism implies concurrency, but the reverse is not true. Almost all operating systems deal with concurrency, which is managed by multiplexing multiple processes or threads on a single processor. A few also deal with parallelism, for example by mapping threads onto physically distinct processors. Typical embedded systems exhibit both concurrency and parallelism, but their context is different from that of general-purpose operating systems in many ways.

In embedded systems, concurrent tasks are often statically defined, largely unchanging during the lifetime of the system. A cellular phone, for example, has relatively few distinct modes of operation (dialing, talking, standby, etc.), and in each mode of operation, a well-defined set of tasks is concurrently active (speech coding, radio modem, dual-tone multi-frequency encoding, etc.). The static structure of the concurrency affords opportunities for much more detailed analysis and optimization than would be possible in a more dynamic environment. This book is about such analysis and optimization.

The ordered transaction strategy, for example, leverages that relatively static nature of embedded software to dramatically reduce the synchronization overhead of communication between processors. It recognizes that embedded software is intrinsically less predictable than hardware and more predictable than general-purpose software. Indeed, minimizing synchronization overhead by leveraging static information about the application is the major theme of this book.

In general-purpose computation, communication is relatively expensive. Consider for example the interface between the audio hardware and the software of a typical personal computer today. Because the transaction costs are extremely high, data is extensively buffered, resulting in extremely long latencies. A path from the microphone of a PC into the software and back out to the speaker typically has latencies of hundreds of milliseconds. This severely limits the utility of the audio hardware of the computer. Embedded systems cannot tolerate such latencies.

A major theme of this book is communication between components. The methods given in the book are firmly rooted in a manipulable and tractable formalism, and yet are directly applied to hardware design. The closely related IPC (interprocessor communication) graph and synchronization graph models, introduced in Chapters 7 and 9, capture the essential properties of this communication. By use of graph-theoretic properties of IPC and synchronization graphs, optimization problems are formulated and solved. For example, the notion of resynchronization, where explicit synchronization operations are minimized through manipulation of the synchronization graph, proves to be an effective optimization tool.

In some ways, embedded software has more in common with hardware than with traditional software. Hardware is highly parallel. Conceptually, hardware is an assemblage of components that operate continuously or discretely in time and interact via synchronous or asynchronous communication. Software is an assemblage of components that trade off use of a CPU, operating sequentially, and communicating by leaving traces of their (past and completed) execution on a stack or in memory.

Hardware is temporal. In the extreme case, analog hardware operates in a continuum, a computational medium that is totally beyond the reach of software. Communication is not just synchronous; it is physical and fluid. Software is sequential and discrete. Concurrency in software is about reconciling sequences. Concurrency in hardware is about reconciling signals. This book examines parallel software from the perspective of signals, and identifies joint hardware/software designs that are particularly well-suited for embedded systems.

The primary abstraction mechanism in software is the procedure (or the method in object-oriented designs). Procedures are terminating computations. The primary abstraction mechanism in hardware is a module that operates in parallel with the other components. These modules represent nonterminating computations. These are very different abstraction mechanisms. Hardware modules do not start, execute, complete, and return. They just are. In embedded systems, software components often have the same property. They do not terminate.

Conceptually, the distinction between hardware and software, from the perspective of computation, has only to do with the degree of concurrency and the role of time. An application with a large amount of concurrency and a heavy temporal content might as well be thought of using the abstractions that have been successful for hardware, regardless of how it is implemented. An application that is sequential and ignores time might as well be thought of using the abstractions that have succeeded for software, regardless of how it is implemented. The key problem becomes one of identifying the appropriate abstractions for representing the design. This book identifies abstractions that work well for the joint design of embedded software and the hardware on which it runs.

The intellectual content in this book is high. While some of the methods it describes are relatively simple, most are quite sophisticated. Yet examples are given that concretely demonstrate how these concepts can be applied in practical hardware architectures. Moreover, there is very little overlap with other books on parallel processing. The focus on application-specific processors and their use in embedded systems leads to a rather different set of techniques. I believe that this book defines a new discipline. It gives a systematic approach to problems that engineers previously have been able to tackle only in an ad hoc manner.

*Edward A. Lee*
*Professor*
*Department of Electrical Engineering and Computer Sciences*
*University of California at Berkeley*
*Berkeley, California*

# PREFACE

Software implementation of compute-intensive multimedia applications such as video conferencing systems, set-top boxes, and wireless mobile terminals and base stations is extremely attractive due to the flexibility, extensibility, and potential portability of programmable implementations. However, the data rates involved in many of these applications tend to be very high, resulting in relatively few processor cycles available per input sample for a reasonable processor clock rate. Employing multiple processors is usually the only means for achieving the requisite compute cycles without moving to a dedicated ASIC solution. With the levels of integration possible today, one can easily place hundreds of digital signal processors on a single die; such an integrated multiprocessor strategy is a promising approach for tackling the complexities associated with future systems-on-a-chip. However, it remains a significant challenge to develop software solutions that can effectively exploit such multiprocessor implementation platforms.

Due to the great complexity of implementing multiprocessor software, and the severe performance constraints of multimedia applications, the development of automatic tools for mapping high level specifications of multimedia applications into efficient multiprocessor realizations has been an active research area for the past several years. Mapping an application onto a multiprocessor system involves three main operations: assigning tasks to processors, ordering tasks on each processor, and determining the time at which each task begins execution. These operations are collectively referred to as *scheduling* the application on the given architecture. A key aspect of the multiprocessor scheduling problem for multimedia system implementation that differs from classical scheduling contexts is the central role of interprocessor communication — the efficient management of data transfer between communicating tasks that are assigned to different processors. Since the overall costs of interprocessor communication can have a dramatic impact on execution speed and power consumption, effective handling of interprocessor communication is crucial to the development of cost-effective multiprocessor implementations.

This books reviews important research in three key areas related to multiprocessor implementation of multimedia systems, and this book also exposes important synergies between efforts related to these areas. Our areas of focus are application modeling techniques for multimedia systems; the incorporation of interprocessor communication costs into multiprocessor scheduling decisions; a modeling methodology, called the "synchronization graph", for multiprocessor system performance analysis; and the application of the synchronization graph model to the development of hardware and software optimizations that can significantly reduce the interprocessor communication overhead of a given schedule.

More specifically, this book reviews various modeling techniques that are relevant to designing embedded multiprocessor systems; reviews several important multiprocessor scheduling strategies that effectively incorporate the consideration of interprocessor communication costs; and highlights the variety of techniques employed in these multiprocessor scheduling strategies to take interprocessor communication into account. The book also reviews a body of research performed by the authors on modeling implementations of multiprocessor schedules, and on the use of these modeling techniques to optimize interprocessor communication costs. A unified framework is then presented for applying arbitrary scheduling strategies in conjunction with the application of alternative optimization algorithms that address specific subproblems associated with implementing a given schedule. We provide several examples of practical applications that demonstrate the relevance of the techniques described in this book.

## AKNOWLEDGMENTS

We are grateful to the Signal Processing Series Editor K. J. Ray Liu for his encouragement of this book; B.J. Clark for his coordination of the first edition of the book; Nora Konopka for her coordination of this second edition; and Amber Donley for her help with the production process. It was a privilege for both of us to be students of Professor Edward A. Lee (University of California at Berkeley). Edward provided a truly inspiring research environment during our doctoral studies, and gave valuable feedback while we were developing many of the concepts that underlie Chapters 7 to 13 in this book.

We are also grateful to the following colleagues with whom we have had valuable interactions that have contributed to this book. Neal Bambha, Celine Badr, Twan Basten, Bishnupriya Bhattacharyya, Jani Boutellier, Joseph Buck, Sek Chai, Nitin Chandrachoodan, Rama Chellappa, Ivan Corretjer, Omkar Dandekar, Ed Deprettere, Alan Gatherer, Marc Geilen, Neil Goldsman, Soohoi Ha, Fiorella Haim, Christian Haubelt, Yashwanth Hemaraj, Chia-Jui Hsu, Shaoxiong Hua, Joern Janneck, Fuat Keceli, Mukul Khandelia, Vida Kianzad, Bart Kienhuis, Branislav Kisacanin, Dong-Ik Ko, Ming-Yung Ko, Jacob Kornerup, Roni Kupershtok, William Levine, Rainer Leupers, Sean Leventhal, Sumit Lohani, Peter Marwedel, Praveen Murthy, Kazuo Nakajima, Martin Peckerar, Newton Petersen, Jose Pino, William Plishker, Gang Qu, Suren Ramasubbu, Christopher

Robbins, Sankalita Saha, Perttu Salmela, Mainak Sen, Shahrooz Shahparnia, Raj Shekhar, Olli Silven, Gary Spivey, Todor Stefanov, Vijay Sundararajan, Jarmo Takala, Juergen Teich, Ankush Varma, Wayne Wolf, Lin Yuan, Claudieu Zissulescu, and Eckart Zitzler.

*Sundararajan Sriram*
*Shuvra S. Bhattacharyya*

May 17, 2008

# CONTENTS

# 1

---

# INTRODUCTION

---

The focus of this book is the exploration of architectures and design methodologies for application-specific parallel systems in the general domain of embedded applications in digital signal processing (DSP). In the DSP domain, such multiprocessors typically consist of one or more central processing units (microcontrollers or programmable digital signal processors), and one or more application-specific hardware components (implemented as custom application specific integrated circuits (ASICs) or reconfigurable logic such as field programmable gate arrays (FPGAs)). Such embedded multiprocessor systems are becoming increasingly common today in applications ranging from digital audio/video equipment to portable devices such as cellular phones and personal digital assistants. With increasing levels of integration, it is now feasible to integrate such heterogeneous systems entirely on a single chip. The design task of such multiprocessor systems-on-a-chip is complex, and the complexity will only increase in the future.

One of the critical issues in the design of embedded multiprocessors is managing communication and synchronization overhead between the heterogeneous processing elements. This book discusses systematic techniques aimed at reducing this overhead in multiprocessors that are designed to be application specific. The scope of this book includes both hardware techniques for minimizing this overhead based on compile time analysis, as well as software techniques for strategically designing synchronization points in multiprocessor implementation with the objective of reducing synchronization overhead. The techniques presented here apply to DSP algorithms that involve predictable control structure; the precise domain of applicability of these techniques will be formally stated shortly.

Applications in signal, image, and video processing require large computing power and have real-time performance requirements. The computing engines in such applications tend to be embedded as opposed to general-purpose. Custom VLSI implementations are usually preferred in such high throughput applications. However, custom approaches have the well-known problems of long design cycles (the advances in high-level VLSI synthesis notwithstanding) and low flexibility in the final implementation. Programmable solutions are attractive in both of these respects: the programmable core needs to be verified for correctness only once, and design changes can be made late in the design cycle by modifying the software program. Although verifying the embedded software to be run on a programmable part is also a hard problem, in most situations changes late in the design cycle (and indeed even after the system design is completed) are much easier and cheaper to make in the case of software than in the case of hardware.

Special processors are available today that employ an architecture and an instruction set tailored towards signal processing. Such software programmable integrated circuits are called **Digital Signal Processors** (**DSP chips** or **DSPs** for short). The special features that these processors employ are discussed extensively by Lapsley, Bier, Shoham, and Lee [LBSL94]. However, a single processor — even DSPs — often cannot deliver the performance requirement of some applications. In these cases, use of multiple processors is an attractive solution, where both the hardware and the software make use of the application-specific nature of the task to be performed.

For a multiprocessor implementation of embedded real-time DSP applications, reducing **interprocessor communication** (**IPC**) costs and synchronization costs becomes particularly important, because there is usually a premium on processor cycles in these situations. For example, consider processing of video images in a video-conferencing application. Video-conferencing typically involves CIF (Common Intermediate Format) images; this format specifies data rates of 30 frames per second, with each frame containing 352 lines and 288 pixels per line. The effective sampling rate of the CIF video signal is 3 Megapixels per second. The highest performance programmable DSP processor available as of this writing (2008) has a cycle time of a little under 1 nanosecond; this allows about 330 instruction cycles per processor for processing each sample of the CIF formatted video signal. In a multiprocessor scenario, IPC can potentially waste these precious processor cycles, negating some of the benefits of using multiple processors. In addition to processor cycles, IPC also wastes power since it involves access to shared resources such as memories and busses. Thus, reducing IPC costs also becomes important from a power consumption perspective for portable devices.

## 1.1    Multiprocessor DSP Systems

Over the past few years several companies have offered boards consisting of multiple DSPs. More recently, semiconductor companies have been offering chips that integrate multiple DSP engines on a single die. Examples of such integrated multiprocessor DSPs include commercially available products such as the Texas Instruments TNETV3020 multi-DSP [Tex08], Philips Trimedia processor [RS98], and the Adaptive Solutions CNAPS processor. The Hydra research at Stanford [HO98] is another example of an effort focussed on single-chip multi-processors. Multiprocessor DSPs are likely to be increasingly popular in the future for a variety of reasons. First, VLSI technology today enables one to "stamp" hundreds of standard DSP processors onto a single die; this trend is certain to continue in the coming years. Such an approach is expected to become increasingly attractive because it reduces the verification and testing time for the increasingly complex VLSI systems of the future.

Second, since such a device is programmable, tooling, testing, and mask costs of building an ASIC (application-specific integrated circuit) for each different application are saved by using such a device for many different applications. This advantage of DSPs is going to be increasingly important as circuit integration levels continue their dramatic ascent, thus enabling a large number of DSP processors to be integrated on a single chip, while the cost of building custom ASICS in newer fabrication technologies continues to grow.

Third, although there has been reluctance in adopting automatic compilers for embedded DSPs, such parallel DSP products make the use of automated tools feasible; with a large number of processors per chip, one can afford to give up some processing power to the inefficiencies in the automatic tools. In addition, new techniques are being researched to make the process of automatically mapping a design onto multiple processors more efficient — the research results discussed in this book are also attempts in that direction. This situation is analogous to how logic designers have embraced automatic logic synthesis tools in recent years — logic synthesis tools and VLSI technology have improved to the point that the chip area saved by manual design over automated design is not worth the extra design time involved: one can afford to "waste" a few gates, just as one can afford to waste a limited amount of processor cycles to compilation inefficiencies in a multiprocessor DSP system.

Finally, a proliferation of telecommunication standards and signal formats, often giving rise to multiple standards for the very same application, makes software implementation extremely attractive. Examples of applications in this category include set-top boxes capable of recognizing a variety of audio/video formats and compression standards, modems supporting multiple standards, multimode cellular phones and base stations that work with multiple cellular standards, multimedia workstations that are required to run a variety of different multimedia software products, and programmable audio/video codecs. Integrated multiprocessor DSP systems provide a very flexible software platform for this rapidly-growing family of applications.

A natural generalization of such fully-programmable, multiprocessor integrated circuits is the class of multiprocessor systems that consists of an arbitrary — possibly heterogeneous — collection of programmable processors as well as a set of zero or more custom hardware elements on a single chip. Mapping applications onto such an architecture is then a hardware/software co-design problem. However, the problems of interprocessor communication and synchronization are, for the most part, identical to those encountered in fully-programmable systems. In this book, when we refer to a "multiprocessor", we will imply an architecture that, as described above, may be comprised of different types of programmable processors, and may include custom hardware elements. Additionally, the multiprocessor systems that we address in this book may be packaged in a single integrated circuit chip, or may be distributed across multiple chips. All of the techniques that we present in this book apply to this general class of parallel processing architectures.

## 1.2    Application-Specific Multiprocessors

Although this book addresses a broad range of parallel architectures, it focuses on the design of such architectures in the context of specific, well-defined families of applications. We focus on application-specific parallel processing instead of applying the ideas in general purpose parallel systems because DSP systems are typically components of embedded applications, and the computational characteristics of embedded applications are fundamentally different from those of general-purpose systems. General purpose parallel computation involves user-programmable computing devices, which can be conveniently configured for a wide variety of purposes, and can be reconfigured any number of times as the users' needs change. Computation in an embedded application, however, is usually one-time programmed by the designer of that embedded system (a digital cellular radio handset, for example) and is not meant to be programmable by the end user.

Also, the computation in embedded systems is specialized (the computation in a cellular radio handset involves specific DSP functions such as speech compression, channel equalization, modulation, etc.), and the designers of embedded multiprocessor hardware typically have specific knowledge of the applications that will be developed on the platforms that they develop. In contrast, architects of general purpose computing systems cannot afford to customize their hardware too heavily for any specific class of applications. Thus, only designers of embedded systems have the opportunity to accurately predict and optimize for the specific application subsystems that will be executing on the hardware that they develop. However, if only general purpose implementation techniques are used in the development of an embedded system, then the designers of that embedded system lose this opportunity.

Furthermore, embedded applications face very different constraints compared to general purpose computation. These constraints include nonrecurring design costs, competitive time-to-market constraints, limitations on the amount and placement of memory, constraints on power consumption, and real-time performance requirements. Thus, for an embedded application, it is critical to apply techniques for design and implementation that exploit the special characteristics of the application in order to optimize for the specific set of constraints that must be satisfied. These techniques are naturally centered around design methodologies that tailor the hardware and software implementation to the particular application.

## 1.3    Exploitation of Parallelism

Parallel computation has of course been a topic of active research in computer science for the past several decades. Whereas parallelism within a single processor has been successfully exploited (instruction-level parallelism), the problem of partitioning a single user program onto multiple such processors is yet to be satisfactorily solved. Although the hardware for the design of multiple processor machines — the memory, interconnection network, input/output subsystems, etc. — has received much attention, efficient partitioning of a general program (written in C, for example) across a given set of processors arranged in a particular configuration is still an open problem. The need to detect parallelism from within the overspecified sequencing in popular imperative languages such as C, the need to manage overhead due to communication and synchronization between processors, and the requirement of dynamic load balancing for some programs (an added source of overhead) complicates the partitioning problem for a general program.

If we turn from general purpose computation to application-specific domains, however, parallelism is often easier to identify and exploit. This is because much more is known about the computational structure of the functionality being implemented. In such cases, we do not have to rely on the limited ability of automated tools to deduce this high-level structure from generic, low-level specifications (for instance, from a general purpose programming language such as C). Instead, it may be possible to employ specialized computational models — such as one of the numerous variants of dataflow and finite state machine models — that expose relevant structure in our targetted applications, and greatly facilitate the manual or automatic derivation of optimized implementations. Such specification models will be unacceptable in a general-purpose context due to their limited applicability, but they present a tremendous opportunity to the designer of embedded applications. The use of specialized computational models — particularly dataflow-based models — is especially prevalent in the DSP domain.

Similarly, focusing on a particular application domain may inspire the discovery of highly streamlined system architectures. For example, one of the most extensively studied family of application-specific parallel processors is the class of *systolic array* architectures [Kun88][Rao85]. These architectures consist of regularly arranged arrays of processors that communicate locally, onto which a certain class of applications, specified in a mathematical form, can be systematically mapped. Systolic arrays are further discussed in Chapter 2.

## 1.4     Dataflow Modeling for DSP Design

The necessary elements in the study of application-specific computer architectures are: 1) a clearly defined set of problems that can be solved using the particular application-specific approach, 2) a formal mechanism for specification of these applications, and 3) a systematic approach for designing hardware and software from such a specification. In this book we focus on embedded signal, image, and video signal processing applications, and a specification model called Synchronous Dataflow that has proven to be very useful for design of such applications.

Dataflow is a well-known programming model in which a program is represented as a set of tasks with data precedences. Figure 1.1 shows an example of a dataflow graph, where computation tasks (**actors**) $A$, $B$, $C$, and $D$ are represented as circles, and arrows (or **arcs**) between actors represent FIFO (first-in-first-out) queues that direct data values from the output of one computation to the input of another. Figure 1.2 shows the semantics of a dataflow graph. Actors consume data (or **tokens**, represented as bullets in Figure 1.2) from their inputs, perform computations on them (**fire**), and produce a certain number of tokens on their outputs. The functions performed by the actors define the overall function of the dataflow graph; for example in Figure 1.1, $A$ and $B$ could be data sources, $C$ could be a simple addition operation, and $D$ could be a data sink. Then the

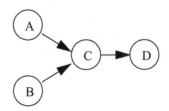

Figure 1.1. An example of a dataflow graph.

function of the dataflow graph would be simply to output the sum of two input tokens.

Dataflow graphs are a very useful specification mechanism for signal processing systems since they capture the intuitive expressivity of block diagrams, flow charts, and signal flow graphs, while providing the formal semantics needed for system design and analysis tools. The applications we focus on are those that can be described by **Synchronous Dataflow (SDF)** [LM87] and its extensions; we will discuss the formal semantics of this computational model in detail in Chapter 3. SDF in its pure form can only represent applications that have no decision making at the task level. Extensions of SDF (such as the **Boolean dataflow (BDF) model** [Lee91][Buc93]) allow control constructs, so that data-dependent control flow can be expressed in such models. These models are significantly more powerful in terms of expressivity, but they give up some of the useful analytical properties possessed the SDF model. For instance, Buck shows that it is possible to simulate any Turing machine in the BDF model [Buc93]. The BDF model can therefore compute all Turing computable functions, whereas this is not possible in the case of the SDF model. We discuss the Boolean dataflow model

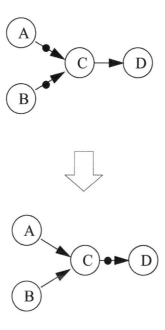

Figure 1.2. Actor "firing."

further in Chapter 4 and Chapter 9.

In exchange for the limited expressivity of an SDF representation, we can efficiently check conditions such as whether a given SDF graph deadlocks, and whether it can be implemented using a finite amount of memory. No such general procedures can be devised for checking the corresponding conditions (deadlock behavior and bounded memory usage) for a computation model that can simulate any given Turing machine. This is because the problems of determining if any given Turing machine halts (the halting problem), and determining whether it will use less than a given amount of memory (or tape) are **undecidable** [LP81]; that is, no general algorithm exists to solve these problems in finite time.

In this work, we first focus on techniques that apply to SDF applications, and we will propose extensions to these techniques for applications that can be specified essentially as SDF, but augmented with a limited number of control constructs (and hence fall into the BDF model).

SDF has proven to be a useful model for representing a significant class of DSP algorithms; several computer-aided design tools for DSP have been developed around SDF and closely related models. Examples of early commercial tools that employed SDF are the Signal Processing Worksystem (SPW) from Cadence [PLN92][BL91], and COSSAP, from Synopsys [RPM92]. More recently-developed commercial tools that use SDF and related models of computation are ADS from Agilent (formerly, from the EEsof division of Hewlett Packard); Cocentric System Studio (formerly called El Greco) from Synopsys [BV00]; LabVIEW from National Instruments [AK98]; and System Canvas from Angeles Design Systems [MCR01]. Tools developed at various research laboratories that use SDF and related models include DESCARTES [RPM92], DIF [HKB05], GRAPE [LEAP94], the Graph Compiler [VPS90], NP-click [SPRK04], PeaCE [SOIH97], PGMT [Ste97], Ptolemy [PHLB95], StreamIt [TKA02], and the Warp compiler [Pri92]. Figure 1.3 shows an example of an FM system specified as a block diagram in Cadence SPW.

The SDF model is popular because it has certain analytical properties that are useful in practice; we will discuss these properties and how they arise in the following section. The most important property of SDF graphs in the context of this book is that it is possible to effectively exploit parallelism in an algorithm specified as an SDF graph by scheduling computations in the SDF graph onto multiple processors at compile or design time rather than at run-time. Given such a schedule that is determined at compile time, we can extract information from it with a view towards optimizing the final implementation. In this book we present techniques for minimizing synchronization and interprocessor communication overhead in statically (i.e., compile time) scheduled multiprocessors in which the program is derived from a dataflow graph specification. The strategy is to model run-time execution of such a multiprocessor to determine how processors communicate and synchronize, and then to use this information to optimize the final implementation.

## 1.5    Utility of Dataflow for DSP

As mentioned before, dataflow models such as SDF (and other closely related models) have proven to be useful for specifying applications in signal processing and communications, with the goal of both simulation of the algorithm at the functional or behavioral level, and for synthesis from such a high level specification to a software description (e.g., a C program) or a hardware description (e.g., VHDL) or a combination thereof. The descriptions thus generated can then be compiled down to the final implementation, e.g., an embedded processor, or an ASIC. In Chapter 4, we provide a broad review of a variety of different dataflow-based modeling techniques for DSP applications.

One of the reasons for the popularity of such dataflow based models is that they provide a formalism for block-diagram based visual programming, which is a very intuitive specification mechanism for DSP; the expressivity of the SDF model sufficiently encompasses a significant class of DSP applications, including multirate applications that involve upsampling and downsampling operations. An equally important reason for employing dataflow is that such a specification exposes parallelism in the program. It is well known that imperative programming styles such as C and FORTRAN tend to over-specify the control structure of a given computation, and compilation of such specifications onto parallel architectures is known to be a hard problem. Dataflow on the other hand

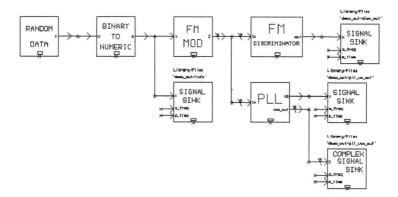

Figure 1.3. A block diagram specification of an FM system in Cadence Signal Processing Worksystem (SPW).

imposes minimal data-dependency constraints in the specification, potentially enabling a compiler to detect parallelism very effectively. The same argument holds for hardware synthesis, where it is also important to be able to specify and exploit concurrency.

The SDF model has also proven to be useful for compiling DSP applications on single processors. Programmable digital signal processing chips tend to have special instructions such as a single cycle multiply-accumulate (for filtering functions), modulo addressing (for managing delay lines), and bit-reversed addressing (for FFT computation). DSP chips also contain built in parallel functional units that are controlled from fields in the instruction (such as parallel moves from memory to registers combined with an ALU operation). It is difficult for automatic compilers to optimally exploit these features; executable code generated by commercially available compilers today utilizes one-and-a-half to two times the program memory that a corresponding hand optimized program requires, and results in two to three times higher execution time compared to hand-optimized code [ZVSM95]. There is, however, a significant body of research that is targeted to narrowing this gap (e.g., see [Leu02]). Moreover, some of the newer DSP architectures such as the Texas Instruments TMS320C6000 are more compiler friendly than past DSP architectures; automatic compilers for these processors often rival hand optimized assembly code for many standard DSP benchmarks.

Block diagram languages based on models such as SDF have proven to be a bridge between automatic compilation and hand coding approaches; a library of reusable blocks in a particular programming language is hand coded, this library then constitutes the set of atomic SDF actors. Since the library blocks are reusable, one can afford to carefully optimize and fine tune them. The atomic blocks are fine to medium grain in size; an atomic actor in the SDF graph may implement anything from a filtering function to a two input addition operation. The final program is then automatically generated by concatenating code corresponding to the blocks in the program according to the sequence prescribed by a schedule. This approach is mature enough that there are commercial tools available today, for example the SPW and COSSAP tools mentioned earlier, that employ this technique. Powerful optimization techniques have been developed for generating sequential programs from SDF graphs that optimize for metrics such as program and data memory usage, the run-time efficiency of buffering code, and context switching overhead between subtasks [BML96].

Scheduling is a fundamental operation that must be performed in order to implement SDF graphs on both uniprocessor as well as multiprocessors. Uniprocessor scheduling simply refers to determining a sequence of execution of actors such that all precedence constraints are met and all the buffers between actors (corresponding to arcs) return to their initial states. Multiprocessor scheduling involves determining the mapping of actors to available processors, in addition to determining of the sequence in which actors execute. We discuss the issues involved in multiprocessor scheduling in subsequent chapters.

## 1.6     Overview

The following chapter describes examples of application specific multi-processors used for signal processing applications. Chapter 3 lays down basic formal notation and definitions that are used in the remainder of this book for modeling scheduling, run-time synchronization, and interprocessor communication. Chapter 4 reviews a variety of different dataflow-oriented models of computation for design and implementation of DSP systems. Chapter 5 describes scheduling models that are commonly employed when scheduling dataflow graphs on multiple processors. Chapter 6 describes scheduling algorithms that attempt to maximize performance while accurately taking interprocessor communication costs into account. Chapters 7 and 8 describe a hardware based technique for minimizing IPC and synchronization costs; the key idea in these chapters is to predict the pattern of processor accesses to shared resources and to enforce this pattern during run-time. We present the hardware design and implementation of a four processor machine — the Ordered Memory Access Architecture (OMA). The OMA is a shared bus multiprocessor that uses shared memory for IPC. The order in which processors access shared memory for the purpose of communication is predetermined at compile time and enforced by a bus controller on the board, resulting in a low-cost IPC mechanism without the need for explicit synchronization. This scheme is termed the Ordered Transactions strategy.

In Chapter 8, we present a graph theoretic scheme for modeling run-time synchronization behavior of multiprocessors using a structure we call the **IPC graph** that takes into account the processor assignment and ordering constraints that a self-timed schedule specifies. We also discuss the effect of run-time variations in execution times of tasks on the performance of a multiprocessor implementation.

In Chapter 9, we discuss ideas for extending the Ordered Transactions strategy to models more powerful than SDF, for example, the Boolean dataflow (BDF) model. The strategy here is to assume we have only a small number of control constructs in the SDF graph and explore techniques for this case. The domain of applicability of compile time optimization techniques can be extended to programs that display some dynamic behavior in this manner, without having to deal with the complexity of tackling the general BDF model.

The ordered memory access approach discussed in Chapters 7 to 9 requires special hardware support. When such support is not available, we can utilize a set of software-based approaches to reduce synchronization overhead.

These techniques for reducing synchronization overhead consist of efficient algorithms that minimize the overall synchronization activity in the implementation of a given self-timed schedule. A straightforward multiprocessor implementation of a dataflow specification often includes **redundant synchronization** points, i.e., the objective of a certain set of synchronizations is guaranteed as a side effect of other synchronization points in the system. Chapter 10 discusses efficient algorithms for detecting and eliminating such redundant synchronization operations. We also discuss a graph transformation called **Convert-to-SC-graph** that allows the use of more efficient synchronization protocols.

It is also possible to reduce the overall synchronization cost of a self-timed implementation by adding synchronization points between processors that were not present in the schedule specified originally. In Chapter 11, we discuss a technique, called **resynchronization**, for systematically manipulating synchronization points in this manner. Resynchronization is performed with the objective of improving throughput of the multiprocessor implementation. Frequently in real-time signal processing systems, latency is also an important issue, and although resynchronization improves the throughput, it generally degrades (increases) the latency.

Chapter 11 addresses the problem of resynchronization under the assumption that an arbitrary increase in latency is acceptable. Such a scenario arises when the computations occur in a feedforward manner, e.g., audio/video decoding for playback from media such as Digital Versatile Disk (DVD), and also for a wide variety of simulation applications. Chapter 12 examines the relationship between resynchronization and latency, and addresses the problem of optimal resynchronization when only a limited increase in latency is tolerable. Such latency constraints are present in interactive applications such as video conferencing and telephony, where beyond a certain point the latency becomes annoying to the user. In voice telephony, for example, the round trip delay of the speech signal is kept below about 100 milliseconds to achieve acceptable quality.

The ordered memory access strategy discussed in Chapters 7 through 9 can be viewed as a hardware approach that optimizes for IPC and synchronization overhead in statically scheduled multiprocessor implementations. The synchronization optimization techniques of Chapter 10 through 13, on the other hand, operate at the level of a scheduled parallel program by altering the synchronization structure of a given schedule to minimize the synchronization overhead in the final implementation. Throughout the book, we illustrate the key concepts by applying them to examples of practical systems.

# 2

# APPLICATION-SPECIFIC
# MULTIPROCESSORS

The extensive research results and techniques developed for general purpose high-performance computation are naturally applicable to signal processing systems. However, in an application-specific domain, it is often possible to simplify the parallel machine architecture and the interconnect structure, thereby potentially achieving the requisite performance in terms of throughput and power consumption at a lower system cost. System costs include not only the dollar cost of manufacturing, but also hardware and software development costs, and testing costs; these costs are particularly crucial for the embedded and consumer applications that are targeted by a majority of current DSP-based systems. In this chapter we discuss some application-specific parallel processing strategies that have been employed for signal processing.

## 2.1    Parallel Architecture Classifications

The classic Flynn categorization of parallel processors as Single Instruction Multiple Data (SIMD) or Multiple Instruction Multiple Data (MIMD) [Fly66] classifies machines according to how they partition control and data among different processing elements. An SIMD machine partitions input data among processors executing identical programs, whereas an MIMD machine partitions input data and allows processors to execute different programs on each data portion. Modern parallel machines may employ a mix of SIMD and MIMD type processing, as we shall see in some of the examples discussed in this section.

Parallelism can be exploited at different levels of granularity: The processing elements making up the parallel machine could either be individual functional units (adders, multipliers, etc.) to achieve fine-grain parallelism, or the elements could themselves be self-contained processors that exploit parallelism within themselves. In the latter case, we can view the parallel program as being split into multiple threads of computation, where each thread is assigned to a processing element. The processing element itself could be a traditional von Neumann-type Central Processing Unit (CPU), sequentially executing instructions fetched from a central instruction storage, or it could employ **instruction level parallelism (ILP)** to realize high performance by executing in parallel multiple instructions in its assigned thread.

The interconnection mechanism between processors is clearly crucial to the performance of the machine on a given application. For fine-grained and instruction level parallelism support, communication often occurs through a simple mechanism such as a multiported register file. For machines composed of more sophisticated processors, a large variety of interconnection mechanism have been employed, ranging from a simple shared bus to 3-dimensional meshes and hyper-trees [Lei92]. Embedded applications often employ simple structures such as hierarchical busses or small cross bars.

The two main flavors of ILP are superscalar and VLIW (Very Long Instruction Word) [PH96]. Superscalar processors (for example the Intel Pentium processor) contain multiple functional units (ALUs, floating point units, etc.); instructions are brought into the machine sequentially and are scheduled dynamically by the processor hardware onto the available functional units. Out-of-order execution of instructions is also supported.

VLIW processors, on the other hand, rely on a compiler to statically schedule instructions onto functional units; the compiler determines exactly what operation each functional unit performs in each instruction cycle. The "long instruction word" arises because the instruction word must specify the control information for all the functional units in the machine. Clearly, a VLIW model is less flexible than a superscalar approach; however, the implementation cost of VLIW is also significantly less because dynamic scheduling need not be supported in hardware. For this reason, several modern DSP processors have adopted the VLIW approach; at the same time, as discussed before, the regular nature of DSP algorithms lend themselves well to the static scheduling approach employed in VLIW machines. We will discuss some of these machines in detail in the following sections.

Given multiple processors capable of executing autonomously, the program threads running on the processors may be tightly or loosely coupled to one another. In a tightly coupled architecture the processors may run in lock step executing the same instructions on different data sets (for example systolic arrays), or they may run in lock step, but operate on different instruction sequences (similar to VLIW). Alternatively, processors may execute their programs independent of one another, only communicating or synchronizing when necessary. Even in this case there is a wide range of how closely processors are coupled, which can range from a shared memory model where the processors may share the same memory address space to a "network of workstations" model where autonomous machines communicate in a coarse-grained manner over a local area network.

In the following sections, we discuss application-specific parallel processors that exemplify the many variations in parallel architectures discussed thus far. We will find that these machines employ tight coupling between processors; these machines also attempt to exploit the predictable run-time nature of the targeted applications, by employing architectural techniques such as VLIW, and employing processor interconnections that reflect the nature of the targeted application set. Also, these architectures rely heavily upon static scheduling techniques for their performance.

## 2.2    Exploiting Instruction Level Parallelism

### 2.2.1    ILP in Programmable DSP Processors

DSP processors have incorporated ILP techniques since inception; the key innovation in the very first DSPs was a single cycle multiply-accumulate unit. In addition, almost all DSP processors today employ an architecture that includes multiple internal busses allowing multiple data fetches in parallel with an instruction fetch in a single instruction cycle; this is also known as a "Harvard" architecture. Figure 2.1 shows an example of a modern DSP processor (Texas Instruments TMS320C54x DSP) containing multiple address and data busses, and parallel address generators.

Since filtering is the key operation in most DSP algorithms, modern programmable DSP architectures provide highly specialized support for this function. For example, a multiply-and-accumulate operation may be performed in parallel with two data fetches from data memory (for fetching the signal sample and the filter coefficient); in addition, an update of two address registers (potentially including modulo operations to support circular buffers and delay lines), and an instruction fetch can also be done in the same cycle. Thus, there are as many as *seven* atomic operations performed in parallel in a single cycle; this allows a finite impulse response (FIR) filter implementation using only one DSP instruction cycle per filter tap. For example, Figure 2.2 shows the assembly code for the inner loop of an FIR filter implementation on a TMS320C54x DSP. The MAC instruction is repeated for each tap in the filter; for each repetition this instruction fetches the coefficient and data pointed to by address registers AR2 and AR3, multiplies and accumulates them into the "A" accumulator, and post-increments the address registers.

Most DSP processors thus have a complex instruction set and follow a philosophy very different from "Reduced Instruction Set Computer" (RISC) architectures, that are prevalent in the general purpose high performance microprocessor domain. The advantages of a complex instruction set are compact object code, and deterministic performance, while the price of supporting a complex instruction set is lower compiler efficiency and lesser portability of the DSP software. The constraint of low power, and high performance-to-cost ratio requirement for embedded DSP applications has resulted in very different evolu-

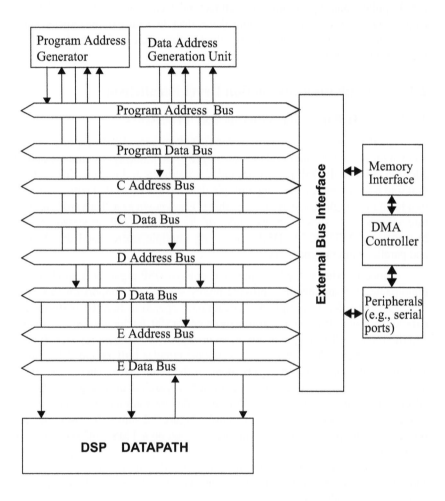

Figure 2.1. A simplified view of the Texas Instruments TMS320C54x, showing multiple on-chip address and data busses.

tion paths for DSP processors compared to general-purpose processors. Whether these paths eventually converge in the future remains to be seen.

### 2.2.2    Subword Parallelism

Subword parallelism refers to the ability to divide a wide ALU into narrower slices so that multiple operations on a smaller data type can be performed on the same datapath in an SIMD fashion (Figure 2.3). Several general purpose microprocessors employ a multimedia enhanced instruction set that exploits subword parallelism to achieve higher performance on multimedia applications that require a smaller precision.

The "MMX Technology"-enhanced Intel Pentium processor [EK97] is perhaps the best known general purpose CPU with an enhanced instruction set to handle throughput intensive "media" processing. The MMX instructions allow a 64-bit ALU to be partitioned into 8-bit slices, providing subword parallelism. The 8-bit ALU slices work in parallel in an SIMD fashion. The MMX enhanced Pentium can perform operations such as addition, subtraction, and logical operations on eight 8-bit samples (for example image pixels) in a single cycle. It also can perform data movement operations such as single cycle swapping of bytes within words, packing smaller sized words into a 64-bit register, etc. More complex operations such as four 8-bit multiplies (with or without saturation), parallel shifts within subwords, and sum of products of subwords, may all be performed in a single cycle.

Similarly enhanced microprocessors have been developed by Sun Microsystems (the "VIS" instruction set for the SPARC processor [TONL96]) and Hewlett-Packard (the "MAX" instructions for the PA RISC processor [Lee96]). The VIS instruction set includes a capability for performing single cycle sum of absolute difference (for image compression applications). The MAX instructions include a subword average, shift and add, and fairly generic permute instructions that change the positions of the subwords within a 64-bit word boundary in a very flexible manner. The permute instructions are especially useful for efficiently aligning data within a 64-bit word before employing an instruction that operates on multiple subwords. DSP processors such as the TMS320C6x and the Philips Trimedia also support subword parallelism.

```
RPT       num_taps - 1
MAC       *AR2+, *AR3+, A
```

Figure 2.2. Inner loop of an FIR filter implemented on a TMS320C54x DSP.

Exploiting subword parallelism clearly requires extensive static or compile time analysis, either manually or by a compiler.

### 2.2.3    VLIW Processors

As discussed before, the lower cost of a compiler-scheduled approach employed in VLIW machines compared to hardware scheduling employed in superscalar processors makes VLIW a good candidate as a DSP architecture. It is therefore no surprise that several semiconductor manufacturers have recently announced VLIW-based signal processor products.

The Philips Trimedia [RS98] processor, for example, is geared towards video signal processing, and employs a VLIW engine. The Trimedia processor also has special I/O hardware for handling various standard video formats. In addition, hardware modules for highly specialized functions such as Variable Length Decoding (used for MPEG video decoding), color and format conversion, are also provided. Trimedia also has instructions that exploit subword parallelism among byte-sized samples within a 32-bit word.

The Chromatics MPACT architecture [Pur97] uses an interesting hardware/software partitioned solution to provide a programmable platform for PC-based multimedia. The target applications are graphics, audio/video processing, and video games. The key idea behind Chromatic's multimedia solution is to use some amount of processing capability in the native x86 CPU, and use the MPACT processor for accelerating certain functions when multiple applications are operated simultaneously (for example when a FAX message arrives while a teleconferencing session is in operation).

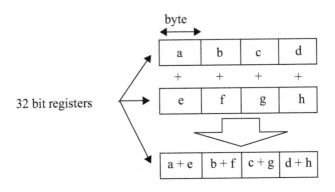

Figure 2.3. Example of subword parallelism: Addition of bytes within a 32 bit register (saturation or truncation could be specified).

Finally, the Texas Instruments TMS320C6x DSP [Tex98] is a high performance, general purpose DSP that employs a VLIW architecture. The C6x processor is designed around eight functional units that are grouped into two identical sets of four functional units each (see Figure 2.4). These functional units are the D unit for memory load/store and add/subtract operations; the M unit for multiplication; the L unit for addition/subtraction, logical and comparison operations; and the S unit for shifts in addition to add/subtract and logical operations. Each set of four functional units has its own register file, and a bypass is provided for accessing each half of the register file by either set of functional units.

Each functional unit is controlled by a 32-bit instruction field; the instruction word for the processor therefore has a length between 32 bits and 256 bits, depending on how many functional units are actually active in a given cycle. Features such as predicated instructions allow conditional execution of instructions; this allows one to avoid branching when possible, a very useful feature considering the deep pipeline of the C6x. The C64x processor, the latest member of the C6x family, is capable of executing four 16x16 multiplies or eight 8x8 multiplies each clock cycle. Support is provided for operating on packet 16-bit and 8-bit data, and up to four 32-bit register loads or stores to memory can be done per

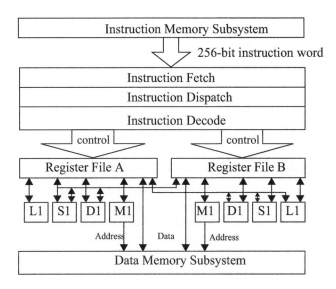

Figure 2.4. The TMS320C6x VLIW architecture.

clock cycle.

## 2.3    Dataflow DSP Architectures

Several multiprocessors geared towards signal processing are based on the dataflow architecture principles introduced by Dennis [Den80]; these machines deviate from the traditional von Neumann model of a computer. Notable among these are Hughes Data Flow Multiprocessor [GB91], the Texas Instruments Data Flow Signal Processor [Gri84], and the AT&T Enhanced Modular Signal Processor [ATT87]. The first two perform the processor assignment step at compile time (i.e., tasks are assigned to processors at compile time) and tasks assigned to a processor are scheduled on it dynamically; the AT&T EMPS performs even the assignment of tasks to processors at run-time. The main steps involved in scheduling tasks on multiple processors are discussed fully in Chapter 5.

Each of these machines employs elaborate hardware to implement dynamic scheduling within processors, and employs expensive communication networks to route tokens generated by actors assigned to one processor to tasks on other processors that require these tokens. In most DSP applications, however, such dynamic scheduling is unnecessary since compile time predictability makes static scheduling techniques viable. Eliminating dynamic scheduling results in much simpler hardware without an undue performance penalty.

Another example of an application-specific dataflow architecture is the NEC μPD7281 [Cha84], which is a single chip processor geared towards image processing. Each chip contains one functional unit; multiple such chips can be connected together to execute programs in a pipelined fashion. The actors are statically assigned to each processor, and actors assigned to a given processor are scheduled on it dynamically. The primitives that this chip supports, convolution, bit manipulations, accumulation, etc., are specifically designed for image processing applications.

## 2.4    Systolic and Wavefront Arrays

Systolic arrays consist of processors that are locally connected and may be arranged in different interconnection topologies: mesh, ring, torus, etc. The term "systolic" arises because all processors in such a machine run in lock-step, alternating between a computation step and a communication step. The model followed is usually SIMD (Single Instruction Multiple Data). Systolic arrays can execute a certain class of problems that can be specified as "Regular Iterative Algorithms (RIA)" [Rao85]; systematic techniques exist for mapping an algorithm specified in a RIA form onto dedicated processor arrays in an optimal fashion. Optimality includes metrics such as processor and communication link utilization, scalability with the problem size, and achieving best possible speedup for a given number of processors. Several numerical computation problems were found to fall into the RIA category: linear algebra, matrix operations, singular value decomposition, etc. (see [Kun88][Lei92] for interesting systolic array implementations of a variety of different numerical problems). Only highly regular computations can be specified in the RIA form; this makes the applicability of systolic arrays somewhat restrictive.

Wavefront arrays are similar to systolic arrays except that processors are not under the control of a global clock [Kun88]. Communication between processors is asynchronous or self-timed; handshake between processors ensures run-time synchronization. Thus, processors in a wavefront array can be complex and the arrays themselves can consist of a large number of processors without incurring the associated problems of clock skew and global synchronization. The flexibility of wavefront arrays over systolic arrays comes at the cost of extra handshaking hardware.

The Warp project at Carnegie Mellon University [A+87] is an example of a programmable systolic array, as opposed to a dedicated array designed for one specific application. Processors are arranged in a linear array and communicate with their neighbors through FIFO queues. Programs are written for this computer in a language called W2 [Lam88]. The Warp project also led to the iWarp design [GW92], which has a more elaborate interprocessor communication mechanism than the Warp machine. An iWarp node is a single VLSI component, composed of a computation engine and a communication engine. The computation agent consists of an integer and logical unit as well as a floating point add and multiply unit. Each unit is capable of running independently, and has access to a multiported register file. The communication agent connects to its neighbors via four bidirectional communication links, and provides the interface to support message passing type communication between cells as well as word-based systolic communication. The iWarp nodes can therefore be connected in various single and two dimensional topologies. Various image processing applications (2-D FFT, image smoothing, computer vision) and matrix algorithms (SVD, QR decomposition) have been reported for this machine [Lou93].

## 2.5    Multiprocessor DSP Architectures

Next, we discuss multiprocessors that make use of multiple off-the-shelf programmable DSP chips. An example of such a system is the SMART architecture [Koh90] that is a reconfigurable bus-based design comprised of AT&T DSP32C processors, and custom VLSI components for routing data between processors. Clusters of processors may be connected onto a common bus, or may form a linear array with neighbor-to-neighbor communication. This allows the multiprocessor to be reconfigured depending on the communication requirement of the particular application being mapped onto it. Scheduling and code generation for this machine is done by an automatic parallelizing compiler [HR92].

The DSP3 multiprocessor [SW92] is comprised of AT&T DSP32C processors connected in a mesh configuration. The mesh interconnect is implemented using custom VLSI components for data routing. Each processor communicates with four of its adjacent neighbors through this router, which consists of input and output queues, and a crossbar that is configurable under program control. Data packets contain headers that indicate the ID of the destination processor.

The Ring Array Processor (RAP) system [M+92] uses TI DSP320C30 processors connected in a ring topology. This system is designed specifically for speech-recognition applications based on artificial neural networks. The RAP system consists of several boards that are attached to a host workstation, and acts as a co-processor for the host. The unidirectional pipelined ring topology employed for interprocessor communication was found to be ideal for the particular algorithms that were to be mapped to this machine. The ring structure is similar to the SMART array, except that no processor ID is included with the data, and processor reads and writes into the ring are scheduled at compile time. The ring is used to broadcast data from one processor to all the others during one phase of the neural network algorithm, and is used to shift data from processor to processor in a pipelined fashion in the second phase.

Several modern off-the-shelf DSP processors provide special support for multiprocessing. Examples include the Texas Instruments TMS320C40 (C40),

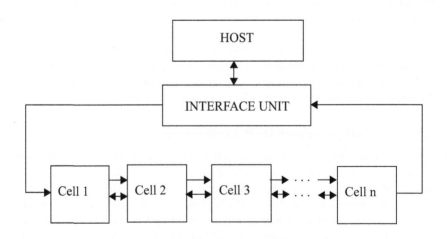

Figure 2.5. Warp array.

Motorola DSP96000, Analog Devices ADSP-21060 "SHARC," as well as the Inmos (now owned by SGS Thompson) Transputer line of processors. The DSP96000 processor is a floating point DSP that supports two independent busses, one of which can be used for local accesses and the other for interprocessor communication. The C40 processor is also a floating point processor with two sets of busses; in addition it has six 8-bit bidirectional ports for interprocessor communication. The ADSP-21060 is a floating point DSP that also provides six bidirectional serial links for interprocessor communication. The Transputer is a CPU with four serial links for interprocessor communications.

Owing to the ease with which these processors can be interconnected, a number of multi-DSP machines have been built around the C40, DSP96000, SHARC, and the Transputer. Examples of multi-DSP machines composed of DSP96000s include MUSIC [G+92] that targets neural network applications as well as the OMA architecture described in Chapter 7; C40 based parallel processors have been designed for beamforming applications [Ger95], and machine vision [DIB96] among others; ADSP-21060 based multiprocessors include speech-recognition applications [T+95], applications in nuclear physics [A+98], and digital music [Sha98]; and machines built around Transputers have targeted applications in scientific computation [Mou96], and robotics [YM96].

## 2.6    Single-Chip Multiprocessors

Olukotun et al. [O+96] present an interesting study that concludes that going to a multiple processor solution is a better path to high performance than going to higher levels of instruction level parallelism (using a superscalar approach, for example). Systolic arrays have been proposed as ideal candidates for application-specific multiprocessor on a chip implementations; however, as pointed out before, the class of application targeted by systolic arrays is limited. Modern VLSI technology enables hundreds of processor "cores" to be placed on a single die, to yield a multiprocessor system-on-a-chip (also called Multiprocessor Systems on Chip, abbreviated as MP-SOC). The achievable clock speeds for such MP-SOCs range from hundreds of Megahertz to more than a Gigahertz.

This trend is increasingly popular, especially in the embedded computing space, resulting in several commercially available single chip processors today. There are three main concerns in designing single chip multiprocessors: 1) Usage of available silicon area, 2) power consumption, and 3) the programming model employed. The available single chip multiprocessors use a spectrum of different approaches to address these concerns.

In terms of silicon area, the variables are the complexity of each processor core; the number of such cores on a single die; the type of interconnect; the amount of on-chip memory, both processor-specific and shared; application specific accelerators if any; and external memory and I/O interfaces. These large number of design variables result in a huge design space for MP-SOCs. The choice of design variables often tends to be application specific.

In terms of power consumption, MP-SOCs offer a natural trade-off between clock speed, operating voltage, and parallelism. The use of multiple processors implies that the clock speed and operating voltage may be reduced, thereby reducing the total energy consumed for accomplishing a given task.

Single chip multiprocessors often provide architectural support for certain types of programming models. Multiprocessors in general are notoriously difficult to program; indeed the utility and widespread use of a multiprocessor architecture is often dictated by the ease of programming and the available tools for software programming and debugging. As we discuss examples of MP-SOCs next, we will also discuss the associated programming paradigms for them.

### 2.6.1     The MathStar MP-SOC

The MathStar MP-SOC [Math07] is an example of a MP-SOC that contains a large number of relatively fine grained processing elements running at speeds of up to a gigahertz. The chip consists of hundreds of field programmable "silicon objects" such as arithmatic logic units, multiply-accumulate units, and register files. The silicon objects consist of building blocks such as ALUs, MACs, and register files, and provide a higer level programming model as compred to an FPGA. The processing elements are 16-bit ALUs that have 32 possible operations. An interconnect fabric provides communication mechanisms between the silicon objects; single cycle communication between up to a distance of three objects is supported. The processing elements and ALUs may be configured using either a graphical block diagram based or using th SystemC language. Preconfigured library of templates are provided for configuring the silicon objects. The end-user is responsible for managing fine grained parallelization of individual arithmatic functions. The MathStar programming model is closely tied to the underlying hardware and resembles hardware design using an FPGA rather than programming a general purpose processor. The targetted applications include video compression/decompression, image recognition, enhancement, 2-D and 3-D rendering.

### 2.6.2     The Ambric Processor

The Ambric processor [Amb08] employs a globally asynchronous locally synchronous architectural model, where each processor is synchronous, but the inter-processor communication mechanism employs self-synchronizing channels that manage communication and synchronization. The chip consists of 360 processors, each of which is a 32-bit RISC core. Ambric employs a heterogeneous processor architecture: half the available processors are single ALU processors for executing control functions, and the remaining half are enhanced to include three ALUs for tackling math intensive DSP functions. Memory on the chip is distributed across processors, with a little over 100 kilobytes available per processor on average. Sending and receiving data from other processors is achieved via blocking reads and writes to channel registers. In terms of programming model, the application parallelism is explicitly specified as a network of interconnected objects, using either a graphical block diagram, or textually using an proprietary language called aStruct. The primitive objects themselves are written in sequential code, either in assembly or in the Java language.

### 2.6.3    The Stream Processor

The Stream processor [Dal07] on the other hand employs a stream oriented programming model, where operations are performed on arrays of data, called streams. The hardware architecture is optimized for processing streams, as found in most signal processing applications. The application is written in StreamC, an extension of C. The application programmer explicitly partitions the program into computation kernels and data streams on which the kernels operate; the kernels are executed on the stream processor in a pipelined manner. The chip architecture employs a large number of registers and a memory hierarchy that is compiler managed to reduce global bandwidth needs, eliminating the need for hardware caches.    Data transfers requirements are statically analyzed by the Stream compiler, and the on-chip DMAs are programmed accordingly at compile time. The architecture exploits ILP at the processor level since the processing element is a VLIW processor, and data parallelism at the stream level where the same compute kernel is applied in parallel to different elements of the stream. The Storm-1 class of processors from Stream Inc. support up to 80 processors. The stream data model lends itself naturally to a data-parallel execution model where elements of the data stream are operated in parallel on multiple processing units. Data-dependent control requires special handling in this model; a specialized switch statement is employed for this purpose.

### 2.6.4    Graphics Processors

Graphics processors also employ stream processing techniques since graphics tends to be a highly data-parallel application. For example the Nvidia GeForce architecture employs 128 stream processors. Each stream processor supports floating point operations and is capable of performing various graphics operations such as vertex processing, pixel shading, rendering, etc. The stream processors support a multithreading model where each processor executes an independent thread; there can thousands of simultaneously executing threads that keep the processors efficiently loaded. Hardware support is provided for managing threads. Although graphics processors are optimized for imaging and graphics application, there have been recent efforts to harness the immense parallel processing capabilities for general purpose scientific computation [O+07].

### 2.6.5     The Sandbridge Sandblaster

Another example of hardware supported multithreading as applied to embedded multiprocessing is the Sandbridge Sandblaster platform, the latest of which is the SB3011 [G+07]. This chip contains four processors; each processor contains hardware support for executing up to eight concurrent threads. Sandbridge does not support simultaneous multithreading where instructions for each thread are issued at every clock cycle. Instead, only one thread issues an instruction per cycle (sometimes called **temporal multithreading**). This allows the processor to be area and power efficient while maintaining some of the key advantages of multithreading. In such an architecture, when cache misses, branches, and other pipeline stalls occur during the execution of one thread, another thread is immediately started. This results in higher processor efficiency because functional units are kept busy for a higher fraction of time as compared to a traditional single threaded architecture. This also translates to lower overall power consumption since fewer processor cycles are required to complete a task, enabling the processor to go into a low power idle mode more often. Also, the processor speed is decoupled from memory speed since the processor performance has tolerance to memory stalls, as long as there are more than one threads executing on the processor. This allows the processor and memory to operate individually at their optimal operating points.

Multiple processors in the Sandbridge architecture are treated as containing a larger number of hardware threads. Automatic multithreading of loops across processors and across hardware threads is supported by the Sandbridge C compiler. Users may also explicitly define threads using Java or the POSIX thread library. Processors are connected via a time-multiplexed unidirectional ring network, and interprocessor communication occurs primarily through shared Level 2 memory. Sandbridge targets applications in mobile multimedia and physical layer DSP processing that require high performance and very low power consumption.

### 2.6.6     picoChip

The picoChip multiprocessor is another commercially available single chip multiprocessor that supports medium to fine grained parallelism utilizing hundreds of processing elements per chip [Pico08]. The PC205 for example contains 248 processing elements. Each processor is a 16-bit VLIW processor capable of executing up to six operations every clock cycle. In the picoChip architecture the amount of memory per processing element is relatively modest; the majority of the processing elements in the PC205 for example contain less than 1 kilobyte of dedicated memory. The picoChip multiprocessor is targetted towards applications in communications, particularly wireless physical layer applications. The processors are connected in a two dimensional mesh of time multiplexed busses. Groups of four processors are connected on each bus, and the bus from each group is connected to the neighboring groups via switches. Data transfer between processors occurs during specific time slots. Scheduling of specific transfers during each time slot and the switch configuration at each time slot is determined statically at compile time; at run time the switches cycle through the predetermined configurations. Multiple pairs of processors can communicate in parallel during each time slot, as long as there are no resource conflicts in terms of bus usage and switch configuration. Processors communicate by issuing "put" and "get" commands; these commands are blocking in that the sending processor halts until the "put" command completes, and similarly the receiving processor halts until the "get" command completes.

In the picoChip programming model the application is specified as a set of processes that communicate via signals. The user specifies the communication bandwidth needed for each signal; the picoChip compiler then works out the details of the specific time slots and switch configurations required. Processes are specified in the C language or assembly, and map one-to-one to the processing elements. The latest picoChip products, for example the PC20x series, also contain hardware accelerators aimed towards specific types of wireless physical layer implementations. This is in recognition of the fact that for well-defined functions, such as channel decoding, FFT, correlation, etc., a targetted hardware co-processor tends to be more area and power efficient than a fully programmable processor. The programmable processing elements can then be used for functions that need the flexibility of a software solution. The drawback is this results in a heterogeneous architecture that is more difficult to program and debug than a fully homogeneous architecture.

### 2.6.7    Single-Chip Multiprocessors from Texas Instruments

In contrast to the picoChip approach of placing hundreds of relatively simple processing elements on chip with a modest amount of memory per processor, it is also possible to place a few powerful processors running at high clock speeds, each with relatively large amounts of memory. This approach is taken by processors from Texas Instruments, such as the TCI6488 [Tex07] and the TNETV3020 [Tex08]. These processors are specifically targetted to the network infrastructure applications; the TCI6488 is optimized for wireless physical layer processing, and the TNETV3020 is optimized for voice channel processing. These applications involve supporting multiple channels (for example voice calls in a cellular network) in parallel. The TNETV3020 integrates six TMS320C64x processors along with over 5 Megabytes of memory, and high speed memory and network interfaces.

The TCI6488 integrates three TMS320C64x processors along with over 3 Megabytes of memory that may be partitioned among the DSPs as Level 2 (L2) memory (Figure 2.6). These multiprocessors naturally employ coarse grained parallelism; application partitioning, interprocessor communications and off-chip communications, are all managed explicitly by the application programmer. Programming each processor is done in the C language or assembly. Instruction level parallelism is exploited within each VLIW processor. On-chip interprocessor communication occurs via explicit transfers between L2 memories of each DSP through a cross-bar like switch fabric. Connectivity between multiprocessor chips is achieved via high speed serial interfaces. Application specific hardware "co-processors" are included for compute intensive functions such as Viterbi and Turbo decoding, and correlation operations for spread spectrum applications.

In the case of the TNETV3020 processor, each voice channel is completely independent of other voice channels; therefore a data parallel programming model where voice packets from each channel are routed to specific processors (based on the channel address for example) is generally used. All processors run the same program in a Single Program Multiple Data fashion. The only run-time task in such a model is one of load balancing.

The physical layer processing applications targetted to the TCI6488 on the other hand require a different approach to partitioning. Although the physical layer processing also must be done for a large number of channels, there is often a significant amount of processing that is common across channels. In other words the processing requirements for separate channels are not completely independent. As a result, a more natural task partitioning involves a functional mapping of tasks across processors, such that a given physical layer function is performed on a particular processor, for all channels. This leads to a heterogeneous partitioning of the application, where each processor executes a different program, and communication occurs at function boundaries. As mentioned earlier, the TCI6488 employs hardware co-processors for compute-intensive tasks such as channel decoding and correlation. Although all processors may access each of the co-processors, restricting access to one processor results in a simpler

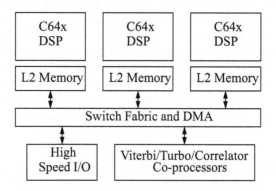

Figure 2.6. TCI6488 multicore DSP.

partition. Communication occurs through explicit data movement in shared memory using an on-chip cross-bar like interconnect. A powerful programmable DMA is configured with memory transfer characteristics. At run-time the DMA is capable of autonomously performing a sequence of memory transfers, and then interrupting one of the processors at the end of the transfers. The DMA can therefore be preconfigured to perform the required interprocessor data movements.

### 2.6.8    The Tilera Processor

An example of a medium-grained single chip multiprocessor is the Tilera processor [Til07]. It consists of 64 processor "tiles" arranged in a 2-D mesh. Each processor tile consists of a processor and a switch to connect the processor to the mesh interconnect. The processor core is a three instruction per cycle VLIW processor, capable of running a general purpose operating system such as Linux. Each processor has it's own Level 1 and Level 2 cache, for a total of 5 Megabytes of on-chip cache memory. Pages cached on one processor are visible to all other processors, enabling one tile to prefetch data for all other tiles. The interconnect hardware maintains cache coherence. The interconnect consists of five meshes, two of which are employed for processor to memory transactions, such as cache miss traffic and DMA. The remaining three meshes are used to move data between processors and I/O. Each processor runs a complete operating system (OS) and processors communicate via OS constructs such as sockets and message passing interface libraries. The applications targetted by Tilera are in networking and video processing.

### 2.6.9    Heterogeneous Multiprocessors

Embedded single-chip multiprocessors may also be composed of heterogeneous processors. For example, many consumer devices today, for example digital cellular phones, modems, hard disk drive controllers, etc., are composed of two processors: one is a DSP to handle signal processing tasks, while the other is a microcontroller such as an ARM processor, for handling control tasks.

Such a two-processor system is increasingly found in embedded applications because of the types of architectural optimization used in each processor. The microcontroller has an efficient interrupt-handling capability, and is more amenable to compilation from a high-level language; however, it lacks the multiply-accumulate performance of a DSP processor. The microcontroller is thus ideal for performing user interface and protocol processing type functions that are somewhat asynchronous in nature, while the DSP is more suited to signal processing tasks that tend to be synchronous and predictable. Even though new DSP processors boasting microcontroller capabilities have been introduced recently (for example the Hitachi SH-DSP and the TI TMS320C27x series) an ARM/DSP two processor solution is expected to be popular for embedded signal processing/control applications in the near future. A good example of such an architecture is described in [Reg94]; this part uses two DSP processors along with a microcontroller to implement audio processing and voice band modem functions in software.

## 2.7     Reconfigurable Computing

Reconfigurable computers are another approach to application-specific computing that has received significant attention lately. Reconfigurable computing is based on implementing a function in hardware using configurable logic (for example a field programmable gate array or FPGA), or higher level building blocks that can be easily configured and reconfigured to provide a range of different functions. Building a dedicated circuit for a given function can result in large speedups; examples of such functions are bit manipulation in applications such as cryptography and compression; bit-field extraction; highly regular computations such as Fourier and Discrete Cosine Transforms; pseudo-random number generation; compact lookup tables, etc. One strategy that has been employed for building configurable computers is to build the machine entirely out of reconfigurable logic; examples of such machines, used for applications such as DNA sequence matching, finite field arithmetic, and encryption, are discussed in [G+91][LM95][GMN96][V+96].

A second and more recent approach to reconfigurable architectures is to augment a programmable processor with configurable logic. In such an architecture, functions best suited to a hardware implementation are mapped to the FPGA to take advantage of the resulting speedup, and functions more suitable to software (for example control dominated applications, and floating point intensive computation) can make use of the programmable processor. The Garp processor [HW97], for example, combines a Sun UltraSPARC core with an FPGA that serves as a reconfigurable functional unit (see Figure 2.7). Special instructions are defined for configuring the FPGA, and for transferring data between the FPGA and the processor. The authors demonstrate a 24x speedup over a Sun UltraSPARC machine, for an encryption application. In [HFHK97] the authors describe a similar architecture, called Chimaera, that augments a RISC processor with an FPGA. In the Chimaera architecture, the reconfigurable unit has access to the processor register file; in the GARP architecture the processor is responsible for directly reading from and writing data to the reconfigurable unit through special instructions that are augmented to the native instruction set of the RISC processor. Both architectures include special instructions in the processor for sending commands to the reconfigurable unit.

Another example of a reconfigurable architecture is Matrix [MD97], which attempts to combine the efficiency of processors on irregular, heavily multiplexed tasks with the efficiency of FPGAs on highly regular tasks. The Matrix architecture allows selection of the granularity according to application needs. It consists of an array of basic functional units (BFUs) that may be configured either as functional units (add, multiply, etc.), or as control for another BFU. Thus one can configure the array into parts that function in SIMD mode under a common control, where each such partition runs an independent thread in an MIMD mode.

In [ASI+98] the authors describe the idea of domain-specific processors that achieve low power dissipation for a small class of applications they are optimized for. These processors augmented with general purpose processors yield a practical trade-off between flexibility, power and performance. The authors estimate that such an approach can reduce the power utilization of speech coding implementations by over an order of magnitude compared to an implementation using only a general purpose DSP processor.

PADDI (Programmable Arithmetic Devices for DIgital signal processing) is another reconfigurable architecture that consists of an array of high performance execution units (EXUs) with localized register files, connected via a flexible interconnect mechanism [CR92]. The EXUs perform arithmetic functions such as add, subtract, shift, compare, accumulate, etc. The entire array is controlled by a hierarchical control structure: A central sequencer broadcasts a global control word, which is then decoded locally by each EXU to determine its

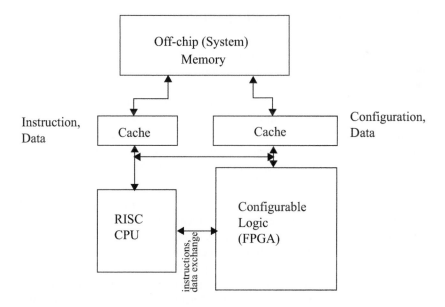

Figure 2.7. A RISC processor augmented with an FPGA-based accelerator [HW97][HFHK97].

action. The local EXU decoder ("nanostore") handles local control, for example the selection of operands and program branching.

Wu and Liu [WLR98] describe a reconfigurable processing unit that can be used as a building block for a variety of video signal processing functions including FIR, IIR, and adaptive filters, and discrete transforms such as DCT. An array of processing units along with an interconnection network is used to implement any one of these functions, yielding throughput comparable to custom ASIC designs but with much higher flexibility and potential for adaptive operation.

Finally, the Stretch architecture [Str08] augments a base processor with reconfigurable fabric for instruction set extension. Software "hot spots" in the application program are converted into single cycle instructions. The instruction set extension fabric consists of 4096 ALUs, each capable of 2x4 multiplies, that may be combined for wider word widths. There are 64 dedicated 8x16 multiplies. The reconfigurable array can be configured at run-time, employing independent configurations optimized for different parts of the code. The application program, specified in C/C++, is analyzed by Stretch tools to indicate potential instruction extensions. Configuration of the reconfigurable fabric and final code generation is done automatically.

## 2.8     Architectures that Exploit Predictable IPC

As we will discuss in Chapter 5, compile time scheduling is very effective for a large class of applications in signal processing and scientific computing. Given such a schedule, we can obtain information about the pattern of interprocessor communication that occurs at run-time. This compile time information can be exploited by the hardware architecture to achieve efficient communication between processors. We exploit this fact in the *ordered transaction* strategy discussed in Chapter 3. In this section we discuss related work in this area of employing compile time information about interprocessor communication coupled with enhancements to the hardware architecture with the objective of reducing IPC and synchronization overhead.

Determining the pattern of processor communications is relatively straightforward in SIMD implementations. Techniques applied to systolic arrays in fact use the regular communication pattern to determine an optimal interconnect topology for a given algorithm. An interesting architecture in this context is the GF11 machine built at IBM [BDW85]. The GF11 is an SIMD machine in which processors are interconnected using a Benes network (Figure 2.8), which allows the GF11 to support a variety of different interprocessor communication topologies rather than a fixed topology.

Benes networks are non-blocking, i.e., they can provide one-to-one connections from all the network inputs to the network outputs simultaneously according to any specified permutation. These networks achieve the functional capability of a full crossbar switch with much simpler hardware. The drawback, however, is that in a Benes network, computing switch settings needed to achieve a particular permutation involves a somewhat complex algorithm [Lei92]. In the GF11, this problem is solved by precomputing the switch settings based on the program to be executed on the array. A central controller is responsible for reconfiguring the Benes network at run-time based on these predetermined switch settings. Interprocessor communication in the GF11 is synchronous with respect to computations in the processors, similar to systolic arrays. The GF11 has been used for scientific computing, for example calculations in quantum physics, finite element analysis, LU decomposition, and other applications.

An example of a mesh connected parallel processor that uses compile time information at the hardware level is the NuMesh system at MIT [SHL+97]. In this system, it is assumed that the communication pattern — source and destination of each message, and the communication bandwidth required — can be extracted from the parallel program specification. Some amount of dynamic execution is also supported by the architecture. Each processing node in the mesh gets a communication schedule which it follows at run-time. If the compile time estimates of bandwidth requirements are accurate, the architecture realizes efficient, hot-spot free, low-overhead communication. Incorrect bandwidth estimates

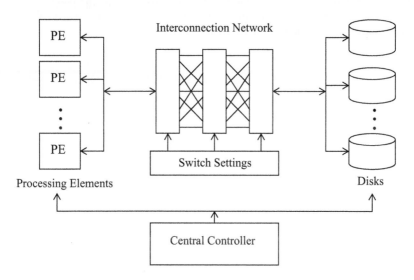

Figure 2.8. The IBM GF11 architecture: an example of statically scheduled communication.

or dynamic execution are not catastrophic, but these do cause lower performance.

The RAW machine [W+97] is another example of a parallel processor where switches are configured statically. The processing elements are tiled in a mesh topology; each element consists of a RISC-like processor, with configurable logic that implements special instructions and configurable data widths. Programmable switches enforce a compile-time determined static communication pattern, allowing dynamic switching when necessary. Implementing the static communication pattern reduces synchronization overhead and network congestion. A compiler is responsible for partitioning the program into threads mapped onto each RAW processor, configuring the reconfigurable logic on each processor, and routing communications statically.

## 2.9    Summary

In this chapter we discussed various types of application-specific multiprocessors employed for signal processing. Although these machines employ standard parallel processing techniques well known in general purpose computing, the predictable nature of the computations allows for simplified system architectures. It is often possible to configure processor interconnects statically to make use of compile time knowledge of inter-processor communication patterns. This allows for low overhead interprocessor communication and synchronization mechanisms that employ a combination of simple hardware support for IPC, and software techniques applied to programs running on the processors. We will explore these ideas further in the following chapters.

# 3

---

# BACKGROUND TERMINOLOGY AND
# NOTATION

---

In this chapter we introduce terminology and definitions used in the remainder of the book, and formalize key dataflow concepts that were introduced intuitively in Chapter 1. We emphasize in this chapter a particular form of dataflow called **homogeneous synchronous dataflow (HSDF)**, which is a specialized form of dataflow that we focus on in later chapters of the book. We also briefly introduce the concept of algorithmic complexity, and discuss various shortest and longest path algorithms in weighted directed graphs along with their associated complexity. These algorithms are used extensively in subsequent chapters.

To start with, we define the difference of two arbitrary sets $S_1$ and $S_2$ by $S_1 - S_2 = \{s \in S_1 | s \notin S_2\}$, and we denote the number of elements in a finite set $S$ by $|S|$. Also, if $r$ is a real number, then we denote the smallest integer that is greater than or equal to $r$ by $\lceil r \rceil$.

## 3.1    Graph Data Structures

A **directed graph** is an ordered pair $(V, E)$, where $V$ is the set of **vertices** and $E$ is the set of **edges**. Each edge is an ordered pair $(v_1, v_2)$ where $v_1, v_2 \in V$. If $e = (v_1, v_2) \in E$, we say that $e$ is directed from $v_1$ to $v_2$; $v_1$ is the **source vertex** of $e$, and $v_2$ is the **sink vertex** of $e$. We also refer to the source and sink vertices of a graph edge $e \in E$ by $src(e)$ and $snk(e)$. In a directed graph we cannot have two or more edges that have identical source *and* sink vertices. A generalization of a directed graph is a **directed multigraph**, in which two or more edges have the same source and sink vertices.

Figure 3.1(a) shows an example of a directed graph, and Figure 3.1(b) shows an example of a directed multigraph. The vertices are represented by circles and the edges are represented by arrows between the circles. Thus, the vertex set of the directed graph of Figure 3.1(a) is $\{A, B, C, D\}$, and the edge set is $\{(A, B), (A, D), (A, C), (D, B), (C, C)\}$.

## 3.2    Dataflow Graphs

A **dataflow graph** is a directed multigraph, where the vertices (actors) represent computation and edges (arcs) represent FIFO (first-in-first-out) queues that direct data values from the output of one computation to the input of another. Edges thus represent data precedences between computations. Recall that actors consume data (or tokens) from their inputs, perform computations on them (fire), and produce certain numbers of tokens on their outputs.

Programs written in high-level functional languages such as pure LISP, and in dataflow languages such as Id and Lucid can be directly converted into dataflow graph representations; such a conversion is possible because these languages are designed to be *free of side-effects*, i.e., programs in these languages are not allowed to contain global variables or data structures, and functions in these languages cannot modify their arguments [Ack82]. Also, since it is possible to simulate any Turing machine in one of these languages, questions such as deadlock (or equivalently, terminating behavior) and determining maximum buffer sizes required to implement edges in the dataflow graph become undecidable. Several models based on dataflow with restricted semantics have been proposed; these models give up the descriptive power of general dataflow in exchange for properties that facilitate formal reasoning about programs specified in these models, and are useful in practice, leading to simpler implementation of the specified computation in hardware or software.

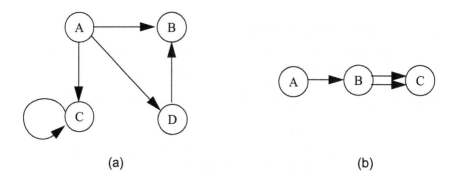

(a)                                                   (b)

Figure 3.1. (a) A directed graph. (b) A directed multigraph.

## 3.3     Computation Graphs

One such restricted model (and in fact one of the earliest graph-based computation models) is the **computation graph** of Karp and Miller [KM66], where the authors establish that their computation graph model is *determinate*, i.e., the sequence of tokens produced on the edges of a given computation graph are unique, and do not depend on the order that the actors in the graph fire, as long as all data dependencies are respected by the firing order. The authors also provide an algorithm that, based on topological and algebraic properties of the graph, determines whether the computation specified by a given computation graph will eventually terminate. Because of the latter property, computation graphs clearly cannot simulate all Turing machines, and hence are not as expressive as a general dataflow language like Lucid or pure LISP. Computation graphs provide some of the theoretical foundations for the SDF model to be discussed in detail in Section 3.5.

## 3.4     Petri Nets

Another model of computation relevant to dataflow is the **Petri net model** [Pet81][Mur89]. A Petri net consists of a set of *transitions*, which are analogous to actors in dataflow, and a set of *places* that are analogous to arcs. Each transition has a certain number of input places and output places connected to it. Places may contain one or more *tokens*. A Petri net has the following semantics: a transition *fires* when all its input places have one or more tokens and, upon firing, it produces a certain number of tokens on each of its output places.

A large number of different kinds of Petri net models have been proposed in the literature for modeling different types of systems. Some of these Petri net models have the same expressive power as Turing machines: for example, if transitions are allowed to possess "inhibit" inputs (if a place corresponding to such an input to a transition contains a token, then that transition is not allowed to fire) then a Petri net can simulate any Turing machine (pp. 201 in [Pet81]). Others (depending on topological restrictions imposed on how places and transitions can be interconnected) are equivalent to finite state machines, and yet others are similar to SDF graphs. Some extended Petri net models allow a notion of time, to model execution times of computations. There is also a body of work on stochastic extensions of timed Petri nets that are useful for modeling uncertainties in computation times. We will touch upon some of these Petri net models again in Chapter 4. Finally, there are Petri nets that distinguish between different classes of tokens in the specification (*colored* Petri nets), so that tokens can have information associated with them. We refer to [Pet81] [Mur89] for details on the extensive variety of Petri nets that have been proposed over the years.

## 3.5     Synchronous Dataflow

The particular restricted dataflow model we are mainly concerned with in this book is the SDF — Synchronous Data Flow — model proposed by Lee and Messerschmitt [LM87]. The SDF model poses restrictions on the firing of actors: the number of tokens produced (consumed) by an actor on each output (input) edge is a fixed number that is known at compile time. The number of tokens produced and consumed by each SDF actor on each of its edges is annotated in illustrations of an SDF graph by numbers at the arc source and sink respectively. In an actual implementation, arcs represent buffers in physical memory. The arcs in an SDF graph may contain **initial tokens**, which we also refer to as **delays**. Arcs with delays can be interpreted as data dependencies across iterations of the graph; this concept will be formalized in the following chapter when we discuss scheduling models. We will represent delays using bullets (•) on the edges of the SDF graph; we indicate more than one delay on an edge by a number alongside the bullet. An example of an SDF graph is illustrated in Figure 3.2.

DSP applications typically represent computations on an indefinitely long data sequence; therefore the SDF graphs we are interested in for the purpose of signal processing must execute in a nonterminating fashion. Consequently, we must be able to obtain periodic schedules for SDF representations, which can then be run as infinite loops using a finite amount of physical memory. Unbounded buffers imply a sample rate inconsistency, and deadlock implies that all actors in the graph cannot be iterated indefinitely. Thus for our purposes, correctly constructed SDF graphs are those that can be scheduled periodically using a finite amount of memory. The main advantage of imposing restrictions on the SDF model (over a general dataflow model) lies precisely in the ability to determine whether or not an arbitrary SDF graph has a periodic schedule that neither deadlocks nor requires unbounded buffer sizes [LM87]. The buffer sizes required

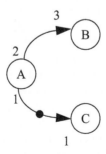

Figure 3.2. An SDF graph.

to implement arcs in SDF graphs can be determined at compile time (recall that this is not possible for a general dataflow model); consequently, buffers can be allocated statically, and run-time overhead associated with dynamic memory allocation is avoided. The existence of a periodic schedule that can be inferred at compile time implies that a correctly constructed SDF graph entails no run-time scheduling overhead.

## 3.6 Analytical Properties of SDF Graphs

This section briefly describes some useful properties of SDF graphs; for a more detailed and rigorous treatment, please refer to the work of Lee and Messerschmitt [LM87][Lee86]. An SDF graph is compactly represented by its **topology matrix**. The topology matrix, referred to henceforth as $\Gamma$, represents the SDF graph structure; this matrix contains one column for each vertex, and one row for each edge in the SDF graph. The $(i, j)$ th entry in the matrix corresponds to the number of tokens produced by the actor numbered $j$ onto the edge numbered $i$. If the $j$ th actor *consumes* tokens from the $i$ th edge, i.e., the $i$ th edge is incident into the $j$ th actor, then the $(i, j)$ th entry is negative. Also, if the $j$ th actor neither produces nor consumes any tokens from the $i$ th edge, then the $(i, j)$ th entry is set to zero.

For example, the topology matrix $\Gamma$ for the SDF graph in Figure 3.2 is:

$$\Gamma = \begin{bmatrix} 2 & -3 & 0 \\ 1 & 0 & -1 \end{bmatrix}, \tag{3-1}$$

where the actors $A$, $B$, and $C$ are numbered 1, 2, and 3 respectively; the edges $(A, B)$ and $(A, C)$, are numbered 1 and 2 respectively. A useful property of $\Gamma$ is stated by the following Theorem.

**Theorem 3.1:** A connected SDF graph with $s$ vertices that has consistent sample rates is guaranteed to have $rank(\Gamma) = s - 1$, which ensures that $\Gamma$ has a null space.

*Proof:* See [LM87].

This can easily be verified for (3-1). This fact is utilized to determine the **repetitions vector** $q$.

**Definition 3.1:** The repetitions vector for an SDF graph with $s$ actors numbered 1 to $s$ is a column vector of length $s$, with the property that if each actor $i$ is invoked a number of times equal to the $i$ th entry of $q$, then the number of tokens on each edge of the SDF graph remains unchanged. Furthermore, $q$ is the smallest integer vector for which this property holds.

Clearly, the repetitions vector is very useful for generating infinite schedules for SDF graphs by indefinitely repeating a finite length schedule, while maintaining small buffer sizes between actors. Also, $q$ will only exist if the SDF graph has consistent sample rates. The conditions for the existence of $q$ is determined by Theorem 3.1 coupled with the following Theorem.

**Theorem 3.2:** The repetitions vector for an SDF graph with consistent sample rates is the smallest integer vector in the nullspace of its topology matrix. That is, $q$ is the smallest integer vector such that $\Gamma q = 0$.

*Proof:* See [LM87].

Note that $q$ may be easily obtained by solving a set of linear equations; these are also called **balance equations**, since they represent the constraint that the number of samples produced and consumed on each edge of the SDF graph be the same after each actor fires a number of times equal to its corresponding entry in the repetitions vector.

For the example of Figure 3.2, from (3-1),

$$
q = \begin{bmatrix} 3 \\ 2 \\ 3 \end{bmatrix} . \tag{3-2}
$$

Clearly, if actors $A$, $B$, and $C$ are invoked $3$, $2$, and $3$ times respectively, the number of tokens on the edges remain unaltered (no token on $(A, B)$, and one token on $(A, C)$). Thus, the repetitions vector in (3-2) brings the SDF graph back to its "initial state."

## 3.7    Converting a General SDF Graph into a Homogeneous SDF Graph

**Definition 3.2:** An SDF graph in which every actor consumes and produces only one token from each of its inputs and outputs is called a **homogeneous SDF graph (HSDFG)**.

An HSDFG actor fires when it has one or more tokens on all its input edges; it consumes one token from each input edge when it fires, and produces one token on all its output edges when it completes execution. We note that an HSDFG is very similar to a **marked graph** in the context of Petri nets [Pet81]; transitions in the marked graph correspond to actors in the HSDFG, places correspond to edges, and initial tokens (or initial marking) of the marked graph correspond to initial tokens (or delays) in HSDFGs.

The repetitions vector defined in the previous section can be used to convert any general SDF graph to an equivalent HSDFG. We outline this transformation next.

Let us denote as $q(A)$ the number of invocations of an actor $A$ in the repetitions vector $q$ of the SDF graph $G$ that is to be transformed. Let us represent by $G_H$ the homogeneous graph obtained as a result of this transformation. For each actor $A$ in $G$, $G_H$ contains $q(A)$ copies (or invocations) of $A$; let us call these invocations $A_i$, $1 \leq i \leq q(A)$. For each edge $(A, B)$ in $G$, let $n_A$ represent the number of tokens produced onto $(A, B)$ when $A$ fires, and let $n_B$ represent the number of tokens consumed by $B$ from the edge $(A, B)$ when it fires. In the homogeneous graph, each $A_i$ produces and consumes only one token from each of its edges. Thus, for each edge $(A, B)$ of which $A$ is a source, the corresponding $A_i$ in $G_H$ must now be the source vertex for $n_A$ edges. Each of these $n_A$ edges corresponds to the $n_A$ samples produced by the actor $A$ onto the edge $(A, B)$ in the original SDF graph $G$. Similarly, each $B_i$ will have $n_B$ edges pointing to it, corresponding to the $n_B$ samples actor $B$ consumes in the original SDF graph.

For notational convenience, let us assume the samples produced by $A_i$ appear from an "output port" on $A_i$. For each edge $(A, B)$, there will be $n_A$ such output ports on each $A_i$; similarly, there will be $n_B$ "input ports" on each $B_i$. Let us call these output and input ports $P_{A_i}^k$, $1 \leq k \leq n_A$, and $P_{B_j}^l$, $1 \leq l \leq n_A$ respectively. The $k$th sample produced by the $i$th invocation of $A$ in $G_H$ is thus produced at the output port $P_{A_i}^k$. Note that even though actor $A$ produces all $n_A$ tokens onto edge $(A, B)$ simultaneously according to the SDF model, the $n_A$ generated tokens have a fixed relative order in which they are generated on the edge $(A, B)$. This is because each edge $(A, B)$ is essentially a first-in-first-out buffer. The superscript $k$ in $P_{A_i}^k$ reflects this ordering. Edges in $G_H$ between input and output ports will be denoted by the usual notation $(P_{A_i}^k, P_{B_j}^l)$, and delays on these edges will be denoted by $delay(P_{A_i}^k, P_{B_j}^l)$. The concept of input and output ports is illustrated in Figure 3.3 for a given edge $(A, B)$. In this example, $n_A = 1$, and $n_B = 3$.

The idea behind the SDF to HSDF transformation is to connect the input and output ports such that the samples produced and consumed by every invocation of each actor in the HSDFG remains identical to that in the original SDF graph. An example of such a transformation is shown in Figure 3.4. The input and output ports are not shown to avoid clutter; the port numbering, however, has to be taken into account when implementing the functionality of each actor in the HSDFG. An algorithm to perform the transformation is shown in Figure 3.6, where $(x \bmod y)$ equals $x$ taken modulo $y$, and $\lfloor x/y \rfloor$ equals the integer quotient obtained when $x$ is divided by $y$.

Based on the ideas discussed above, Figure 3.6 outlines an algorithm to convert an SDF graph into a homogeneous SDF graph.

## 3.8      Acyclic Precedence Expansion Graph

As discussed in the previous section, a general, consistent SDF graph that is not an HSDFG can always be converted into an equivalent HSDFG [Lee86]. The resulting HSDFG has a larger number of actors than the original SDF graph. It in fact has a number of actors equal to the sum of the entries in the repetitions vector. In the worst case, the SDF to HSDFG transformation may result in an exponential increase in the number of actors — e.g., see [PBL95] for an example of a family of SDF graphs in which this kind of exponential increase occurs. Such a technique, however, appears to be necessary when constructing periodic multiprocessor schedules from multirate SDF graphs in a way that takes into account all of the available parallelism among different actor invocations (distinct firings within a periodic schedule). However, for SDF scheduling techniques that provide better scalability in terms of the time and memory required to compute schedules, it is useful to explore methods that bypass construction of the HSDFG, and derive schedules directly from the SDF graph or through alternative

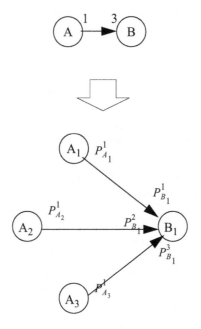

Figure 3.3. Expansion of an edge in an SDF graph $G$ into multiple edges in the equivalent HSDFG $G_H$. Note the input and output ports on the vertices of $G_H$.

intermediate representations. For examples of efforts in this direction, we refer the reader to [PBL95, SBGC07]. In addition to avoiding HSDFG construction, the technique of [SBGC07] is useful in that it can handle multiple SDF graphs, and take into account differing performance requirements across the set of input graphs.

An SDF graph converted into an HSDFG for the purposes of multiprocessor scheduling can be further converted into an **Acyclic Precedence Expansion Graph (APEG)** by removing from the HSDFG arcs that contain initial tokens (delays). Recall that arcs with initial tokens on them represent dependencies between successive iterations of the dataflow graph. An APEG is therefore useful for constructing multiprocessor schedules that, for algorithmic simplicity, do not attempt to overlap multiple iterations of the dataflow graph by exploiting precedence constraints across iterations. Figure 3.5 shows an example of an APEG. Note that the precedence constraints present in the original HSDFG of Figure 3.4 are maintained by this APEG, as long as each iteration of the graph is completed before the next iteration begins.

Since we are concerned with multiprocessor schedules, we assume that we start with an application represented as a homogeneous SDF graph henceforth, unless we state otherwise. This of course results in no loss of generality because a general SDF graph is converted into a homogeneous graph for the purposes of multiprocessor scheduling anyway. In Chapter 9 we discuss how the ideas that apply to HSDF graphs can be extended to graphs containing actors that display data-dependent behavior (i.e., *dynamic* actors).

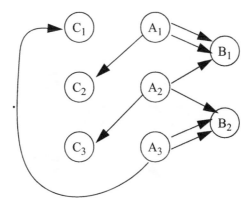

Figure 3.4. HSDFG obtained by expanding the SDF graph in Figure 3.2.

## 3.9     Application Graph

**Definition 3.3:**  An HSDFG representation of an algorithm (for example, a speech coding system, a filter bank, or a Fast Fourier Transform) is called an **application graph**.

For example, Figure 3.7(a) shows an SDF representation of a two-channel multirate filter bank that consists of a pair of analysis filters followed by a pair of synthesis filters. This graph can be transformed into an equivalent HSDFG, which represents the application graph for the two-channel filter bank, as shown in Figure 3.7(b). Algorithms that map applications specified as SDF graphs on to single and multiple processors take the equivalent application graph as input. Such algorithms will be discussed in Chapters 5 and 6. Chapter 8 will discuss how the performance of a multiprocessor system — after scheduling is completed — can be effectively modeled by another HSDFG called the **interprocessor communication graph**, or IPC graph. The IPC graph is derived from the original application graph, and the given parallel schedule. Furthermore, Chapters 10 to 12 will discuss how a third HSDFG, called the **synchronization graph**, can be used to analyze and optimize the synchronization structure of a multiprocessor system. The full interaction of the application graph, IPC graph, and synchronization graphs, and also the formal definitions of these graphs will then be further elaborated in Chapters 8 through 12.

## 3.10     Synchronous Languages

SDF should not be confused with **synchronous languages** [Hal93][BB91] (e.g., LUSTRE, SIGNAL, and ESTEREL), which have very different semantics

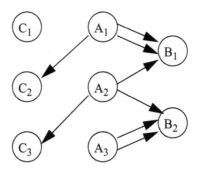

Figure 3.5. APEG obtained from the HSDFG in Figure 3.4.

from SDF. Synchronous languages have been proposed for formally specifying and modeling reactive systems, i.e., systems that constantly react to stimuli from a given physical environment. Signal processing systems fall into the reactive category, and so do control and monitoring systems, communication protocols, man-machine interfaces, etc. In synchronous languages, variables are possibly infinite sequences of data of a certain type. Associated with each such sequence

**Function** ConvertToHSDFG
**Input:** SDF graph $G$
**Output:** Equivalent HSDFG $G_H = (E, V)$

1. Compute the repetitions vector $q$; for each actor $A$ in $G$, let $q(A)$ be the corresponding entry in $q$
2. **For** each actor $A$ in $G$
    Add $q(A)$ vertices $A_1, A_2, ..., A_{q(A)}$ to vertex set $V$ of $G_H$
**Endfor**
3. **For** each edge $(A, B)$ in $G$
    Let $A$ produce $n_A$ tokens on $(A, B)$, and $B$ consume $n_B$ tokens
    Denote the number of delays on $(A, B)$ by $d$
    Add $n_A$ output ports on each $A_i$, $1 \le i \le q(A)$
    Add $n_B$ input ports on each $B_j$, $1 \le j \le n_B$
    **For** each $i$ such that $1 \le i \le q(A)$
        **For** each $k$ such that $1 \le k \le n_A$
            Let $l = [(d + (i-1)n_A + k - 1) \bmod n_B q(B)] \bmod + 1$
            Let $j = \left\lfloor \dfrac{(d + (i-1)n_A + k - 1) \bmod n_B q(B)}{n_B} \right\rfloor + 1$
            Add edge $(P_{A_i}^k, P_{B_j}^l)$ to edge set $E$ of $G_H$
            Set $delay(P_{A_i}^k, P_{B_j}^l) = \left\lfloor \dfrac{d + (i-1)n_A + k - 1}{n_B q(B)} \right\rfloor$
        **Endfor**
    **Endfor**
**Endfor**

Figure 3.6. Algorithm to convert an SDF graph into a homogeneous SDF graph.

is a conceptual (and sometimes explicit) notion of a *clock signal*. In LUSTRE, each variable is explicitly associated with a clock, which determines the instants at which the value of that variable is defined. SIGNAL and ESTEREL do not have an explicit notion of a clock. The clock signal in LUSTRE is a sequence of Boolean values, and a variable in a LUSTRE program assumes its $n$th value when its corresponding clock takes its $n$th TRUE value. Thus, we may relate one variable with another by means of their clocks. In ESTEREL, on the other hand, clock ticks are implicitly defined in terms of instants when the reactive system corresponding to an ESTEREL program receives (and reacts to) external events. All computations in synchronous language are defined with respect to these clocks.

In contrast, the term "synchronous" in the SDF context refers to the fact that SDF actors produce and consume fixed number of tokens, and these numbers are known at compile time. This allows us to obtain periodic schedules for SDF graphs such that the average rates of firing of actors are fixed relative to one

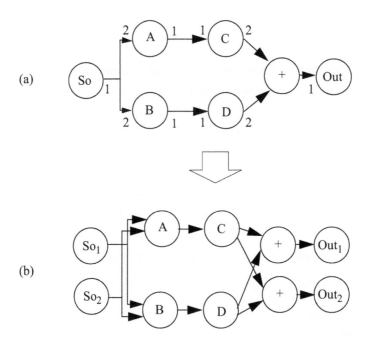

Figure 3.7. (a) SDF graph representing a two-channel filter bank. (b) Application graph.

another. We will not be concerned with synchronous languages, although these languages have a close and interesting relationship with dataflow models used for specification of signal processing algorithms [LP95].

## 3.11    HSDFG Concepts and Notations

A **homogeneous SDF graph (HSDFG)** is a directed multigraph $(V, E)$, and we denote the delay (or the number of initial tokens) on $e$ by $delay(e)$. We say that $e$ is an **output edge** of $src(e)$, and that $e$ is an **input edge** of $snk(e)$. We will also use the notation $(v_i, v_j)$, $v_i, v_j \in V$, for an edge directed from $v_i$ to $v_j$. The delay on the edge is denoted by $delay((v_i, v_j))$ or simply $delay(v_i, v_j)$.

A **path** in $(V, E)$ is a finite, nonempty sequence $(e_1, e_2, ..., e_n)$, where each $e_i$ is a member of $E$, and $snk(e_1) = src(e_2)$, $snk(e_2) = src(e_3)$, ..., $snk(e_{n-1}) = src(e_n)$. We say that the path $p = (e_1, e_2, ..., e_n)$ **contains** each $e_i$ and each subsequence of $(e_1, e_2, ..., e_n)$; $p$ is **directed from** $src(e_1)$ **to** $snk(e_n)$; and each member of $\{ src(e_1), src(e_2), ..., src(e_n), snk(e_n) \}$ is **on** $p$. We say that the path $p$ **originates** at vertex $src(e_1)$ and **terminates** at vertex $snk(e_n)$. A **dead-end path** is a path that terminates at a vertex that has no successors. That is, $p = (e_1, e_2, ..., e_n)$ is a dead-end path such that for all $e \in E$, $src(e) \neq snk(e_n)$. A path that is directed from a vertex to itself is called a **cycle**, and a **fundamental cycle** is a cycle of which no proper subsequence is a cycle.

If $(p_1, p_2, ..., p_k)$ is a finite sequence of paths such that $p_i = (e_{i,1}, e_{i,2}, ..., e_{i,n_i})$, for $1 \leq i \leq k$, and $snk(e_{i,n_i}) = src(e_{i+1,1})$, for $1 \leq i \leq (k-1)$, then we define the **concatenation** of $(p_1, p_2, ..., p_k)$, denoted $\langle (p_1, p_2, ..., p_k) \rangle$, by

$$\langle (p_1, p_2, ..., p_k) \rangle \equiv (e_{1,1}, ..., e_{1,n_1}, e_{2,1}, ..., e_{2,n_2}, ..., e_{k,1}, ..., e_{k,n_k}).$$

Clearly, $\langle (p_1, p_2, ..., p_k) \rangle$ is a path from $src(e_{1,1})$ to $snk(e_{k,n_k})$.

If $p = (e_1, e_2, ..., e_n)$ is a path in an HSDFG, then we define the **path delay** of $p$, denoted $Delay(p)$, by

$$Delay(p) = \sum_{i=1}^{n} delay(e_i). \tag{3-3}$$

Since the delays on all HSDFG edges are restricted to be nonnegative, it is easily seen that between any two vertices $x, y \in V$, either there is no path directed from $x$ to $y$, or there exists a (not necessarily unique) **minimum-delay path** between $x$ and $y$. Given an HSDFG $G$, and vertices $x, y$ in $G$, we define $\rho_G(x, y)$ to be equal to the path delay of a minimum-delay path from $x$ to $y$ if there exist one or more paths from $x$ to $y$, and equal to $\infty$ if there is no path from $x$ to $y$. If $G$ is understood, then we may drop the subscript and simply write "$\rho$" in place of "$\rho_G$." It is easily seen that minimum delay path lengths satisfy the following *triangle inequality*

$$\rho_G(x, z) + \rho_G(z, y) \geq \rho_G(x, y), \text{ for any } x, y, z \text{ in } G. \tag{3-4}$$

By a **subgraph** of $(V, E)$, we mean the directed graph formed by any $V' \subseteq V$ together with the set of edges $\{e \in E | src(e), snk(e) \in V'\}$. We denote the subgraph associated with the vertex-subset $V'$ by $subgraph(V')$.

We say that $(V, E)$ is **strongly connected** if for each pair of distinct vertices $x, y$, there is a path directed from $x$ to $y$ and there is a path directed from $y$ to $x$. We say that a subset $V' \subseteq V$ is strongly connected if $subgraph(V')$ is strongly connected. A **strongly connected component (SCC)** of $(V, E)$ is a strongly connected subset $V' \subseteq V$ such that no strongly connected subset of $V$ properly contains $V'$. If $V'$ is an SCC, then when there is no ambiguity, we may also say that $subgraph(V')$ is an SCC. If $C_1$ and $C_2$ are distinct SCCs in $(V, E)$, we say that $C_1$ is a **predecessor SCC** of $C_2$ if there is an edge directed from some vertex in $C_1$ to some vertex in $C_2$; $C_1$ is a **successor SCC** of $C_2$ if $C_2$ is a predecessor SCC of $C_1$. An SCC is a **source SCC** if it has no predecessor SCC; and an SCC is a **sink SCC** if it has no successor SCC. An edge $e$ is a **feedforward** edge of $(V, E)$ if it is not contained in an SCC, or equivalently, if it is not contained in a cycle; an edge that is contained in at least one cycle is called a **feedback** edge.

A sequence of vertices $(v_1, v_2 ..., v_k)$ is a *chain* that joins $v_1$ and $v_k$ if $v_{i+1}$ is adjacent to $v_i$ for $i = 1, 2, ..., (k-1)$. We say that a directed multigraph is a **connected graph** if for any pair of distinct members $A$, $B$ of $Z$, there is a chain that joins $A$ and $B$. Given a directed multigraph $G = (V, E)$, there is a unique partition (unique up to a reordering of the members of the partition) $V_1, V_2, ..., V_n$ such that for $1 \leq i \leq n$, $subgraph(V_i)$ is connected; and for each $e \in E$, $src(e), snk(e) \in V_j$ for some $j$. Thus, each $V_i$ can be viewed as a maximal connected subset of $V$, and we refer to each $V_i$ as a **connected component** of $G$.

A **topological sort** of an acyclic directed multigraph $(V, E)$ is an ordering $v_1, v_2, ..., v_{|V|}$ of the members of $V$ such that for each $e \in E$,

$$((src(e) = v_i) \text{ and } (snk(e) = v_j) \Rightarrow (i < j));$$

that is, the source vertex of each edge occurs earlier in the ordering than the sink vertex. An acyclic directed multigraph is said to be **well-ordered** if it has only one topological sort, and we say that an $n$-vertex well-ordered directed multigraph is **chain-structured** if it has $(n-1)$ edges.

For elaboration on any of the graph-theoretic concepts presented in this section, we refer the reader to Cormen, Leiserson, and Rivest [CLR92].

## 3.12    Complexity of Algorithms

To discuss the run-time complexity of algorithms described in this book, we use the standard $O$-notation that describes an asymptotic upper bound on complexity. Formally, a function $f(n)$ is said to be $O(g(n))$ if there exist constants $c$ and $N$ such that $cg(n) \geq f(n)$, $\forall n \geq N$. We say an algorithm is **polynomial time** if it has running time $O(g(n))$ where $g(n)$ is a polynomial, and $n$ is the size of the input. Thus, for an efficient algorithm $g(n)$ will be a polynomial of low degree.

While the goal of designing an efficient algorithm is to achieve low running time, there are some classes of well-defined problems for which no polynomial time algorithms are known. One of the more important subclasses of problems in this class is the set of **NP-complete problems**. A wide variety of optimization problems, including some of the problems formulated in this book, fall into this set. We will briefly discuss the concept of NP-complete problems informally. For a formal discussion of complexity classes such as NP, the reader is referred to the book by Garey and Johnson [GJ79], which provides an excellent treatment of the subject of NP-completeness, and includes a large set of documented NP-complete problems.

All of the problems within the set of NP-complete problems are **polynomially equivalent** in terms of complexity — an algorithm for one NP-complete problem can be converted into an algorithm for any other NP-complete problem, and this conversion itself can be done in polynomial time. This implies that an efficient algorithm for any one NP-complete problem would result in an efficient solution for all other problems in the set. However, polynomial time algorithms for NP-complete problems are not known, and it is widely believed that no efficient algorithms exist for these problems. Lack of a polynomial time algorithm implies that as the input size of the problem increases, the time required to compute a solution becomes impractically large. Given an optimization problem, to show that the problem is NP-complete, or is at least as complex as any documented NP-complete problem, is very useful, for it shows that the existence of an efficient algorithm for that problem is highly unlikely. In such a case we may resort to efficient suboptimal heuristics that yield good solutions for practical instances of the problem.

In order to show that a given problem is at least as complex as one of these problems, we use the concept of **polynomial transformability**. To prove that a given problem "A" is at least as complex as an NP-complete problem, it suffices to show that an algorithm to solve any problem in the set of NP-complete problems (say, "B") can be transformed into an algorithm to solve A, and that this transformation itself can be done in polynomial time. Such a conversion procedure is called *polynomial transformation*. The existence of a polynomial transformation from "B" to "A" implies that a polynomial time algorithm to solve "A" can be used to solve "B" in polynomial time, and if "B" is NP-complete then the transformation implies that "A" is at least as complex as any NP-complete problem. Such a problem is called **NP-hard** [GJ79].

We illustrate this concept with a simple example. Consider the **set covering problem,** where we are given a collection of subsets $C$ of a finite set $S$, and a positive integer $I \leq |C'|$. The problem is to find out if there is a subset $C' \subseteq C$ such that $|C'| \leq I$ and each element of $S$ belongs to at least one set in $C'$.

By finding a polynomial transformation from a known NP-complete problem to the set-covering problem we can prove that the set cover problem is NP-hard. For this purpose, we choose the **vertex cover** problem, where we are given a graph $G = (V, E)$ and a positive integer $I \leq |V|$, and the problem is to determine if there exists a subset of vertices $V' \subseteq V$ such that $|V'| \leq I$ and for each edge $e \in E$ either $src(e) \in V'$ or $snk(e) \in V'$. The subset $V'$ is said to be a *cover* of the set of vertices $V$. The vertex cover problem is known to be NP-complete, and by transforming it to the set covering problem in polynomial time, we can show that the set covering problem is NP-hard.

Given an instance of vertex cover, we can convert it into an instance of set-covering by first letting $S$ be the set of edges $E$. Then for each vertex $v \in V$, we construct the subset of edges $T_v = \{e \in E \mid v = src(e) \text{ or } v = snk(e)\}$. The set $\{T_v | v \in V\}$ forms the collection $C'$. Clearly, this transformation can be done in time at most linear in the number of edges of the input graph, and the resulting $C'$ has size equal to $|V|$. Our transformation ensures that $V'$ is a vertex cover for $G$ if and only if $\{T_v | v \in V\}$ is a set cover for the set of edges $E$. Now, we may use a solution of set cover to solve the transformed problem, since a vertex cover $|V'| \leq I$ exists if and only if a corresponding set cover $|C'| \leq I$ exists for $E$. Thus, the existence of a polynomial time algorithm for set cover implies the existence of a polynomial time algorithm for vertex cover. This proves that set cover is NP-hard.

It can easily be shown that the set cover problem is also NP-complete by showing that it belongs to the class NP. However, since a formal discussion of complexity classes is beyond the scope of this book, we will refer the interested reader to [GJ79] for a comprehensive discussion of complexity classes and the definition of the class NP.

In summary, by finding a polynomial transformation from a problem that is known to be NP-complete to a given problem, we can prove that the given problem is NP-hard. This implies that a polynomial time algorithm to solve the given problem in all likelihood does not exist, and if such an algorithm does exist, a major breakthrough in complexity theory would be required to find it. This provides a justification for solving such problems using suboptimal polynomial time heuristics. It should be pointed out that a polynomial transformation of an NP-complete problem to a given problem, if it exists, is often quite involved, and is not necessarily as straightforward as in the case of the set-covering example discussed here.

In Chapter 11, we use the concepts outlined in this section to show that a particular synchronization optimization problem is NP-hard by reducing the set-covering problem to the synchronization optimization problem. We then discuss efficient heuristics to solve that problem.

## 3.13    Shortest and Longest Paths in Graphs

There is a rich history of work on shortest path algorithms and there are many variants and special cases of these problems (depending, for example, on the topology of the graph, or on the values of the edge weights) for which efficient algorithms have been proposed. In what follows we focus on the most general, and from the point of view of this book, most useful shortest path algorithms.

Consider a weighted, directed graph $G = (V, E)$, with real valued edge weights $w(u, v)$ for each edge $(u, v) \in E$. The single-source shortest path problem finds a path with minimum weight (defined as the sum of the weights of the edges on the path) from a given vertex $v_o \in V$ to all other vertices $u \in V$, $u \neq v$, whenever at least one path from $v_o$ to $u$ exists. If no such path exists, then the shortest path weight is set to $\infty$.

The two best known algorithms for the single-source shortest path algorithm are Dijkstra's algorithm and the Bellman-Ford algorithm. Dijkstra's algorithm is applicable to graphs with nonnegative weights ($w(u, v) \geq 0$). The running time of this algorithm is $O(|V|^2)$. The Bellman-Ford algorithm solves the single-source shortest path problem for graphs that may have negative edge weights; the Bellman-Ford algorithm detects the existence of negative weight cycles reachable from $v_o$ and, if such cycles are detected, it reports that no solution to the shortest path problem exists. If a negative weight cycle is reachable from $v_o$, then clearly we can reduce the weight of any path by traversing this negative cycle one or more times. Thus, no finite solution to the shortest path problem exists in this case.

An interesting fact to note is that for graphs containing negative cycles, the problem of determining the weight of the shortest *simple* path between two vertices is NP-hard [GJ79]. A simple path is defined as one that does not visit the same vertex twice, i.e., a simple path does not include any cycles.

The all-pairs shortest path problem computes the shortest path between all pairs of vertices in a graph. Clearly, the single-source problem can be applied repeatedly to solve the all-pairs problem. However, a more efficient algorithm based on dynamic programming — the Floyd-Warshall algorithm — may be used to solve the all-pairs shortest path problem in $O(|V|^3)$ time. This algorithm solves the all-pairs problem in the absence of negative cycles.

The corresponding longest path problems may be solved using the shortest path algorithms. One straightforward way to do this is to simply negate all edge weights (i.e., use the edge weights $w'(u, v) = -w(u, v)$) and employ the Bellman-Ford algorithm for the single-source problem and Floyd-Warshall for the all-pairs problem. If all the edge weights are positive, determining the weight of the longest simple path becomes NP-hard for a general graph containing cycles reachable from the source vertex.

In the following sections, where we briefly describe the shortest path algorithms discussed thus far. We describe the algorithms in pseudo-code, and assume we only need the weight of the longest or shortest path; these algorithms also yield the actual path, but we do not need this information for the purposes of this book. Also, we will not delve into the correctness proofs for these algorithms, but will refer the reader to texts such as [CLR92] and [AHU87] for detailed discussion of these graph algorithms.

### 3.13.1    Dijkstra's Algorithm

The pseudo-code for the algorithm is shown in Figure 3.8. The *While* loop is run $|V|$ times, the total time spent in the *For* loop is $O(|E|)$, and a straightforward implementation of extracting the minimum element from an array takes time at most $O(|V|)$ for each iteration of the *While* loop. Dijkstra's algorithm can therefore be implemented in time $O(|V|^2 + |E|)$. If the input graph is sparse, a more clever implementation of the minimum extraction step leads to a modified implementation of the algorithm with running time of $O(|V|\log(|V|) + |E|)$ [CLR92].

### 3.13.2    The Bellman-Ford Algorithm

As discussed before, the Bellman-Ford algorithm solves the single-source shortest path problem even for the case where the edge weights are negative, provided there are no negative cycles in the graph that are reachable from the designated source. The algorithm detects such negative cycles when these are present. The algorithm pseudo-code is listed in Figure 3.9.

The nested *For* loop in Step 4 determines the complexity of the algorithm; this step clearly takes $O(|V||E|)$ time, which is the complexity of Bellman-Ford. This algorithm is based on the *dynamic-programming* technique.

### 3.13.3    The Floyd-Warshall Algorithm

Next, consider the all-pairs shortest path problem. One simple method of solving this is to apply the single-source problem to all vertices in the graph; this takes $O(|V|^2|E|)$ time using the Bellman-Ford algorithm. The Floyd-Warshall algorithm improves upon this. A pseudo-code specification of this algorithm is given in Figure 3.10.

The triply nested *For* loop in this algorithm clearly implies a complexity of $O(|V|^3)$. This algorithm is also based upon dynamic programming: At the $k$th iteration of the outermost *For* loop, the shortest path from the vertex numbered $i$ to vertex $j$ is determined among all paths that do not visit any vertex numbered $k$ or higher. Again, we leave it to texts such as [CLR92][AHU87] for a formal proof of correctness.

**Function** SingleSourceShortestPath
**Input**: Weighted directed graph $G = (V, E)$, with nonnegative edge
weights $w(e)$ for each $e \in E$, and a source vertex $s \in V$.
**Output**: $d(v)$, the weight of the shortest path from $s$ to each vertex
$v \in V$

1. Initialize $d(s) = 0$, and $d(v) = \infty$ for all other vertices
2. $V_S \leftarrow \varnothing$
3. $V_Q \leftarrow V$
4. **While** $V_Q \neq \varnothing$ do
    Extract $u \in V_Q$ such that $d(u) = min(d(v)|v \in V_Q)$
    $V_S \leftarrow u$
    **For** each $e \in E$ such that $u = src(e)$
        Set $t = snk(e)$
        $d(t) \leftarrow min(d(t), d(u) + w(e))$
    **Endfor**
**Endwhile**

Figure 3.8. Dijkstra's algorithm.

## 3.14     Solving Difference Constraints Using Shortest Paths

As discussed in subsequent chapters, a feasible schedule for an HSDFG is obtained as a solution of a system of **difference constraints**. Difference constraints are of the form

$$x_i - x_j \leq c_{ij}, \tag{3-5}$$

where $x_i$ are unknowns to be determined, and $c_{ij}$ are given; this problem is a

**Function** SingleSourceShortestPath

**Input:** Weighted directed graph $G = (V, E)$, with edgeweight $w(e)$ for each $e \in E$, and a source vertex $s \in V$.

**Output:** $d(v)$, the weight of the shortest path from $s$ to each vertex $v \in V$, or else a Boolean indicating the presence of negative cycles reachable from $s$

1. Initialize $d(s) = 0$, and $d(v) = \infty$ for all other vertices
2. $V_S \leftarrow \varnothing$
3. $V_Q \leftarrow V$
4. **For** $i \leftarrow 1$ to $|V| - 1$
     **For** each $(u, v) \in E$
         $d(v) \leftarrow min(d(v), d(u) + w(u, v))$
     **Endfor**
**Endfor**
5. Set NegativeCyclesExist = FALSE
6. **For** each $(u, v) \in E$
     **If** $d(v) > d(u) + w(u, v)$
         Set NegativeCyclesExist = TRUE
     **Endif**
**Endfor**
**Endfor**

Figure 3.9. The Bellman-Ford algorithm.

special case of linear programming. The data precedence constraints between actors in a dataflow graph often lead to a system of difference constraints, as we shall see later. Such a system of inequalities can be solved using shortest path algorithms, by transforming the difference constraints into a *constraint graph*. This graph consists of a number of vertices equal to the number of variables $x_i$, and for each difference constraint $x_i - x_j \leq c_{ij}$, the graph contains an edge $(v_j, v_i)$, with edge weight $w(v_j, v_i) = c_{ij}$. An additional vertex $v_0$ is also included, with zero weight edges directed from $v_0$ to all other vertices in the graph.

**Function** AllPairsShortestPath

**Input:** Weighted directed graph $G = (V, E)$, with edgeweight $w(e)$ for each $e \in E$.

**Output:** $d(u, v)$, the weight of the shortest path from $s$ to each vertex $v \in V$.

1. Let $|V| = n$; number the vertices $1, 2, ..., n$.

2. Let $A$ be an $n \times n$ matrix, set $A(i, j)$ as the weight of the edge from the vertex numbered $i$ to the vertex $j$. If no such edge exists, $A(i, j) = \infty$. Also, $A(i, i) = 0$.

3. **For** $k \leftarrow 1$ to $n$

    **For** $i \leftarrow 1$ to $n$

        **For** $j \leftarrow 1$ to $n$

            $A(i, j) \leftarrow min(A(i, j), A(i, k) + A(k, j))$

        **Endfor**

    **Endfor**

**Endfor**

4. For vertices $u, v \in V$ with enumeration $u \leftarrow i$ and $v \leftarrow j$, set
$$d(u, v) = A(i, j)$$

Figure 3.10. The Floyd-Warshall algorithm.

The solution to the system of difference constraints is then simply given by the weights of the shortest path from $v_0$ to all other vertices in the graph [CLR92]. That is, setting each $x_i$ to be the weight of the shortest path from $v_0$ to $x_i$ results in a feasible solution to the set of difference constraints. A feasible solution exists if, and only if, there are no negative cycles in the constraint graph. The difference constraints can therefore be solved using the Bellman-Ford algorithm. The reason for adding $v_0$ is to ensure that negative cycles in the graph, if present, are reachable from the source vertex. This in turn ensures that given $v_0$ as the source vertex, the Bellman-Ford algorithm will determine the existence of a feasible solution.

For example, consider the following set of inequalities in three variables:

$$x_1 - x_2 \leq 2$$
$$x_2 - x_3 \leq -3$$
$$x_3 - x_1 \leq 1$$

The constraint graph obtained from these inequalities is shown in Figure 3.11. A feasible solution is obtained by computing the shortest paths from $v_0$ to each $x_i$; thus $x_1 = -1$, $x_2 = -3$, and $x_3 = 0$, is a feasible solution. Clearly, given such a feasible solution $\{x_1, x_2, x_3\}$, if we add the same constant to each $x_i$, we obtain another feasible solution.

We make use of such a solution of difference constraints in Chapter 8.

## 3.15    Maximum Cycle Mean

The *maximum cycle mean* for an HSDF graph $G$ is defined as:

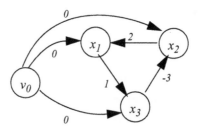

Figure 3.11. Constraint graph.

$$MCM(G) = \max_{\text{cycle } C \text{ in } G} \left\{ \frac{\sum\limits_{v \text{ is on } C} t(v)}{Delay(C)} \right\}. \tag{3-6}$$

As we shall see in subsequent chapters, the maximum cycle mean is related to the maximum achievable throughput for a given HSDFG.

Dasdan and Gupta [DG98] provide a comprehensive overview of all known algorithms for computing the maximum cycle mean. Out of these, the algorithm due to Hartmann and Orlin [HO93] appears to have the most efficient algorithm, with a running time of $O(|E|\Gamma)$, where $\Gamma$ is the sum of $t(v)$ over all actors in the HSDFG.

## 3.16   Summary

In this chapter, we covered background relating to dataflow models of computation and the SDF model. We discussed conversion of a general SDF graph into a homogeneous SDF (HSDFG), and generation of an Acyclic Precedence Expansion Graph (APEG). We briefly described asymptotic notation for computational complexity, and the concept of NP-complete problems. We then described some useful shortest path algorithms that are used extensively in the following chapters, and defined the maximum cycle mean. This background will be used extensively in the remainder of this book.

# 4

# DSP-ORIENTED DATAFLOW MODELS OF COMPUTATION

In Chapters 1 and 3, we discussed the utility of dataflow models of computation for design and implementation of embedded multiprocessor systems, and we described in detail certain aspects of the synchronous dataflow (SDF) model. SDF is the first dataflow-based model of computation to gain broad acceptance in DSP design tools, and most of the techniques developed in this book are developed in the context of SDF.

The expressive power of SDF is restricted, which means that not all applications can be described by SDF alone. This limitation arises directly from the restriction in SDF that production and consumption rates on dataflow edges cannot vary dynamically. Various researchers have proposed alternatives or extensions to SDF for dataflow-based design of signal processing systems. Many of these newer modeling techniques have been developed to provide more programming flexibility, thereby trading off some of the predictability and potential for rigorous analysis in SDF in favor of broader applicability.

In this chapter, we review some of the key DSP-oriented dataflow modeling techniques that have been developed to provide useful features that are not available in or difficult to derive from pure SDF representations. We focus in this chapter on several representative models of computation that are geared towards DSP system design, and employ or closely relate to dataflow modeling in significant ways. We also draw connections in this chapter to other related forms of dataflow that have emerged in recent years.

Techniques for analysis and optimization of SDF-based and HSDF-based representations, such as those developed in the later chapters of this book, are very much applicable in the context of the more general dataflow models reviewed in this chapter. In particular, when more general dataflow models are applied to represent DSP applications, significant subsystems often adhere to

SDF restrictions, or more broadly, to restrictions that enable derivation of equivalent homogeneous SDF (HSDF) representations, using methods related to those discussed in Section 3.7. Thus, by preprocessing more general dataflow representations to detect SDF subgraphs or to derive equivalent HSDF representations of selected subsystems, one has the potential to apply SDF- or HSDF-based techniques to key parts of the overall design.

Note that not all of the modeling techniques described in this chapter can be viewed as pure dataflow techniques. However, all of these models build upon or otherwise relate to dataflow concepts in significant ways.

## 4.1    Scalable Synchronous Dataflow

The **scalable synchronous dataflow** (**SSDF**) model is an extension of SDF that enables software synthesis of implementations that employ actor-level *vectorization*, which is also referred to in this context as **block processing**. This form of implementation exploits opportunities to process blocks of data in an efficient fashion within dataflow actors.

As a simple example of a vectorized actor, consider an addition actor that consumes one token per invocation on each of two inputs and outputs the sum of the values of the two input tokens on its single output. A vectorized version of this actor would operate on blocks of $N$ successive input tokens in an uninterrupted fashion, where $N$ is a positive-integer *vectorization factor* that can be associated with the actor by the designer or by a design tool. Thus, the vectorized form of this actor would effectively consume input in groups of $N$-element blocks of tokens and similarly output the corresponding sums in units of $N$-element blocks. Although a scheduler can schedule an "unvectorized" (scalar) version of the addition actor to execute $N$ times in succession (assuming that such a scheduling decision does not introduce deadlock [BML96]), the internal code for a vectorized actor can be more efficient because the actor programmer effectively has access to the enclosing "loop" of $N$ iterations. For example, when developing a vectorized actor, an actor programmer may be able to distribute the enclosing loop throughout the actor code, sometimes with significant flexibility.

To enable this kind of optimization for actor implementations, Ritz, Pankert, and Meyr developed the SSDF form of dataflow. In SSDF, production and consumption rates for all graph edges are required to be constant, and statically known, as in SDF; however, each actor $A$ in an SSDF-based design is specified in terms of a vectorization factor $f(A)$ that provides for a parameterized scaling of the associated production and consumption rates based on the desired block sizes for block processing within $A$. Such scaling is done directly by the designer when the actor is instantiated or edited in a graph, or by a design tool, when the graph is scheduled for synthesis or simulation. In SSDF, the scaling must be done prior to execution so that the production and consumption rates of all actors after vectorization are known statically. However, SSDF can be integrated with dynamic parameter control — for example, through the framework of parameterized dataflow, which is described in Section 4.4. Such a dynamic approach to vectorization is a useful direction for further investigation.

Vectorization for DSP-oriented dataflow models was first studied as part of the COSSAP project at the Aachen University of Technology [RPM92, RPM93]. COSSAP was commercialized by Cadis GmbH, which was acquired in 1994 by Synopsys. Berkeley Design Technology presents an industry perspective on this acquisition and on related forms of integration that occurred in commercial DSP tools around the same time period [BDTI94].

Vectorized implementation of DSP-oriented actors leads to increased regularity in the code in a form that reduces overhead associated with address generation and inter-actor context switching. Such advantages can result in significant improvements in throughput and energy efficiency. On the negative side, vectorization generally leads to increased latency and increased buffer memory requirements. Key advantages associated with vectorization have been motivated in detail by Ritz, Pankert, and Meyr [RPM93]. An experimental study of trade-offs associated with vectorization has been presented by Ko, Shen, and Bhattacharyya [KSB08].

A number of useful scheduling algorithms related to SSDF representations has been developed. Ritz, Pankert and Meyr have developed techniques to maximize the impact of vectorization for single appearance schedules of general SSDF representations. Single appearance schedules are a class of schedules that provide for minimal code size implementation of dataflow graphs, and have been studied extensively in the context of SDF (e.g., see [BML96]). This is generally a broad class of schedules with significant potential to streamline towards secondary optimization objectives while maintaining the minimal code size property. The vectorization methods developed by Ritz attempt to jointly maximize vectorization while adhering to the single appearance schedule constraint.

A number of approaches have explored retiming techniques that can be applied to SSDF graphs or to restricted classes of SSDF graphs to improve vectorization. Retiming involves the rearrangement of delays in a dataflow graph in ways that do not violate key functional properties of the graphs. Intuitively, retiming can be beneficial for vectorization because it has the potential to concentrate delays in graphs on isolated edges, which effectively provides for blocks of initial data to initialize vectorized executions. Retiming methods that are relevant to SSDF implementation include those by Zivojnovic, Ritz, and Meyr [ZRM94], and by Lalgudi, Papaefthymiou, and Potkonjak [LPP00].

Optimized vectorization of SSDF graphs under buffer memory constraints has been explored by Ritz, Willems, and Meyr [RWM95], and by Ko, Shen, and Bhattacharyya [KSB08].

## 4.2      Cyclo-Static Dataflow

Cyclo-static dataflow [BELP96] is a generalization of SDF in which token production and consumption rates take the form of periodic sequences of constant values. Thus, the amount of data produced and consumed by each actor invocation is fully deterministic, although some variation can occur across different invocations of the actor. CSDF is useful largely because it allows for a finer granularity of dataflow representation, thereby providing greater potential for dataflow-graph-level optimization, while maintaining key decidability features of SDF — in particular, CSDF, like SDF, allows for decidable verification of bounded memory and deadlock-free operation.

In a CSDF graph, each actor is viewed as executing through a periodically-repeating sequence of phases, where the production and consumption behavior of the actor is constant (as in SDF) for a given phase but may vary across different phases. For example, consider a downsample-by-two actor that has a single input port and a single output port, and discards every alternate token on its input stream, while passing the remaining input tokens to its output stream. An SDF version of this actor could be represented as shown in Figure 4.1(a), where the actor consumes 2 tokens and produces 1 token on each invocation. However, in CSDF, the actor can be viewed as executing through an alternating sequence of two phases, where the first phase consumes and produces 1 token, while the second phase consumes 1 token but does not produce any token (i.e., has a production rate of 0). The second phase corresponds to the discarding of

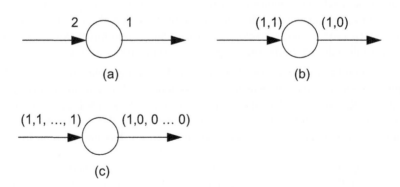

Figure 4.1. Different forms of a downsampler actor as a basic illustration of cyclo-static dataflow.

alternative tokens by the downsampler. This CSDF form of a downsample-by-two actor is illustrated in Figure 4.1(b), where the input and output ports of the actor are annotated with ordered pairs that specify the production and consumption corresponding to the two phases through which the actor continually alternates. Figure 4.1(c) shows a generalization of this CSDF actor to handle downsampling by an arbitrary factor $N > 2$. Here, the actor has $N$ phases, and the production and consumption rates are represented as $N$-element tuples. Note that the phase associated with the downsampling actor can be adjusted by changing the position of the 1 -valued entry in the production rate tuple.

A CSDF actor $A$ is characterized by a fundamental period $\tau(A)$, which gives the number of distinct phases associated with the actor. Given an execution of a CSDF graph that contains $A$, each $i$ th invocation of $A$ in $G$ is based on the $P(j)$ th phase, where

$$P(j) = ((i - 1) \bmod \tau(A)) + 1 . \tag{4-1}$$

Here, actor invocations and actor phases are assumed to be numbered starting at 1 . The sequence of actor invocations increases without any general, prespecified bound depending on how long the iterative execution of the graph continues; whereas the actor phases cycle periodically through the sequence $\{1, 2, ..., \tau(A)\}$.

Each edge $e$ in a CSDF graph has an associated *production tuple* $(p_1, p_2, ..., p_{\tau(src(e))})$, where each $p_i$ is a nonnegative integer that gives the number of tokens produced during phase $i$ of actor $src(e)$. Similarly, each edge $e$ has an associated consumption tuple $(c_1, c_2, ..., c_{\tau(snk(e))})$ that specifies the numbers of tokens consumed during the different phases of actor $snk(e)$. For example, the edges incident to the CSDF actors in Figure 4.1(b) and Figure 4.1(c) are annotated with the corresponding production tuples (at the output ports), and consumption tuples (at the input ports).

An equivalent HSDF graph can be constructed from a CSDF graph in a manner that extends naturally from the SDF-to-HSDF conversion techniques discussed in Section 3.7. Here, we outline a specific process for performing CSDF-to-HSDF conversion, and refer the reader to [BELP96] for further details.

## 4.2.1    Lumped SDF Representations

Suppose that we are given a CSDF graph $G = (V, E)$, where $V = \{a_1, a_2, ..., a_n\}$ is the set of actors in $G$. An initial step in constructing an equivalent HSDF graph from $G$ (if one exists) is constructing a *lumped SDF* representation $G_L$ of $G$, where each CSDF actor $a_i$ is converted into an SDF actor $L(a_i)$ that corresponds to $\tau(a_i)$ successive phase executions (i.e., a fundamental period) of $a_i$. Each edge $e$ in $G$ is then converted into an SDF edge $M(e)$ in $G_L$ with $src(M(e)) = L(src(e))$; $snk(M(e)) = L(snk(e))$;

$$p_1(M(e)) = \sum_{i=1}^{\tau(src(e))} p_i(e) \text{; and } c_1(M(e)) = \sum_{i=1}^{\tau(snk(e))} c_i(e). \qquad (4\text{-}2)$$

Here, $p_i(x)$ represents the number of tokens produced onto CSDF edge $x$ during the $i$th phase of actor $src(x)$, and similarly, $c_i(x)$ represents the number of tokens consumed from $x$ during the $i$th phase of $snk(x)$. Note that in this notation, we view each SDF actor as a CSDF actor that has a single phase, and hence the production and consumption rates associated with each SDF edge $x$ can be represented by the ordered pair $(p_1(x), c_1(x))$, as shown in the left hand sides of the equations in (4-2).

Figure 4.2 shows a CSDF graph (Figure 4.2(a)) and its corresponding lumped SDF graph (Figure 4.2(b)), and Figure 4.3 illustrates the execution of a periodic schedule for the CSDF graph of Figure 4.2(a). Recall from Section 3.5 that a periodic schedule is a schedule that does not deadlock, and that produces no net change in the number of tokens that are queued on any of the graph edges. The specific periodic schedule illustrated in Figure 4.3 is the schedule

$$S = (B, A, B, A, A, B, B, B, A, A, A, B), \qquad (4\text{-}3)$$

which consists of 3 fundamental periods of $A$ and 2 fundamental periods of $B$, or equivalently 6 phases each of $A$ and $B$. Figure 4.3 illustrates how the sequence of buffer states (the numbers of tokens queued on the graph edges) changes as the periodic schedule of (4-3) executes. The initial buffer state shows one token queued on the top edge, corresponding to the unit delay on the corresponding edge in Figure 4.2. Each arrow between buffer state diagrams in Figure 4.3 is annotated with a string of the form $F[n]$, which indicates that the $n$th execution of actor $F$ is the next actor invocation in the periodic schedule. The last buffer state diagram is equivalent to the first buffer state diagram, which confirms that the schedule of (4-3) is indeed a periodic schedule.

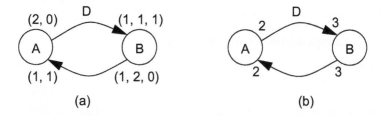

(a)                          (b)

Figure 4.2. An example of a CSDF graph and its corresponding lumped SDF graph.

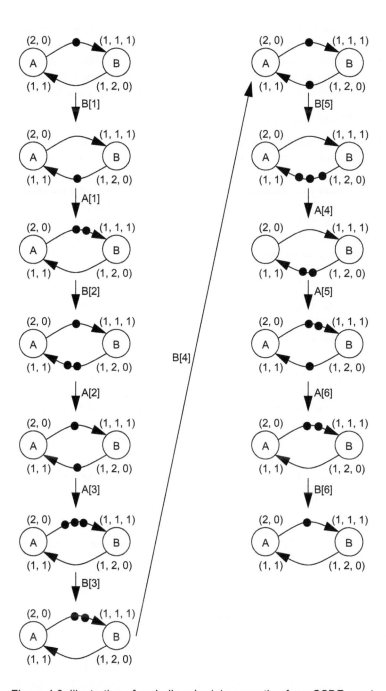

Figure 4.3. Illustration of periodic schedule execution for a CSDF graph.

Observe that while the CSDF graph of Figure 4.2(a) has a periodic schedule, as demonstrated by Figure 4.3, the lumped SDF version of this CSDF is deadlocked, and therefore does not have a periodic schedule. This illustrates one of the advantages — more flexibility in avoiding deadlock — that is provided by the finer granularity of actor interface behavior that can be expressed by CSDF.

### 4.2.2 Phase Repetitions Vectors

Once a corresponding lumped SDF representation $G_L$ has been derived from a CSDF graph $G$, one can attempt to derive the *phase repetitions vector*, which (if it exists) gives the number of CSDF actor invocations (phase executions) for each actor in a minimal periodic schedule for $G$. Using a minor abuse of notation, we represent the phase repetitions vector by $p$. The phase repetitions vector can be derived by first computing the repetitions vector $q$ for the SDF graph $G$ (see section 3.6 for elaboration on repetitions vectors for SDF graphs), and then computing each component of $p$ by

$$p(a_i) = \tau(a_i) \times q(L(a_i)) \text{ for each } a_i \in V. \tag{4-4}$$

In other words, the phase repetitions vector component for each CSDF graph actor $a_i$ is obtained by multiplying the corresponding component $q(L(a_i))$ of the SDF repetitions vector with the number of phases in a fundamental period of $a_i$.

If the lumped SDF graph $G_L$ does not have a repetitions vector — i.e., if its sample rates are not consistent — then the CSDF graph $G$ does not have a phase repetitions vector, and hence, $G$ cannot be implemented with a periodic schedule. Otherwise, if $G_L$ does have a repetitions vector, then we can compute (4-4), and a periodic schedule can be derived for $G$ provided that $G$ is not deadlocked.

For example, consider the example of Figure 4.2. The repetitions vector of the lumped SDF graph (Figure 4.2(b)) is given by

$$q(A) = 3, \text{ and } q(B) = 2. \tag{4-5}$$

Combining (4-4) and (4-5) gives the phase repetitions vector for the CSDF graph of Figure 4.2(a):

$$p(A) = 6, \text{ and } p(B) = 6. \tag{4-6}$$

### 4.2.3 Equivalent HSDF Graphs

From the phase repetitions vector $p$, one can construct an equivalent HSDF graph for a CSDF graph by instantiating an HSDF graph actor corresponding to each actor phase represented in $p$, and then instantiating edges and inserting edge delays in a manner analogous to the approach for SDF graphs that was covered in Section 3.7.

From (4-6) we see that the phase repetitions vector for the CSDF graph of Figure 4.2(a) represents 12 actor phases — that is, 6 phases for actor $A$, and another 6 phases for actor $B$. The equivalent HSDF graph for this example is shown in Figure 4.4.

For further details on this process of "CSDF-to-HSDF edge expansion," we refer the reader to [BELP96].

Conversion of a CSDF graph to its equivalent HSDF graph can reveal computations (actors or subgraphs) whose results are not needed, and therefore can be removed during synthesis as a form of dead code elimination. For example, if actor $A$ is not associated with any system output, then the executions $A[4]$ and $A6]$ in Figure 4.4 need not be implemented because their results are not used by any other actor invocations. In general, complete subgraphs can be removed from consideration if their computations do not influence any of the system outputs. Parks, Pino, and Lee have shown that such "dead subgraphs" can be detected using a simple, efficient algorithm [PPL95].

### 4.2.4    Analysis Techniques

Various techniques have been developed to analyze CSDF graphs for help in deriving efficient implementations from them. Denolf et al. develop methods for buffer optimization in CSDF graphs with emphasis on CSDF representations of multimedia applications [DBC+07]. Fischaber, Woods, and McAllister develop methods for deriving efficient memory hierarchies when mapping CSDF representations of image processing algorithms onto system-on-chip platforms [FWM07]. Saha, Puthenpurayil, and Bhattacharyya generalize the process of converting CSDF graphs to lumped SDF graphs to allow one to control the granularity of the resulting SDF representation, and therefore have greater flexibility when trying to apply SDF-based analysis techniques [SPB06].

## 4.3    Multidimensional Synchronous Dataflow

When dataflow graphs execute iteratively, as in DSP-oriented dataflow models, graph edges correspond to streams of data that are exchanged between the source and sink actors as the graph executes. In SDF, these streams are assumed to be one-dimensional — that is, the index space associated with the data stream for each edge has a single dimension. This is the simplest way to view streams of data and contributes to the intuitive simplicity of the SDF model; however, it is not always the most descriptive way to model multidimensional signal processing systems, such as systems for processing image and video signals.

The multidimensional synchronous dataflow (MDSDF) model of compu-

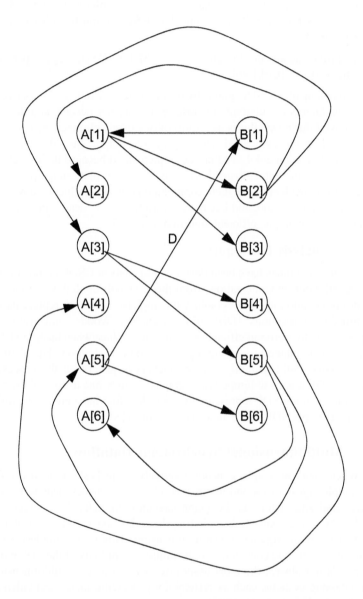

Figure 4.4. An illustration of the equivalent HSDF graph for the CSDF graph of Figure 4.2(a).

tation extends synchronous dataflow so that edges can have multidimensional index spaces associated with their data streams [Lee93]. The constant-integer production and consumption rates associated with graph edges are generalized to tuples of the form $(x_1, x_2, ..., x_n)$, where each $x_i$ is a nonnegative integer, and $n$ gives the dimensionality of the index space associated with the edge. When the production and consumption rates for an edge have differing dimensionalities, the dimensions of the lower-dimensionality rate $r_l$ are placed in correspondence with the lower-index dimensions of the higher dimensionality rate $r_h$, and the remaining dimensions of $r_l$ are assumed to be 1 . This allows for flexibility in connecting actors that have been designed with different dimensionalities at their interfaces.

The "repetitions count" (repetitions vector component) $q(A)$ for an SDF actor is generalized in MDSDF to a repetitions space $s(A)$ . The repetitions space can be specified as a vector $s(A) = (r_1, r_2, ..., r_n)$ that has the same dimensionality as the edges that are connected to $A$ , and each component $r_i$ is a positive integer that gives the size of the $i$ th dimension in the index space of $A$ within a periodic schedule for the enclosing graph $G$ . The total number of invocations of $A$ in a periodic schedule for $G$ is therefore given by the product

$$\prod_{i=1}^{n} r_i . \qquad (4\text{-}7)$$

For each input edge of $A$ , the correspondence between the indices associated with invocations of $A$ and the indices in the index space $e$ must be maintained through appropriate buffer management. This correspondence must be maintained to ensure that each invocation of $A$ consumes tokens from the appropriate invocation or invocations of $src(e)$ . Similarly, for each output edge of $A$ , appropriate bookkeeping is needed to ensure that the tokens produced by a given invocation of $A$ are consumed by the appropriate invocation or invocations of $snk(e)$ . Thus, when implementing MDSDF, one must extend to multiple dimensions the buffer management considerations of SDF in terms of appropriately "lining up" tokens between corresponding producer and consumer actors.

An example of an MDSDF-based specification, which is adapted from an example in [Lee93], is shown in Figure 4.5. Here each invocation of actor $A$ produces a $40 \times 48$ image (e.g., through an interface with some kind of image sensor); actor $B$ computes an $8 \times 8$ discrete cosine transform (DCT); and actor $C$ stores its input tokens sequentially in a file for later use. The dimensionality of the index space referenced by the input port of $C$ is expanded implicitly, and the resulting consumption rate tuple of $C$ becomes $(1, 1)$ . Such adjustment is not needed for the edge $(A, B)$ since the production and consumption rates for that edge have the same dimensionalities in their original forms.

As in SDF and CSDF, correctly-constructed MDSDF graphs can be scheduled statically through periodic schedules. However, static scheduling of MDSDF specifications differs from SDF and CSDF in that actor invocations have multidimensional index spaces, corresponding to the index spaces of the incident edges, and therefore, we replace the scalar "repetitions counts" (repetitions vector components) of individual SDF and CSDF actors with multidimensional "repetitions count vectors." These vectors give the dimensions of the index spaces associated with the corresponding actor invocations, and the products of their components, as given by (4-7), give the total number of invocations for the corresponding actors in a minimal periodic schedule for the overall MDSDF graph.

For example, the repetitions count vectors for the actors in Figure 4.5 are given by

$$Q(A) = (1, 1), Q(B) = (5, 6), \text{ and } Q(C) = (40, 48). \tag{4-8}$$

Thus, each actor in this example can be viewed as having two separate components of its "repetitions specification" — the number of "horizontal" repetitions, and the number of "vertical" repetitions. For example, actor $B$ has 5 horizontal repetitions and 6 vertical repetitions for a total of $5 \times 6 = 30$ invocations within a periodic schedule for the graph.

Equivalent HSDF graphs can also be constructed for MDSDF using extensions of the techniques for SDF and CSDF graphs. Again, the process involves instantiating an HSDF actor corresponding to each periodic schedule invocation of each actor, and then connecting the HSDF graph actors by tracking the corresponding "producer" and "consumer" invocations associated with each token that is transferred across an edge in the MDSDF graph during a periodic schedule. Following this approach, the example of Figure 4.5 results in an equivalent HSDF graph consisting of 1 actor corresponding to the MDSDF actor $A$, 30 actors corresponding to $B$, and $40 \times 48 = 1920$ actors corresponding to $C$.

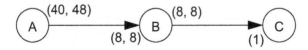

Figure 4.5. An example of an MDSDF graph (adapted from [Lee93]) for a simple image processing application.

One limitation of MDSDF in describing multirate signal processing systems is its restriction to streams that are sampled on standard rectangular sampling lattices. This is problematic for handling some kinds of multidimensional signals, such as interlaced video signals, that are sampled along nonrectangular lattices. Murthy and Lee have extended MDSDF to a generalized form that handles both rectangular and non-rectangular lattices in the same framework [ML02].

MDSDF has been implemented in the Ptolemy design environment, where it provides for single-processor simulation of multidimensional signal processing systems [CL94]. Baumstark, Guler, and Wills apply an extended form of MDSDF as a target model of computation for deriving data parallel programs from image processing algorithms that are specified originally as sequential programs [BGW03].

Keinert, Haubelt, and Teich have integrated considerations of multidimensional block-based and sliding window operations with the deterministic, multidimensional production and consumption rates of MDSDF [KHT06, KFHT07]. The resulting model of computation, called **windowed synchronous dataflow**, provides for useful modeling detail that can be exploited to specify multidimensional signal processing systems more completely (at the dataflow graph level), and optimize implementation aspects, such as buffering efficiency [KHT07, KHT08].

## 4.4    **Parameterized Dataflow**

Parameterized dataflow is a modeling technique that allows parameters of dataflow actors and edges to be changed dynamically [BB01]. A major motivation of the parameterized dataflow approach is support for dynamic parameter changes in dataflow graphs through the construction of parameterized schedules. In these parameterized schedules, a significant part of the schedule structure can be fixed at compile time with, and symbolic placeholders can be used within the schedules to represent variables that can change parts of the schedules dynamically. For example, loops (iterative constructs) across sequences of actor invocations can be parameterized with such symbolic variables so that the actual numbers of times these loops execute dynamically are determined at run time, based on the current value of the variables or symbolic expressions that provide the iteration counts for the loops. Such adjustments through the symbolic placeholders in a schedule can be made to adapt the schedule based on dynamic properties of dataflow graph execution, such as data-dependent production and consumption rates in one or more graph actors. In connection with this kind of "quasi-static" scheduling, Ko et al. develop and apply a general form of dataflow graph schedules that are in terms of nested loops with dynamically parameterized iteration counts [KZP+07].

Parameterized dataflow is a meta-modeling technique for DSP-oriented dataflow graphs: instead of defining a specific dataflow-based model of computation, parameterized dataflow provides a general mechanism through which a variety of different kinds of dataflow models can be augmented with dynamic parameters, and associated quasi-static scheduling capabilities. More specifically, parameterized dataflow can be applied to any underlying dataflow model of computation that has or can be associated with a well-defined concept of a *graph iteration*. When parameterized dataflow is applied to such a model $X$, the resulting model of computation can be viewed as a dynamically parameterized augmentation of $X$, which we refer to as the "parameterized $X$" model of computation. In the context of parameterized $X$, the model $X$ is referred to as the **base model**.

For example, when parameterized dataflow is applied to SDF as the base model, we obtain an extended form of SDF called **parameterized SDF (PSDF)**. In PSDF, actors and edges are specified largely in terms of SDF concepts. However, the actors and edges can have parameters that are changed dynamically, as a result of the computations associated with other PSDF subsystems in a given PSDF specification. Any actor parameter can be changed in this way, even if the parameter effects the production or consumption rates associated with one or more of the actor ports. Similarly, edge parameters, such as the amount of delay on the edge or the values of the initial tokens that correspond to the edge delays, can be changed dynamically. Scheduling techniques for the PSDF model have been developed in [BB00, BB01]. Parameterized CSDF, the application of parameterized dataflow with CSDF as the base model, has been explored in [SPB06].

The concept of graph iterations is important in parameterized dataflow because changes to dynamic parameters are allowed to take effect at certain points in time during execution, and these points are defined in terms of iterations of specific graphs that are associated with specific subsystems in a parameterized dataflow specification. In PSDF and PCSDF, graph iterations can be naturally defined in terms of repetitions vectors and phase repetitions vectors, respectively. Enhanced potential for using quasi-static, parameterized schedules results from structural restrictions imposed by parameterized dataflow on how parameter values are changed, along with restrictions that changes to parameter values can occur only between iterations of certain graphs or subsystems that "enclose" the parameterized actors.

Parameterized dataflow can be viewed as a modeling technique that supports dynamically reconfigurable dataflow graphs — dataflow graphs that are defined in terms of possible graph configurations such that the specific configuration through which the graph executes can be changed dynamically. Another form of reconfigurable dataflow graph modeling is the reactive process networks

(RPN) model of computation, which is discussed in Section 4.5. Reconfigurable dataflow graphs provide useful abstractions because they allow designers to capture important kinds of application dynamics while working in the context of familiar, well-studied formal models — e.g., SDF in the case of PSDF and Kahn process networks [Kah74] in the case of RPN. General techniques for analyzing reconfigurable dataflow graph specifications are developed by Neuendorffer and Lee [NL04]. This approach develops a behavioral type theory for dataflow graph reconfiguration, and applies this theory to check whether or not a given reconfigurable dataflow specification can reach certain kinds of invalid states as a result of dynamic changes to parameter values.

## 4.5 Reactive Process Networks

The **reactive process networks** model (**RPN**) of computation, developed at the Eindhoven University of Technology, is geared towards supporting an integrated, three-level view of signal processing applications involving the layers of control, dynamic stream processing, and signal processing kernels (low level, relatively regularly-structured functions) [GB04].

Dataflow actors and their interconnections are referred to in RPN terminology as processes and channels, respectively. An RPN specification has a fixed interface to the environment, and an underlying universal set of processes (candidate actors). Arbitrary subsets of the universal actor set along with arbitrary subsets of channels can be activated in a given configuration. Configurations also specify all other parameterized aspects of actor behavior so that once a network is configured, it is ready to be executed in a determinate way. An RPN is specified along with an initial configuration; however, changes to an RPN configuration can be made dynamically. Such changes can include changes to the set of active actors and channels, which can be used to adapt the structure of the overall processing network.

Changes in configuration are done in response to *events*, which arrive in ways that may not be predictable at compile time. To improve the predictability of event handling and associated changes in dataflow graph configuration, the RPN model employs a concept of **maximal streaming transactions** (**MSTs**). MSTs define the boundaries of RPN processing when handling of new events is allowed. Although new events may arrive while an MST is being processed, the next new event cannot be processed until the current MST has executed to completion. Intuitively, this constraint ensures that before processing any new event, the processing of input that has arrived before the event is carried out as far as possible.

To handle both event-based and dataflow-based computations, the RPN model supports two distinct modes of computation — stream processing and event processing. Stream processing is handled in terms of general Kahn process networks [Kah74], a model of computation that is closely related to dataflow with the main distinction being that in Kahn process networks, the executions associated with graph vertices ("processes") do not necessarily decompose into discrete units of execution, whereas dataflow actors execute in terms of distinct, indivisible actor invocations or "firings." Lee and Parks explore in more detail the relationships between KPNs and DSP-oriented dataflow models [LP95], and develop a formal foundation that unifies key principles that are common to this class of dataflow models. In the context of general DSP-oriented computation, KPNs and dataflow graphs are so closely related that the terms "KPN" and "dataflow" are sometimes used interchangeably.

Event processing in RPNs involves computations that are driven by inputs (events) that arrive unpredictably, in general, and that can trigger changes in processing state, such as changes to parameter values or reorganization of network structure, which can in turn change the characteristics of subsequent stream processing functionality. The integration of event and stream processing in RPNs is done in a flexible way that can in general lead to nondeterministic results — that is, the output streams produced by an RPN may be dependent not only on the input streams and initial state specifications, but also on run-time scheduling decisions across event and stream processing functions.

In practice, restrictions can be imposed on RPNs, such as the use of event handling priorities, to constrain run-time execution so that it provides deterministic operation or other forms of enhanced predictability. Such restrictions are not prescribed by the RPN model and are left instead for the user to specify in application-specific ways. For example, one could limit the stream processing aspect of RPN to SDF. A computational model in line with this idea is the Scenario-Aware DataFlow (SADF) model developed by Theelen et al. [TGB+06].

An implementation of the RPN model has been developed based on YAPI, which is an application programming interface for modeling signal processing applications in terms of KPNs [DSV+00].

## 4.6      Integrating Dataflow and State Machine Models

A variety of other modeling techniques have been developed in recent years that are related to the general, three-level, signal processing modeling hierarchy described in Section 4.5. A key objective in this kind of modeling hierarchy is to promote more flexibility in the high-level control of dataflow graph behavior. Various efforts in this direction have focused on methods for integrating finite state machine (FSM) representations or related state-based representations with dataflow graphs. Several of these efforts are described in the remainder of this section.

### 4.6.1    CAL

CAL is a textual language for specifying the functionality of dataflow graph actors [EJ03a]. CAL is based on an underlying dataflow model of computation in which each actor is characterized by a set of valid states; transitions that map states and input tokens into next states and output tokens; and priorities, which take the form of a partial order relation on the transitions, and help to select among multiple transitions that may be enabled at a given point in time — that is, multiple transitions that match the current actor state and input tokens at any given time [EJ03b]. An invocation of an actor in CAL (called a **step**) is associated with a specific transition, and can be carried out whenever the actor state and input tokens match the corresponding components of the transition, and no other matching transition exists that has higher priority.

In CAL, transitions are specified intensionally through the use of **actions** and **guards**. This is arguably a more intuitive and compact style of specification for many kinds of actors compared to a more direct specification of the underlying transition relation. Actions in CAL correspond to different modes in which the actor can execute, and have, in general, different computations associated with them. Formally, actions correspond to subsets of transitions in the transition relation. An actor can have any number of actions, and a guard is a logical expression associated with an action that specifies when the corresponding subset of transitions matches the actor state and input tokens. Actor invocations are mapped to specific actions when the associated guard expressions are satisfied, and no higher-priority actions have guards that are satisfied.

A distinguishing characteristic of CAL is its support for analyzing the internal behavior of actors, and integrating detailed intra-actor analysis with graph-level analysis of dataflow representations. This is facilitated by the formal semantics of the underlying actor model in terms of state transitions. An example of the flexibility offered through CAL-based analysis is the ability to "project" actors onto subsets of ports that may satisfy certain restrictions, such as the SDF restrictions of constant production and consumption rates [EJ03b]. Such projections of an actor can be combined with similar projections of neighboring actors, and collections of ports that are grouped in this way can be analyzed with specialized techniques that are based on the restrictions that were used when formulating the projections. This feature of CAL provides a foundation for integrating different kinds of specialized, dataflow-based analysis techniques in a way that is fully transparent to the programmer.

CAL is being used at Xilinx to model applications for efficient implementation on FPGAs. CAL is also used as a part of a novel methodology for reconfigurable video coding (RVM) that is under standardization by MPEG [LMKJ07, TJM+07]. In this RVM approach, the media content (bit stream) is transmitted along with the functionality required to decode the content. This functionality is encoded in the form of a network of CAL actors, drawing from a standard library of CAL actors, along with the parameter settings for the actors that are used. Since the library of actors is standardized, the encoding of functionality does not include the actors themselves, just their interconnections and configurations. Different networks can be used to activate different encoding standards, and different CAL actor specifications can be used for the same network to select different module implementations within the same overall standard.

While CAL is useful for determining whether or not a given actor specification or dataflow graph subsystem conforms to different kinds of restricted dataflow models, such as SDF, the underlying model of computation is highly expressive, and can be used to express nondeterministic behaviors as well as a wide range of dynamic dataflow behaviors.

### 4.6.2     FunState

**FunState** (functions as state machines) is an intermediate representation for embedded system implementation that allows for explicit representation of FSM-based dataflow graph control [TSZ+99]. In addition to modeling the behavior of such mixed control and dataflow representations, FunState allows for modeling of scheduling methods, and therefore can be used to analyze alternative mapping methods for the same design [STZ+99]. Specialized dataflow models, such as HSDF, SDF, CSDF, and Boolean dataflow (see Section 4.7.1), are shown to have equivalent representations in the FunState formalism. Like RPNs and CAL, FunState supports both deterministic and nondeterministic stream processing behaviors.

SysteMoC is a library for model-based design of DSP that is based on extending concepts from FunState and integrating them into a SystemC-based design environment [HFK+07]. The associated SystemCoDesigner toolset provides automated mapping from SysteMoC representations into FPGA implementations. SystemCoDesigner applies sophisticated, multiobjective evolutionary algorithms to optimize the synthesis of derived implementation in terms of latency, throughput and FPGA resource utilization.

### 4.6.3     Starcharts and Heterochronous Dataflow

**Starcharts** provides a general framework for hierarchically combining finite state machines with different kinds of concurrency models; when an appropriate form of dataflow is chosen as the underlying concurrency model, the Starcharts approach can be used to integrate high level control with stream processing capabilities [GLL99]. In the Starcharts approach, one specific form of hierarchical, FSM-dataflow integration that is developed in depth is the integration of FSMs and SDF graphs. This allows designers to provide hierarchical refinements of SDF actors as FSMs, and conversely, to provide hierarchical refinements of FSM states as SDF graphs.

To facilitate the latter form of hierarchical integration — refinement of FSM states as SDF graphs — a form of dataflow called **heterochronous dataflow (HDF)** is defined. A dataflow actor $A$ with $m$ input ports $(I_1, I_2, ..., I_m)$ and $n$ output ports $(O_1, O_2, ..., O_n)$ is an HDF actor if there is a finite set of pairs $S(A) = \{(C_i, P_i)|(1 \le i \le k)\}$ (for some positive integer $k$), where each $C_i$ is an $m$-component vector of positive integers; each $P_i$ is an $n$-component vector of positive integers; and for any invocation (firing) $F$ of $A$, there exists some $(C_j, P_j) \in S(A)$ such that each from each input port $I_i$, $A$ consumes exactly $C_j[i]$ tokens from $I_i$ throughout execution of $F$, and to each output port $O_i$, $A$ produces exactly $P_j[i]$ tokens to $O_j$ throughout execution of $F$.

In other words, an HDF actor can be viewed a representing a finite set of SDF behaviors that it can "draw upon," and the specific SDF behavior that is used can vary arbitrarily between distinct invocations of the HDF actor. Thus, HDF provides for a form of dynamic dataflow modeling where the dynamics is represented as variations across finite, predefined sets of candidate SDF behaviors.

If an FSM has different states that have refinements as SDF graphs with different numbers of tokens produced and consumed at their interfaces, then HDF provides a useful abstraction for embedding the FSM as a dataflow actor in a higher-level dataflow subsystem. The HDF representation of such an embedding captures the dynamic dataflow possibilities of the associated FSM-SDF combination while also representing how the dynamics draws from a specific, finite set of possible SDF behaviors. If the dataflow graph in which the FSM-SDF combination is itself assumed to be an HDF graph $G$, and every HDF actor $A$ in $G$ is assumed to draw from the same candidate SDF behavior from $S(A)$ (i.e., to conform to the same $(P_j, C_j)$) within an appropriately-defined iteration of $G$, then the execution of $G$ can be viewed as the execution of a sequence of SDF graphs. This restricted form of HDF is chosen as the HDF semantics for Starcharts. The restriction allows for significant use of SDF properties for SDF graphs that are embedded within an HDF graph through FSMs. For example, the restriction allows an HDF graph to be implemented through a finite set of periodic schedules, where schedules are repeatedly selected from this set and executed to completion (a single iteration) before the next schedule is selected.

### 4.6.4    DF-STAR

**DF-STAR** (Data Flow combined with STAte machine Reconfiguration model) integrates support for dynamic parameter control, dynamic changes in dataflow graph structure, asynchronous event handling, and nondeterminism. DF-STAR was originally introduced by Cossement, Lauwereins, and Catthoor [CLC00], and subsequently extended to a more powerful form by Mostert et al. [MCVL01]. DF-STAR builds on CSDF principles to allow for specification of and dynamic selection among different modes ("code segments") of actor behavior. The specific mode that is selected at any given time is determined by a *block controller*, which has a general form resembling that of a nondeterministic finite state machine.

### 4.6.5      DFCharts

The DFCharts model of computation integrates hierarchical, concurrent finite state machines with SDF graphs [RSR06]. In the DFCharts approach, SDF actors are represented as FSMs so that a given DFCharts specification can be translated into an equivalent *flat FSM*. While flat FSMs are useful for formal verification, the implementation of DFCharts allows for separate realizations of the SDF and FSM based subsystems, which allows SDF-based optimization techniques to be imported into the framework.

## 4.7      Controlled Dataflow Actors

Another form of control and dataflow integration in DSP-oriented modeling techniques has focused on specifying control conditions for dataflow actors as a more integrated part of the actor definition process, without hierarchical organization of FSMs or other kinds of established, heterogeneous modeling styles.

### 4.7.1      Boolean Dataflow

The **Boolean dataflow (BDF)** model was proposed by Lee [Lee91] and was further developed by Buck [Buc93] for extending the SDF model to allow data-dependent control actors in dataflow graphs. BDF actors are allowed to contain a *control* input, and the number of tokens consumed and produced on the edges of a BDF actor can be a two-valued function of a token consumed at the control input. Actors that follow SDF semantics, i.e., that consume and produce fixed numbers of tokens on their edges, clearly form a subset of the set of allowed BDF actors (SDF actors simply do not have any control inputs).

Two basic dynamic actors in the BDF model are the SWITCH and SELECT actors shown in Figure 4.6. The switch actor consumes one Boolean-valued control token and another input token; if the control token is TRUE, the input token is copied to the output labelled $T$, otherwise it is copied to the output labelled $F$. The SELECT actor performs the complementary operation; it reads an input token from its $T$ input if the control token is TRUE, otherwise it reads from its $F$ input. In either case, it copies the token to its output. Constructs such as conditionals and data-dependent iterations can easily be represented in a BDF graph, as illustrated in Figure 4.7. The actors labeled $A$, $B$, $C$, $D$, and $E$ in Figure 4.7 need not be atomic actors; they can also be arbitrary SDF subgraphs. A BDF graph allows SWITCH and SELECT actors to be connected in arbitrary topologies. Buck [Buc93] in fact shows that any Turing machine can be expressed as a BDF graph, and therefore, the problems of determining whether such a graph deadlocks and whether it uses bounded memory are undecidable. Buck proposes heuristic solutions to these problems based on extensions of certain techniques for SDF graphs to the BDF model.

The principles of Boolean dataflow have been extended by Buck to general, integer-valued control signals [Buck94].

## 4.7.2    Stream-Based Functions

Stream-based functions is a form of dataflow that has been developed at Leiden University [KD01]. Each actor in an SBF graph can select dynamically among a set of possible functions, and consume input tokens, perform computations, and produce output tokens based on the selected function. The formulation of SBF in terms of mathematical (side-effect-free) functions promotes compile time analysis, while the capabilities for dynamically selecting among different functions provide for enhanced modeling flexibility and expressive power.

Associated with each SBF actor $a$ is a set $R(a)$ of functions, called the **function repertoire**, a data space, and a controller. The controller in turn consists of a controller state space; a state transition function, which determines the next controller state given the current data and controller state; and a binding function, which is a mapping from the controller state space into the function repertoire. An SBF actor executes through successive executions of its controller and the associated function from $R(A)$ that is enabled, based on the binding function, by the most recent controller execution. Only the controller can change the controller state of an SBF actor, and only a function from the function repertoire can change the data state. Furthermore, as each function from the function repertoire executes, it consumes one or more inputs from the input edges of the SBF actor, and produces one or more outputs on the output edges.

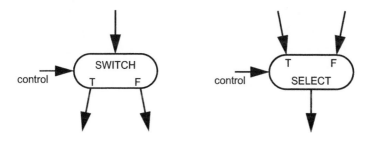

Figure 4.6. BDF actors SWITCH and SELECT.

More specifically, each function $f$ from the function repertoire has a fixed set $f_I$ of input edges from which it reads inputs, and fixed set $f_O$ of output edges onto which it produces outputs. Either $f_I$ or $f_O$ can be empty for a given $f$, and the sets $f_I$ and $f_O$ can be different across different functions $f$. During each execution of a function $f \in R(A)$, exactly one token is consumed from each input edge $e_i \in f_I$, and exactly one token is produced onto each output edge $e_o \in f_O$. The consumption of inputs from the input edges is performed using blocking reads, which means that to read a token, an actor must suspend execution until the token is available. In general, a function can process data from its associated input ports as well as private *data state* that is associated with the enclosing SBF actor. Similarly, a function can write to actor data state in addition to its associ-

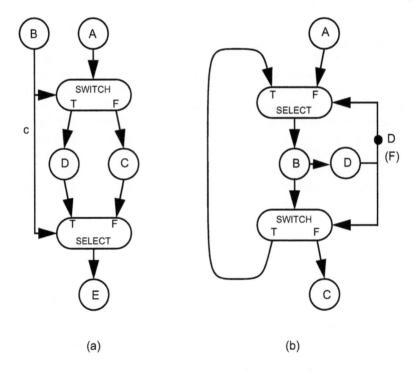

(a)                        (b)

Figure 4.7. (a) Conditional (if-then-else) dataflow graph. The branch outcome is determined at run-time by actor B. (b) Graph representing data-dependent iteration. The termination condition for the loop is determined by actor D.

ated output ports.

Intuitively, an SBF actor $A$ can be viewed as a collection of homogenous synchronous dataflow actors — corresponding to the collection of functions in the function repertoire — that operate across a common set of input and output edges, and are enabled one at a time by the controller of $A$. Each invocation of the SBF actor corresponds to an invocation of the controller, followed by an execution of the function that is enabled by the controller based on the controller state and the binding function. Since each actor in the function repertoire can be viewed as an HSDF actor whose inputs and outputs map to subsets of the input and output ports of the enclosing SBF actor, each SBF actor invocation results in binary-valued production and consumption volumes from the input and output ports of the SBF actor. In other words, the number of tokens produced or consumed by an SBF actor per invocation on its incident ports is 0 or 1. Therefore, SBF can be viewed as a form of dynamic dataflow — that is, a form of dataflow in which the production and consumption rates on dataflow graph edges are not always statically determined. More specifically, dynamic properties of SBF arise because the transition from one function state to another during execution can be data-dependent. Furthermore, SBF actors have associated initialization functions, which determine the initial function states of the actors, and these initialization functions can consume and operate based on data from the actor input ports.

Note that there is no limitation in the general approach of SBF that fundamentally limits the underlying functions to operate based on HSDF semantics. SBF-style networks can also be constructed by building on top of actors that have, for example, general SDF semantics. The resulting modeling approach can be viewed as a natural extension of the original SBF model.

Figure 4.8 illustrates an example of an SBF actor and its mapping from actor ports to function inputs. This actor has two input ports $p_{i,1}$ and $p_{i,2}$; one output port $p_o$; and a function repertoire that consists of three functions $f_1$, $f_2$, and $f_3$. The function $f_1$ consumes no data from either of the actor input ports, and reads only from the data state of the actor; the function $f_2$ consumes one token from each of the input ports; and the function $f_3$ consumes one token from $p_{1,2}$ and does not consume data from $p_{i,1}$. Functions $f_1$ and $f_2$ each produce a single token on the single actor output port, whereas $f_3$ does not produce any actor output tokens, and writes results only to the data state. During any given invocation of the SBF actor, at most one of the three functions $f_1$, $f_2$, and $f_3$ are enabled for execution.

Although SBF is a dynamic dataflow model of computation, specifications in the SBF model are determinate — that is, the output streams are determined uniquely by the input streams, and the initial state. The restricted way in which SBF actors can access their input ports (through blocking reads) helps to ensure determinacy.

SBF has been developed in the context of the Compaan project at Leiden University [SZT+04]. Compaan is a tool for compiling MATLAB programs to embedded processing platforms through dataflow-based intermediate representations. The SBF model helps to provide for the distributed control, distributed memory style of implementations that are targeted by the Compaan tool chain. Particular emphasis in the Compaan project has been placed on mapping to FPGA implementations, and powerful optimization techniques have been developed in this context to analyze the dataflow-based intermediate representations in Compaan. For example, Turjan, Kienhuis, and Deprettere have developed methods to streamline the implementation of buffers between actors based on information derived about the reordering of on along dataflow graph edges, along with multiplicity properties associated with dataflow tokens [TKD04]. Zissulescu, Kienhuis, and Deprettere develop methods for synthesizing expressions to implement parameterized, distributed control structures in Compaan [ZKD05]. Deprettere et al. explore the relationship between the cyclo-static dataflow model of computation and the intermediate representations used in Compaan [DSBS06]. Verdoolaege, Nikolov, and Stefanov present a recently-developed, improved version of Compaan called *pn* [VNS07]. A parameterized version of the SBF model is presented by Nikolov and Deprettere [ND08].

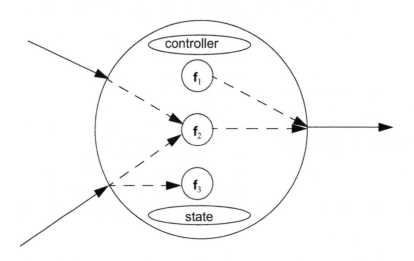

Figure 4.8. An example of an SBF actor and its mapping from actor ports to inputs and outputs of functions in the function repertoire.

### 4.7.3 Enable-Invoke Dataflow

Related to SBF in some ways is the **enable-invoke dataflow (EIDF)** [PSK+08] model of computation. In EIDF, actor behavior is also specified in terms of distinct modes of operation, which are analogous to the distinct functions in an SBF function repertoire. Additionally, the specification of each actor is separated into an *enable* function and an *invoke* function. The enable function determines the set of modes that are ready (enabled) for execution based on the amount of data available on the actor input ports. At any given time, this set of enabled modes can be empty, which means that the actor must remain in a suspended state; can contain a single element, which means that there is a unique next mode of execution; or can contain multiple elements, which means that the next mode of execution is to be selected based on design choices within the invoke function. The support for multiple-element return values of the enable function allows for nondeterministic execution. Also, because the enable function is dedicated to checking for the presence of sufficient input data, the invoke function can consume input data on demand (i.e., without the need for blocking reads). This strengthened separation between actor mode checking and actor execution is useful in efficient prototyping and implementation of dynamic scheduling techniques.

EIDF has been implemented in the **dataflow interchange format (DIF)**, which is a textual programming language and associated design tool for working with DSP-oriented dataflow models [HKB05, HCK+07].

## 4.8 Summary

This chapter has reviewed a variety of DSP-oriented modeling techniques that have been explored in various years. Various qualities are relevant in understanding the utility of a specific model for a specific class of applications. These include the expressive power of the model; any formal properties that are known about the model; the degree to which the model can be presented to designers through an intuitive programming interface; and the kind of tool support that is available for analyzing, simulating, or compiling the model.

The relationships among some of the models along some dimensions are relatively clear. For example, it can easily be argued that HSDF (homogenous synchronous dataflow), SDF (synchronous dataflow), CSDF (cyclo-static dataflow), PCSDF (parameterized cyclo-static dataflow), and BDF (Boolean dataflow) form a sequence of dataflow models that have increasing expressive power. In other cases, the relationships (e.g., those involving tool support or available formal properties) are functions not only of inherent properties of the models, but also of the kind of effort that has been devoted to understanding or applying the models. For example, SDF is perhaps the most mature form of DSP-oriented dataflow, and the range of techniques that is available for SDF is powerful not just because of the inherent properties of SDF, but because the model has been under active investigation for a relatively long time. Because of the complex trade-offs among the dimensions that are relevant to DSP-oriented dataflow models, the further development of individual models, the integration of alternative models, and the exploration of new models are all active areas of ongoing research.

# 5

---

# MULTIPROCESSOR SCHEDULING
# MODELS

---

This chapter discusses parallel scheduling of application graphs. The performance metric of interest for evaluating schedules is the **average iteration period** $T$: the average time it takes for all the actors in the graph to be executed once. Equivalently, we could use the throughput $T^{-1}$ (i.e., the number of iterations of the graph executed per unit time) as a performance metric. Thus, an optimal schedule is one that minimizes $T$.

## 5.1    Task-Level Parallelism and Data Parallelism

In the execution of a dataflow graph, actors fire when sufficient number of tokens are present at their inputs. A dataflow graph therefore lends itself naturally to **functional** or **task-level parallelism**, where the problem is to assign tasks with data dependencies to multiple processors. Systolic and wavefront arrays (Section 2.4) on the other hand exploit **data parallelism**, where the data set is partitioned among multiple processors executing the same program. Ideally, we would like to exploit data parallelism along with functional parallelism within the same parallel programming framework. Such a combined framework is currently an active research topic; several parallel languages have been proposed recently that allow a programmer to specify both data as well as functional parallelism. For example, Ramaswamy et al. propose a hierarchical macro dataflow graph representation of programs written in FORTRAN [RSB97]. Atomic nodes at the lowest level of the hierarchy represent tasks that are run in a data parallel fashion on a specified number of processors. The nodes themselves are run concurrently, utilizing functional parallelism. The work of Printz [Pri91] on geometric scheduling, and the Multidimensional SDF model proposed by Lee in [Lee93], are two other promising approaches for combining data and functional parallelism.

In this book we focus on task-level parallelism, and assume that the application graph implicitly specifies data-parallelism. This could be done, for example, by actors in the application graph that explicitly copy and partition data into multiple blocks, and other actors that collect processed data blocks. Determining how the data must be partitioned to optimally exploit data-parallelism is beyond the scope of this book; the interested reader is referred to [Kun88], and [RSB97].

Scheduling an application graph to exploit task-level parallelism involves assigning actors in the graph to processors (the **processor assignment** step), ordering execution of these actors on each processor (the **actor ordering** step), and determining when each actor fires such that all data precedence constraints are met. Each of these three tasks may be performed either at run-time (a dynamic strategy) or at compile time (static strategy). We restrict ourselves to **nonpreemptive schedules**, i.e., schedules where an actor executing on a processor cannot be interrupted in the middle of its execution to allow another task to be executed. This is because preemption entails a significant implementation overhead and is therefore of limited use in embedded, time-critical applications.

## 5.2    Static versus Dynamic Scheduling Strategies

Lee and Ha [LH89] propose a scheduling taxonomy based on which of the scheduling tasks are performed at compile time and which at run-time; we use the same terminology here. To reduce run-time computation costs, it is advantageous to perform as many of the three scheduling tasks as possible at compile time, especially in the context of algorithms that have hard real-time constraints. Which of these can be effectively performed at compile time depends on the information available about the execution time of each actor in the HSDFG.

In [LH89], a scheduling approach that performs all scheduling tasks at compile time is called a **fully-static** strategy. Relaxing this approach to an approach where the assignment and ordering are performed at compile time but the determination of start times is done at run-time gives rise to the **self-timed** strategy. A strategy that performs only the assignment step at compile time is called **static allocation** scheduling. Finally, an approach that makes all scheduling decision at run-time is deemed a **fully dynamic** scheduling strategy. As we move towards a fully dynamic strategy, the run-time overhead increases; however, the more run-time decisions we make, the more generally applicable the scheduling strategy becomes. Static techniques depend on how much compile time information is available about the application; this makes them applicable to a narrower application domain.

A **quasi-static** scheduling technique tackles data-dependent execution by localizing run-time decision making to specific types of data dependent constructs such as conditionals, and data dependent iterations. An **ordered transactions** strategy is essentially a self-timed approach where the order in which processors communicate is determined at compile time, and the target hardware enforces the predetermined transaction order during run-time. Such a strategy leads to a low overhead interprocessor communication mechanism. We will discuss this model in greater detail in the following two chapters. The trade-off between generality of the applications that can be targeted by a particular scheduling model, and the run-time overhead and implementation complexity entailed by that model is shown in Figure 5.1.

We discuss these scheduling strategies in detail in the following sections.

## 5.3    Fully-Static Schedules

In the **fully-static (FS)** strategy, the exact firing time of each actor is assumed to be known at compile time. Such a scheduling style is used in conjunction with systolic array architectures discussed in Section 2.4, for scheduling VLIW processors discussed in 2.2.3, and also in high-level VLSI synthesis of applications that consist only of operations with guaranteed worst-case execution times [De 94]. Under a fully-static schedule, all processors run in lock step; the

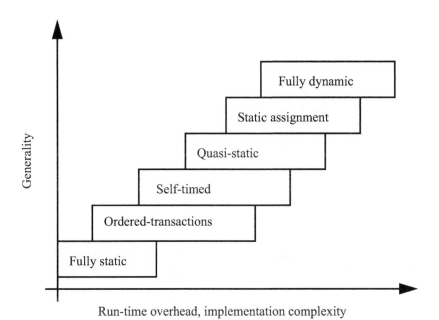

Figure 5.1. Trade-off of generality against run-time overhead and implementation complexity.

operation each processor performs on each clock cycle is predetermined at compile time and is enforced at run-time either implicitly (by the program each processor executes, perhaps augmented with "nop"s or idle cycles for correct timing) or explicitly (by means of a program sequencer, for example).

A fully-static schedule of a simple HSDFG $G$ is illustrated in Figure 5.2.

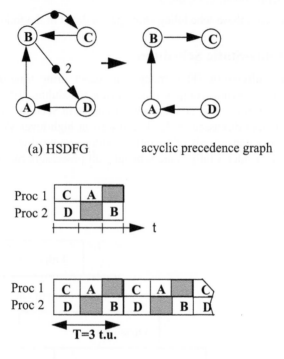

(a) HSDFG      acyclic precedence graph

(b) blocked schedule

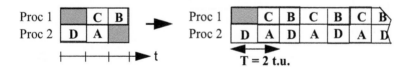

(c) overlapped schedule

Figure 5.2. Fully static schedule.

The fully-static schedule is schematically represented as a **Gantt chart**, which indicates the processors along the vertical axis, and time along the horizontal axis. The actors are represented as rectangles with horizontal length equal to the execution time of the actor. The left edge of each rectangle in the Gantt chart corresponds to the starting time of the corresponding actor. The Gantt chart can be viewed as a processor-time plane; scheduling can then be viewed as a mechanism to tile this plane while minimizing total schedule length, or equivalently minimizing idle time ("empty spaces" in the tiling process). Clearly, the fully-static strategy is viable only if actor execution time estimates are accurate and data-independent, or if tight worst-case estimates are available for these execution times.

As shown in Figure 5.2, two different types of fully-static schedules arise, depending on how successive iterations of the HSDFG are treated. Execution times of all actors are assumed to be one time unit (t.u.) in this example. The fully-static schedule in Figure 5.2(b) represents a **blocked schedule**: successive iterations of the HSDFG in a blocked schedule are treated separately so that each iteration is completed before the next one begins. A more elaborate blocked schedule on five processors is shown in Figure 5.3. The HSDFG is scheduled as if it executes for only one iteration, i.e., interiteration dependencies are ignored; this schedule is then repeated to get an infinite periodic schedule for the HSDFG. The length of the blocked schedule determines the average iteration period $T$. The scheduling problem is then to obtain a schedule that minimizes $T$ (which is also called the **makespan** of the schedule). A lower bound on $T$ for a blocked schedule is simply the length of the **critical path** of the graph, which is the longest delay-free path in the graph.

Ignoring the inter-iteration dependencies when scheduling an application graph is equivalent to the classical multiprocessor scheduling problem for an Acyclic Precedence Expansion Graph (APEG). As discussed in Section 3.8, the APEG is obtained from the given application graph by eliminating all edges with delays on them (edges with delays represent dependencies across iterations) and replacing multiple edges that are directed between the same two vertices in the same direction with a single edge. This replacement is done because such multiple edges represent identical precedence constraints; these edges are taken into account individually during buffer assignment, however. Optimal multiprocessor scheduling of an acyclic graph is known to be NP-hard [GJ79], and a number of heuristics have been proposed for this problem. One of the earliest, and still popular, solutions to this problem is *list-scheduling*, first proposed by Hu [Hu61]. List-scheduling is a greedy approach: whenever a task is ready to run, it is scheduled as soon as a processor is available to run it. Tasks are assigned priorities, and among the tasks that are ready to run at any instant, the task with the highest priority is executed first. Various researchers have proposed different priority mechanisms for list-scheduling [ACD74], some of which use critical-path-based

(CPM) methods [RCG72][Koh75][Bla87] ([Bla87] summarizes a large number of CPM based heuristics for scheduling). We discuss list-scheduling in detail in Chapter 6.

The heuristics mentioned above ignore communication costs between processors, which is often inappropriate in actual multiprocessor implementations.

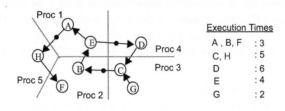

(a) HSDFG "G"

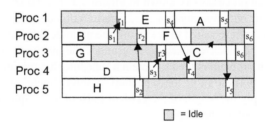

(b) Static schedule

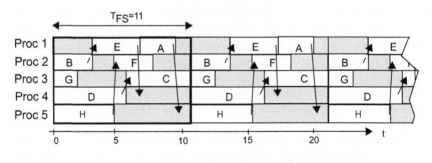

(c) Fully-static execution

Figure 5.3. Fully-static schedule on five processors.

An edge of the HSDFG that crosses processor boundaries after the processor assignment step represents **interprocessor communication (IPC)** (illustrated in Figure 5.4(a)). These communication points are usually implemented using *send* and *receive* primitives that make use of the processor interconnect hardware. These primitives then have an execution cost associated with them that depends on the multiprocessor architecture and hardware being employed. Fully-static scheduling heuristics that take communication costs into account include [Sar89], [Sih91], and [Pri91].

Computations in the HSDFG, however, are iterated essentially infinitely. The blocked scheduling strategies discussed thus far ignore this fact, and thus pay a penalty in the quality of the schedule they obtain. Two techniques that enable blocked schedules to exploit interiteration parallelism are **unfolding** and **retiming**. The unfolding strategy schedules $N$ iterations of the HSDFG together, where $N$ is called the **blocking factor**. Thus, the schedule in Figure 5.2(b) has $N = 1$. Unfolding often leads to improved blocked schedules (pp. 78-100 [Lee86], [PM91]), but it also implies a factor of $N$ increase in program memory size and also in the size of the scheduling problem. The construction of an unfolded HSDFG is described in detail in Section 5.8.

Retiming involves manipulating delays in the HSDFG to reduce the critical path in the graph. This technique has been explored in the context of maximizing clock rates in synchronous digital circuits [LS91], and has been proposed for improving blocked schedules for HSDFGs ("cutset transformations" in [Lee86], and [Hoa92]).

Figure 5.2(c) illustrates an example of an **overlapped schedule**. Such a schedule is explicitly designed such that successive iterations in the HSDFG overlap. Obviously, overlapped schedules often achieve a lower iteration period than blocked schedules. In Figure 5.2, for example, the iteration period for the blocked schedule is 3 units whereas it is 2 units for the overlapped schedule. One might wonder whether overlapped schedules are fundamentally superior to blocked schedules with the unfolding and retiming operations allowed. This question is settled in the affirmative by Parhi and Messerschmitt [PM91]; the authors provide an example of an HSDFG for which no blocked schedule can be found, even allowing unfolding and retiming, that has a lower or equal iteration period than the overlapped schedule they propose.

Optimal resource constrained overlapped scheduling is of course NP-hard, although a periodic overlapped schedule in the absence of processor constraints can be computed efficiently and optimally in polynomial time [PM91][GS92].

Overlapped scheduling heuristics have not been as extensively studied as blocked schedules. The main work in this area is by Lam [Lam88], and deGroot [dGH92], who propose a modified list-scheduling heuristic that explicitly constructs an overlapped schedule. Another work related to overlapped scheduling is the "cyclo-static scheduling" approach proposed by Schwartz. This approach attempts to optimally tile the processor-time plane to obtain the best possible schedule. The search involved in this process has a worst-case complexity that is exponential in the size of the input graph, although it appears that the complexity is manageable in practice, at least for small examples [SI85].

## 5.4    Self-Timed Schedules

The fully-static approach introduced in the previous section cannot be used when actors have variable execution times; the fully-static approach requires precise knowledge of actor execution times to guarantee sender-receiver synchronization. It is possible to use worst-case execution times and still employ a fully-static strategy, but this requires tight worst-case execution time estimates that may not be available to us. An obvious strategy for solving this problem is to introduce explicit synchronization whenever processors communicate. This leads to the **self-timed scheduling (ST)** strategy in the scheduling taxonomy of Lee and Ha [LH89]. In this strategy we first obtain a fully-static schedule using techniques that will be discussed in Chapter 6, making use of the execution time estimates. After computing the fully-static schedule (Figure 5.4 (b)), we simply discard the timing information that is not required, and only retain the processor assignment and the ordering of actors on each processor as specified by the fully-static schedule (Figure 5.4(c)). Each processor is assigned a sequential list of actors, some of which are *send* and *receive* actors, which it executes in an infinite loop. When a processor executes a communication actor, it synchronizes with the processor(s) it communicates with. Exactly when a processor executes each actor depends on when, at run-time, all input data for that actor is available, unlike the fully-static case where no such run-time check is needed. Conceptually, the processor sending data writes data into a FIFO buffer, and blocks when that buffer is full; the receiver, on the other hand, blocks when the buffer it reads from is empty. Thus flow control is performed at run-time. The buffers may be implemented using shared memory, or using hardware FIFOs between processors. In a self-timed strategy, processors run sequential programs and communicate when they execute the communication primitives embedded in their programs, as shown schematically in Figure 5.4 (c).

All of the multiprocessor DSP platforms that we discussed in Section 2.5 employ some form of self-timed scheduling. Clearly, general purpose parallel machines can also be programmed using the self-timed scheduling style, since these machines provide mechanisms for run-time synchronization and flow control.

A self-timed scheduling strategy is robust with respect to changes in execution times of actors, because sender-receiver synchronization is performed at run-time. Such a strategy, however, implies higher IPC costs compared to the fully-static strategy because of the need for synchronization (e.g., using sema-phore management). In addition the self-timed scheduling strategy faces arbitra-tion costs: the fully-static schedule guarantees mutually exclusive access of shared communication resources, whereas shared resources need to be arbitrated at run-time in the self-timed schedule. Consequently, whereas IPC in the fully-static schedule simply involves reading and writing from shared memory (no synchronization or arbitration needed), implying a cost of a few processor cycles for IPC, the self-timed scheduling strategy requires of the order of tens of proces-sor cycles, unless special hardware is employed for run-time flow control.

Run-time flow control allows variations in execution times of tasks; in addition, it also simplifies the compiler software, since the compiler no longer needs to perform detailed timing analysis and does not need to adjust the execu-tion of processors relative to one another in order to ensure correct sender-receiver synchronization. Multiprocessor designs, such as the Warp array [Ann87][Lam88], that could potentially use fully-static scheduling, still choose to implement such run-time flow control (at the expense of additional hardware) for the resulting software simplicity. Lam presents an interesting discussion on

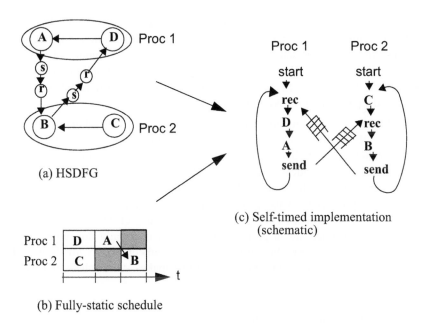

(a) HSDFG

(b) Fully-static schedule

(c) Self-timed implementation (schematic)

Figure 5.4. Steps in a self-timed scheduling strategy.

the trade-off involved between hardware complexity and ease of compilation that ensues when we consider dynamic flow control implemented in hardware versus static flow control enforced by a compiler (pp. 50-68 of [Lam89]).

## 5.5    Dynamic Schedules

In a fully dynamic approach, all scheduling decisions are determined at run-time. Such an approach is most general in terms of applicability, and will handle highly dynamic program behavior by changing the order in which tasks are run, and/or by adjusting processor loads during run-time. However, the cost of such run-time scheduling decisions is very high. Moreover, since scheduling decisions are made during run-time, it is not practical to make globally optimal scheduling decisions; a dynamic scheduler may be forced to make "greedy" locally optimal decisions. In fact, in the presence of compile-time information about task execution times and precedence constraints, a static scheduling approach may lead to better performance [SL93a, SL93b].

Dynamic scheduling may be done by special purpose hardware, or it may be performed by a kernel running on one or more processors. Examples of a hardware solution are superscalar processors, discussed in Section 2.1, where an instruction dispatch unit, implemented in hardware, schedules instructions in the program to multiple parallel functional units at run-time. The classic dataflow computer architectures and the dataflow DSP architectures discussed in Section 2.3 also employ dynamic scheduling. For example, dataflow computers first pioneered by Dennis [Den80] perform the assignment step at compile time (**static assignment**), but employ special hardware to determine, at run-time, when actors assigned to a particular processor are ready to fire.

Dynamic scheduling techniques are often used in distributed, coarsegrained parallel systems such as workstations over a network [D+96]. A distributed algorithm is used to balance the load between the workstations; each workstation periodically broadcasts the number of tasks in its task queue, and when this number exceeds a threshold, a decision is made to move some of the tasks over to other processors. This interaction is clearly complex and lengthy, and hence is done infrequently as compared to the frequency of interprocessor communication. Embedded signal processing systems will usually not require this type of scheduling owing to the run-time overhead and complexity involved, and the availability of compile time information that makes static scheduling techniques practical.

## 5.6    Quasi-Static Schedules

Actors that exhibit data dependent execution time usually do so because they include one or more data-dependent control structures, for example conditionals and data-dependent iterations. In such a case, if we have some knowledge about the statistics of the control variables (number of iterations a loop will go through or the boolean value of the control input to an if-then-else type construct), it is possible to obtain a static schedule that optimizes the average execution time of the overall computation. The key idea here is to define an **execution profile** for each actor in the dataflow graph. An execution profile for a dynamic construct consists of the number of processors assigned to it, and a local schedule of that construct on the assigned processors; the profile essentially defines the shape that a dynamic actor takes in the processor-time plane. In case the actor execution is data-dependent, an exact profile cannot be predetermined at compile time. In such a case, the profile is chosen by making use of statistical information about the actor, e.g., average execution time, probability distribution of control variables, etc. Such an approach is called **quasi-static scheduling** [Lee88b]. Figure 5.5 shows a quasi-static strategy applied to a conditional construct (adapted from [Lee88b]).

Ha [HL97] has applied the quasi-static approach to dataflow constructs representing data-dependent iteration, recursion, and conditionals, where optimal profiles are computed assuming the knowledge of the probability density functions of data-dependent variables that influence the profile. The data-dependent constructs must be identified in a given dataflow graph, either manually or automatically, before Ha's techniques can be applied. These techniques make the simplifying assumption that the control tokens for different dynamic actors are independent of one another, and that each control stream consists of tokens that take TRUE or FALSE values randomly and are independent and identically distributed (i.i.d.) according to statistics known at compile time.

Ha's quasi-static approach constructs a blocked schedule for one iteration of the dataflow graph. The dynamic constructs are scheduled in a hierarchical fashion; each dynamic construct is scheduled on a certain number of processors, and is then converted into a single node in the graph and is assigned a certain execution profile. When scheduling the remainder of the graph, the dynamic construct is treated as an atomic block, and its execution profile is used to determine how to schedule the remaining actors around it; the profile helps tiling actors in the processor-time plane with the objective of minimizing the overall schedule length. Such a hierarchical scheme effectively handles nested control constructs, e.g., nested conditionals. The locally optimal decisions made for the dynamic constructs are shown to be effective when the variability in a dynamic construct is small. We will return to quasi-static schedules again in Chapter 9.

## 5.7    Schedule Notation

To model execution times of actors (and to perform static scheduling), we associate an execution time $t(v) \in Z^+$ (nonnegative integer) with each actor $v$ in the HSDFG; $t(v)$ assigns execution time to each actor $v$ (the actual execution time can be interpreted as $t(v)$ cycles of a base clock). Interprocessor communication costs are represented by assigning execution times to the *send* and *receive* actors. The values $t(v)$ may be set equal to execution time *estimates* when exact execution times are not available, in which case results of the computations that

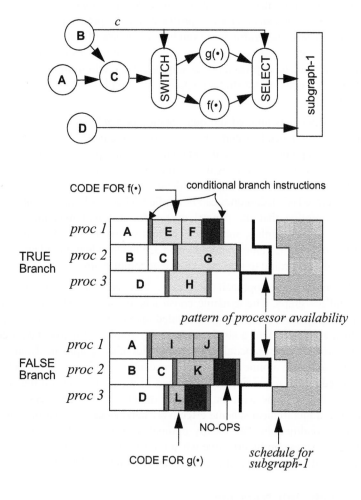

Figure 5.5. A quasi-static schedule for a conditional construct (adapted from [Lee88b]).

make use of these values (e.g., the iteration period $T$) are compile-time estimates.

Recall that actors in an HSDFG are executed essentially infinitely. Each firing of an actor is called an **invocation** of that actor. An **iteration** of the HSDFG corresponds to one invocation of every actor in the HSDFG. A schedule specifies processor assignment, actor ordering and firing times of actors, and these may be done at compile-time or at run-time, depending on the scheduling strategy being employed. To specify firing times, we let the function $start(v, k) \in Z^+$ represent the time at which the $k$th invocation of the actor $v$ starts. Correspondingly, the function $end(v, k) \in Z^+$ represents the time at which the $k$th execution of the actor $v$ completes, at which point $v$ produces data tokens at its output edges. Since we are interested in the $k$th execution of each actor for $k = 0, 1, 2, 3, \ldots$, we set $start(v, k) = 0$ and $end(v, k) = 0$ for $k < 0$ as the "initial conditions". If the $k$th invocation of an actor $v_j$ takes *exactly* $t(v_j)$ time units to complete for all $k$, then we can claim:

$$end(v_j, k) = start(v_j, k) + t(v_j).$$

Recall that a fully-static schedule specifies a processor assignment, actor ordering on each processor, and also the precise firing times of actors. We use the following notation for a fully-static schedule:

**Definition 5.1:** A fully-static schedule $S$ (for $P$ processors) specifies a triple:

$$S = \{\sigma_p(v), \sigma_t(v), T_{FS}\},$$

where $\sigma_p(v) \rightarrow \{1, 2, \ldots, P\}$ is the processor assignment, and $T_{FS}$ is the iteration period. A fully-static schedule specifies the firing times $start(v, k)$ of all actors, and since we want a finite representation for an infinite schedule, a fully-static schedule is constrained to be periodic:

$$start(v, k) = \sigma_t(v) + kT_{FS},$$

$\sigma_t(v)$ is thus the starting time of the first execution of actor $v$ (i.e., $start(v, 0) = \sigma_t(v)$). Clearly, the throughput for such a schedule is $T_{FS}^{-1}$.

The $\sigma_p(v)$ function and the $\sigma_t(v)$ values are chosen so that all data precedence constraints and resource constraints are met. We define precedence constraints as follows:

**Definition 5.2:** An edge $(v_j, v_i) \in E$ in an HSDFG $(V, E)$ represents the (data) **precedence constraint**:

$$start(v_i, k) \geq end(v_j, k - delay(v_j, v_i)), \text{ for all } k \geq delay(v_j, v_i). \quad (5\text{-}1)$$

The above definition arises because each actor consumes one token from each of its input edges when it fires. Since there are already $delay(e)$ tokens on each incoming edge $e$ of actor $v$, another $(k - delay(e) - 1)$ tokens must be produced on $e$ before the $k$th execution of $v$ can begin. Thus, the actor $src(e)$ must have completed its $(k - delay(e) - 1)$th execution before $v$ can begin its $k$th execution. The "$-1$ s" arise because we define $start(v, k)$ for $k \geq 0$ rather than $k > 0$. This is done for notational convenience.

Any schedule that satisfies all the precedence constraints specified by edges in an HSDFG $G$ is also called an **admissible schedule** for $G$ [Rei68]. A **valid execution** of an HSDFG corresponds to a set of firing times $\{ start(v_i, k) \}$ that correspond to an admissible schedule. That is, a valid execution respects all data precedences specified by the HSDFG.

For the purposes of the techniques presented in this book, we are only interested in the precedence relationships between actors in the HSDF graph. In a general HSDFG one or more pairs of vertices can have multiple edges connecting them in the same "direction"; in other words a general HSDFG is a *directed multigraph* (Section 3.1). Such a multigraph often arises when a multirate SDF graph is converted into an HSDFG. Multiple edges between the same pair of vertices in the same direction are redundant as far as precedence relationships are concerned. Suppose there are multiple edges from vertex $v_i$ to $v_j$, and amongst these edges, the minimum edge delay is equal to $d_{min}$. Then, if we replace all of these edges by a single edge with delay equal to $d_{min}$, it is easy to verify that this single edge maintains the precedence constraints for *all* of the edges that were directed from $v_i$ to $v_j$. Thus, a general HSDFG may be preprocessed into a form where the source and sink vertices uniquely identify an edge in the graph, and we may represent an edge $e \in E$ by the ordered pair $(src(e), snk(e))$. That is, an HSDFG that is a directed multigraph may be transformed into an HSDFG that is a directed graph such that the precedence constraints of the original HSDFG are maintained by the transformation. Such a transformation is illustrated in Figure 5.6. The multiple edges are taken into account individually when buffers are assigned to the arcs in the graph. We perform such a transformation to avoid needless clutter in analyzing HSDFGs, and to reduce the running time of algo-

Figure 5.6. Transforming an HSDFG that is a directed multigraph into one that is a directed graph while maintaining precedence constraints.

rithms that operate on HSDFGs.

In a self-timed scheduling strategy, we determine a fully-static schedule, $\{\sigma_p(v), \sigma_t(v), T_{FS}\}$ using the execution time estimates, but we retain only the processor assignment $\sigma_p$ and the ordering of actors on each processor as specified by $\sigma_t$, and we discard the precise timing information specified in the fully-static schedule. Although we may start out with setting $start(v, 0) = \sigma_t(v)$, the subsequent $start(v, k)$ values are determined at run-time based on the availability of data at the input of each actor. The average iteration period of a self-timed schedule is represented by $T_{ST}$. We analyze the evolution of a self-timed schedule further in Chapter 8.

## 5.8    Unfolding HSDF Graphs

As we discussed in Section 5.3, in some cases it is advantageous to *unfold* a graph by a certain unfolding factor, say $N$, and schedule $N$ iterations of the graph together in order to exploit interiteration parallelism more effectively. In this section, we describe the unfolding transformation.

An HSDFG $G = (V, E)$ unfolded $N$ times represents $N$ iterations of the original graph $G$; the unfolding transformation therefore results in another HSDFG that contains $N$ copies of each of the vertices of $G$. We denote the unfolded graph by $G^N = (V^N, E^N)$. For each vertex $v_i \in V$, and the $N$ copies of $v_i$ in $G^N$ by $v_i^0, v_i^1, ..., v_i^{N-1} \in V^N$. Clearly, $|V^N| = N|V|$. From the definition of $v_i^l$, it is obvious that:

$$start(v_i^l, m) = start(v_i, mN + l), \text{ and} \qquad (5\text{-}2)$$

$$end(v_i^l, m) = end(v_i, mN + l), \text{ for all } m \geq 0, \text{ and } 0 \leq l < N. \qquad (5\text{-}3)$$

Also, $G^N$ maintains exactly the same precedence constraints as $G$, and therefore the edges $E^N$ must reflect the same inter and intraiteration precedence constraints as the edges $E$. For the precedence constraint in $G$ represented by the edge $(v_j, v_i) \in E$, there will be a set of one or more edges in $E^N$ that represents the same precedence constraint in $G^N$. The construction of $E^N$ is as follows.

From (5-1), an edge $(v_j, v_i) \in E$ represents the precedence constraint:

$start(v_i, k) \geq end(v_j, k - delay(v_j, v_i))$, for all $k \geq delay(v_j, v_i)$.

Now, we can let $k = mN + l$, and write $delay(v_j, v_i)$ as

$$delay(v_j, v_i) = \left\lfloor \frac{delay(v_j, v_i)}{N} \right\rfloor N + delay(v_j, v_i) \bmod N, \qquad (5\text{-}4)$$

where $(x \bmod y)$ equals the value of $x$ taken modulo $y$, and $\lfloor x/y \rfloor$ equals the quotient obtained when $x$ is divided by $y$. Then, (5-1) can be written as:

$$start(v_i, mN + l) \geq end\left(v_j, \left(m - \left\lfloor \frac{delay(v_j, v_i)}{N} \right\rfloor\right)N + (l - delay(v_j, v_i) \bmod N)\right)$$

(5-5)

We now consider two cases:

1. If $delay(v_j, v_i) \bmod N \leq l < N$, then (5-5) may be combined with (5-2) and (5-3) to yield:

$$start(v_i^l, m) \geq end\left(v_j^{l - delay(v_j, v_i) \bmod N}, m - \left\lfloor \frac{delay(v_j, v_i)}{N} \right\rfloor\right).$$

(5-6)

2. If $0 \leq l < delay(v_j, v_i) \bmod N$, then (5-5) may be combined with (5-2) and (5-3) to yield:

$$start(v_i^l, m) \geq end\left(v_j^{N + l - delay(v_j, v_i) \bmod N}, m - \left\lfloor \frac{delay(v_j, v_i)}{N} \right\rfloor - 1\right).$$

(5-7)

Equations (5-6) and (5-7) are summarized as follows. For each edge $(v_j, v_i) \in E$, there are a set of $N$ edges in $E^N$, which is the edge set of $G^N$. In particular, $E^N$ contains edges from $v_j^{l - delay(v_j, v_i) \bmod N}$ to $v_i^l$, each with delay $\lfloor delay(v_j, v_i)/N \rfloor$, for values of $l$ such that $delay(v_j, v_i) \bmod N \leq l < N$. In addition, $E^N$ contains edges from $v_j^{N + l - delay(v_j, v_i) \bmod N}$ to $v_i^l$, each with delay $\lfloor delay(v_j, v_i)/N \rfloor + 1$, for the values of $l$ such that $0 \leq l < delay(v_j, v_i) \bmod N$. Note that if $delay(v_j, v_i) = 0$, then there is a zero-delay edge from each $v_j^l$ to $v_i^l$ for $0 \leq l < N$. Figure 5.7 shows an example of the unfolding transformation. Figure 5.8 lists an algorithm that may be used for unfolding. Note that this algorithm has a complexity of $O(N|E|)$.

When constructing a schedule for an unfolded graph, the processor assignment $\sigma_p$ and the actor starting times $\sigma_t$ are defined for all vertices of the *unfolded* graph (i.e., $\sigma_p$ and $\sigma_t$ are defined for $N$ invocations of each actor); $T_{FS}$ is the iteration period for the unfolded graph, and the average iteration period for the original graph is then $(T_{FS}/N)$. In the remainder of this book, we assume we are dealing with the unfolded graph and we refer only to the iteration period and throughput of the unfolded graph, if unfolding is in fact employed, with the understanding that these quantities can be scaled by the unfolding factor to obtain the corresponding quantities for the original graph.

## 5.9     **Execution Time Estimates and Static Schedules**

We assume that we have reasonably good estimates of actor execution times available to us at compile time to enable us to exploit static scheduling techniques; however, these estimates need not be exact, and execution times of actors may even be data-dependent. Thus we allow actors that have different execution times from one iteration of the HSDFG to the next, as long as these variations are small or rare. This is typically the case when estimates are available for the task execution times, and actual execution times are close to the corresponding estimates with high probability, but deviations from the estimates of (effectively) arbitrary magnitude occasionally occur due to phenomena such as cache misses, interrupts, user inputs, or error handling. Consequently, tight worst-case execution time bounds cannot generally be determined for such operations; however, reasonably good execution time estimates can in fact be obtained for these operations, so that static assignment and ordering techniques are viable. For such applications self-timed scheduling is ideal, because the performance penalty due

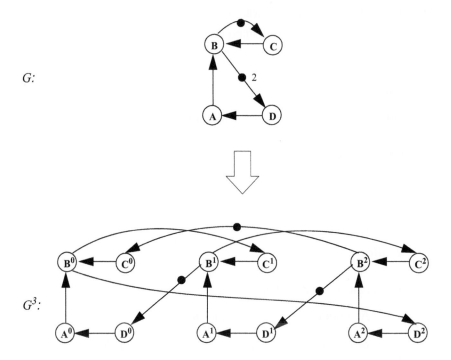

Figure 5.7. Example of an unfolding transformation: HSDFG $G$ is unfolded by a factor of 3 to obtain the unfolded HSDFG $G^3$.

to lack of dynamic load balancing is overcome by the much smaller run-time scheduling overhead involved when static assignment and ordering is employed.

The estimates for execution times of actors can be obtained by several different mechanisms. The most straightforward method is for the programmer to provide these estimates while developing the library of primitive blocks (actors). In this method, the programmer specifies the execution time estimates for each actor as a mathematical function of the parameters associated with that actor (e.g., number of filter taps for an FIR filter, or the block size of a block operation such as an FFT). This strategy is used in the Ptolemy system [Pto98] for example, and is especially effective for libraries in which the primitives are written in the assembly language of the target processor. The programmer can provide a good estimate for blocks written in such a low-level library by counting the number of processor cycles each instruction consumes, or by profiling the block on an instruction-set simulator.

**Function** Unfold HSDFG

**Input:** HSDFG $G = (V, E)$, with delays $delay(v_j, v_i)$ for each edge $(v_j, v_i) \in E$, and an integer unfolding factor $N$.

**Output:** Unfolded HSDFG $G^N = (V^N, E^N)$, with delays $delay(v_j^l, v_i^l)$.

1. **For each** $v_i \in V$

   Add $N$ vertices $v_i^0, v_i^1, ..., v_i^{N-1}$ to the set $V^N$
   
**Endfor**

2. **For each** $(v_j, v_i) \in E$

   **For** $0 \le l < delay(v_j, v_i) \bmod N$

   Add edge $(v_j^{N+l-delay(v_j, v_i) \bmod N}, v_i^l)$ to $E^N$

   Set $delay(v_j^{N+l-delay(v_j, v_i) \bmod N}, v_i^l) = \lfloor delay(v_j, v_i)/N \rfloor + 1$

   **Endfor**

   **For** $delay(v_j, v_i) \bmod N \le l < N$

   Add edge $(v_j^{l-delay(v_j, v_i) \bmod N}, v_i^l)$ to $E^N$

   Set $delay(v_j^{l-delay(v_j, v_i) \bmod N}, v_i^l) = \lfloor delay(v_j, v_i)/N \rfloor$

   **Endfor**

**Endfor**

Figure 5.8. An algorithm for unfolding HSDFGs.

It is more difficult to estimate execution times for blocks that contain control constructs such as data-dependent iterations and conditionals within their body, and when the target processor employs pipelining and caching. Also, it is difficult, if not impossible, for the programmer to provide reasonably accurate estimates of execution times for blocks written in a high-level language (as in the C code generation library in Ptolemy). The solution adopted in the GRAPE system [LEP90] is to automatically estimate these execution times by compiling the block (if necessary) and running it by itself in a loop on an instruction-set simulator for the target processor. To take into account data-dependent execution behavior, different input data sets can be provided for the block during simulation. Either the worst-case or the average-case execution time is used as the final estimate.

The estimation procedure employed by GRAPE is obviously time consuming; in fact, estimation turns out to be the most time-consuming step in the GRAPE design flow. Analytical techniques can be used instead to reduce this estimation time; for example, Li and Malik [LM95] have proposed algorithms for estimating the execution time of embedded software. Their estimation technique, which forms a part of a tool called **cinderella**, consists of two components: 1) determining the sequence of instructions in the program that results in maximum execution time (program path analysis) and 2) modeling the target processor to determine how much time the worst case sequence determined in step 1 takes to execute (micro-architecture modeling). The target processor model also takes the effect of instruction pipelines and cache activity into account. The input to the tool is a generic C program with annotations that specify the loop bounds (i.e., the maximum number of iterations for which a loop runs). Although the problem is formulated as an integer linear program (ILP), the claim is that practical inputs to the tool can be efficiently analyzed using a standard ILP solver. The advantage of this approach, therefore, is the efficient manner in which estimates are obtained as compared to simulation.

It should be noted that the program path analysis component of the Li and Malik technique is, in general, an undecidable problem; therefore, for these techniques to function, the programmer must ensure that his or her program does not contain pointer references, dynamic data structures, recursion, etc., and must provide bounds on all loops. Li and Malik's technique also depends on the accuracy of the processor model, although one can expect good models to eventually evolve for DSP chips and microcontrollers that are popular in the market.

The problem of estimating execution times of blocks is central for us to be able to effectively employ compile time design techniques. This problem is an important area of research in itself, and the strategies employed in Ptolemy and GRAPE, and those proposed by Li and Malik are useful techniques, and we expect better estimation techniques to be developed in the future.

## 5.10     Summary

In this chapter, we discussed various scheduling models for dataflow graphs on multiprocessor architectures that differ in whether scheduling decisions are made at compile time or at run-time. The scheduling decisions are actor assignment, actor ordering, and determination of exact firing times of each actor. A fully-static strategy lies at one extreme, in which all of the scheduling decisions are made at compile time, whereas a dynamic strategy makes all scheduling decisions at run-time. The trade-off involved is the low complexity of static techniques against the greater generality and tolerance to data dependent behavior in dynamic strategies.

For dataflow-oriented signal processing applications, the availability of compile time information makes static techniques very attractive. A self-timed strategy is commonly employed for such applications, where actor assignment and ordering is fixed at compile time (or system design time) but the exact firing time of each actor is determined at run-time, in a data driven fashion. Such a strategy is easily implemented in practical systems through sender-receiver synchronization during interprocessor communication.

# 6

---

# IPC-CONSCIOUS SCHEDULING
# ALGORITHMS

---

In this chapter, we focus on techniques that are used in self-timed scheduling algorithms to handle IPC costs. Since a tremendous variety of scheduling algorithms have been developed to date, it is not possible here to provide comprehensive coverage of the field. Instead, we highlight some of the most fundamental developments to date in IPC-conscious multiprocessor scheduling strategies for HSDFGs.

## 6.1    Problem Description

To date, most of the research on scheduling HSDFGs has focused on the problem of minimizing the schedule *makespan* $\mu$, which is the time required to execute all actors in the HSDFG once. When a schedule is executed repeatedly — for example by being encapsulated within an infinite loop, as would typically be the case for a DSP application — the resulting throughput is equal to the reciprocal $(1/\mu)$ of the schedule makespan if all processors synchronize (perform a "barrier synchronization" as described later in Section 10.1) at the end of each schedule iteration. The throughput can often be improved beyond $(1/\mu)$ by abandoning a global barrier synchronization, and implementing a self-timed execution of the schedule, as described in Section 5.4 [Sri95]. To further exploit parallelism between graph iterations, one may employ the technique of unfolding, which is discussed in Section 5.8.

To model the transit time of interprocessor communication data in a multiprocessor system (for example, the time to write and read data values to and from a shared memory), HSDF edges are typically weighted by an estimate of the delay to transmit and receive the associated data if the source and sink actors of the edge are assigned to different processors. Such **IPC cost** estimates are similar to the execution time estimates that we use to model the run-time of individual dataflow actors, as discussed in Section 5.9.

In this chapter, we are concerned primarily with efficient scheduling of HSDFGs in which an IPC cost is associated with each edge. We refer to the problem of constructing a minimum makespan schedule for such an HSDFG and for a given target multiprocessor architecture as the **scheduling problem**. As this definition suggests, solutions to the scheduling problem are heavily dependent on the underlying target architecture. In an attempt to decompose this problem and separate target-specific aspects from aspects of the problem that are fundamental to the structure of the input HSDFG, some researchers have applied a two-phased approach, pioneered by Sarkar [Sar89], to addressing the scheduling problem. The first phase involves scheduling the input HSDFG onto an **infinite-resource multiprocessor architecture (IRMA)**, which consists of an infinite number of processors that are interconnected by a **fully-connected** interconnection network. In a fully-connected network, any number of processors can perform interprocessor communication simultaneously. The second phase of the decomposed scheduling process involves mapping the IRMA schedule that is derived in the first phase onto the given (resource-constrained) target multiprocessor architecture.

Clearly, the complexity of this second phase of the scheduling process is heavily dependent on the target architecture. For certain degenerate cases, such as a uniprocessor target or a chain-structured processor interconnection topology, optimal solutions can be derived easily; however, in general, this phase is highly complex. The first phase, scheduling an HSDFG onto an IRMA, is NP-complete even for some very restricted special cases [Pra87, Chr89, PY90]. Thus, research on IPC-conscious algorithms to address the scheduling problem has typically focused on the derivation of effective heuristic rather than exact algorithms.

Since the interiteration dependencies represented by SDF delays are not relevant in the context of minimum-makespan scheduling, HSDF edges that have delays are typically ignored in the scheduling problem. Thus, the input to an algorithm for the scheduling problem is usually a delayless HSDFG. Such a graph is often referred to as a **task graph**. More precisely, a task graph is an HSDFG application graph specification $(V, E)$ such that for each $e \in E$, we have $delay(e) = 0$. In the situation that the original application is specified as an SDF graph, the task graph is identical to the APEG (Section 3.8) obtained by expanding the SDF graph and removing edges with non-zero delays.

## 6.2    Stone's Assignment Algorithm

A classic algorithm for computing an assignment of actors to processors based on network flow principles was developed by Stone [Sto77]. This algorithm is designed for heterogeneous multiprocessor systems, and its goal is to map actors to processors so that the sum of computation time and time spent on IPC is minimized. More specifically, suppose that we are given a target multiprocessor architecture consisting of $n$ (possibly heterogeneous) processors $P_1, P_2, ..., P_n$; a set of actors $A_1, A_2, ..., A_m$; a set of actor execution times $\{t_i(A_j)\}$, where for each $i \in \{1, 2, ..., n\}$ and each $j \in \{1, 2, ..., m\}$, $t_i(A_j)$ gives the execution time of actor $A_j$ on processor $P_i$; and a set of inter-actor communication costs $\{C_{ij}\}$, where $C_{ij} = 0$ if actors $A_i$ and $A_j$ do not exchange data, and otherwise, $C_{ij}$ gives the cost of exchanging data between $A_i$ and $A_j$ if $A_i$ and $A_j$ are assigned to different processors. The goal of Stone's assignment algorithm is to compute an assignment

$$F : \{A_1, A_2, ..., A_m\} \rightarrow \{P_1, P_2, ..., P_n\} \tag{6-1}$$

such that the net computation and communication cost

$$\text{cost}(F) = \left( \sum_{j=1}^{m} t_{F(A_j)}(A_j) \right) + \left( \sum_{F(A_i) \neq F(A_j)} C_{ij} \right) \tag{6-2}$$

is minimized. Note that minimizing (6-2) is not equivalent to minimizing the makespan. For example, if the set of target processors is homogeneous, then an optimal solution with respect to (6-2) results from simply assigning all actors to a single processor.

The core of the algorithm is an elegant approach for transforming a given instance $I$ of the assignment problem into an instance $Z(I) = (V(I), E(I))$ of the minimum-weight cutset problem. For example, for a two-processor system, two vertices $p_1$ and $p_2$ are created in $Z(I)$ corresponding to the two heterogeneous target processors, and a vertex $a_i$ is created for each actor $A_i$. For each $A_i$, an undirected edge in $Z(I)$ is instantiated between $a_i$ and $p_1$, and the weight of this edge is set to the execution time $t_2(a_i)$ of $a_i$ on $p_2$. This edge models the execution time cost of actor $A_i$ that results if $a_i$ and $p_1$ lie on opposite sides of the cutset that is computed for $Z(I)$. Similarly an edge $(a_i, p_2)$ is instantiated with weight $w((a_i, p_2)) = t_1(a_i)$. Finally, for each pair of vertices $a_i$ and $a_j$ such that $C_{ij} \neq 0$, an edge $(a_i, a_j)$ in $Z(I)$ is instantiated with weight $w((a_i, a_j)) = C_{ij}$.

From a minimum-weight cutset in $Z(I)$ that separates $p_1$ and $p_2$, an optimal solution to the heterogeneous processor assignment problem can easily be derived. Specifically, if $R \subseteq E(I)$ is a minimum-weight cutset that separates $p_1$ and $p_2$, an optimal assignment can be derived from $R$ by:

$$F(A_i) = \begin{cases} P_1 \text{ if } ((a_i, p_2) \in R) \\ P_2 \text{ if } ((a_i, p_1) \in R) \end{cases} \text{ for } i = 1, 2, ..., m. \tag{6-3}$$

The net computation and communication cost of this assignment is simply

$$\text{cost}(F) = \sum_{e \in R(I)} w(e). \tag{6-4}$$

An illustration of Stone's Algorithm is shown in Figures 6.1 and 6.2. Figure 6.2(a) shows the actor interaction structure of the input application; Figure 6.2(b) specifies the actor execution times; and Figure 6.2(c) gives all nonzero communication costs $C_{ij} \neq 0$. The associated instance $Z(I)$ of the minimum-weight cutset problem that is derived by Stone's Algorithm is depicted in Figure 6.1(a), and a minimum-weight cutset

$$R = \{(a_2, p_1), (a_3, p_1), (a_4, p_2), (a_1, p_2)\} \tag{6-5}$$

for $Z(I)$ is shown in Figure 6.1(b). The optimal assignment that results from this cutset is given by

$$F(A_2) = F(A_3) = P_1, \text{ and } F(A_1) = F(A_4) = P_2. \tag{6-6}$$

For the two-processor case ($n = 2$), a variety of efficient algorithms can be used to derive a minimum weight cutset for Stone's construction $C(I)$ [CLR92]. When the target architecture contains more than two processors ($n \geq 3$), the weight of each edge $(a_i, p_j)$ is set to a weighted sum of the values $\{t_j(A_i) | (1 \leq j \leq n)\}$, and an optimal assignment $F$ is derived by computing a minimum $n$-way cutset in $C(I)$. When $n \geq 4$, Stone's approach becomes com-

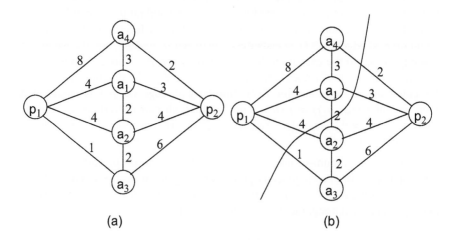

(a)                                    (b)

Figure 6.1. (a) The instance of the minimum-weight cutset problem that is derived from the example of Figure 6.2. (b) An illustration of a solution to this instance of the minimum-weight cutset problem.

(a)

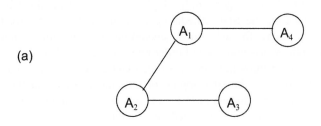

(b)

| $i$ | $t_1(A_i)$ | $t_2(A_i)$ |
|---|---|---|
| 1 | 3 | 4 |
| 2 | 4 | 4 |
| 3 | 6 | 1 |
| 4 | 2 | 8 |

(c)

| $i$ | $j$ | $C_{ij} = C_{ji}$ |
|---|---|---|
| 1 | 4 | 3 |
| 1 | 2 | 2 |
| 2 | 3 | 2 |

Figure 6.2. An example that is used to illustrate Stone's Algorithm for computing heterogeneous processor assignments.

putationally intractable.

Although Stone's algorithm has high intuitive appeal and has had considerable influence on the SDF scheduling community, the most effective algorithms known today for self-timed scheduling of SDF graphs have jointly considered both the assignment and ordering subproblems. The approaches used in these joint algorithms fall into two broad categories — approaches that are driven by iterative, list-based mapping of individual tasks, and those that are based on constructing clusters of tasks that are to be executed on the same processor. These two categories of scheduling techniques are discussed in the following three sections.

## 6.3     List Scheduling Algorithms

In classical **list scheduling**, a *priority list* $L$ of actors in constructed; a global time clock $c_G$ is maintained; and each task $T$ is eventually mapped into a time interval $[x_T, y_T]$ on some processor (the time intervals for two distinct actors assigned to the same processor cannot overlap). The priority list $L$ is a linear ordering $(v_1, v_2, ..., v_{|V|})$ of the actors in the input task graph $G = (V, E)$ ($V = \{v_1, v_2, ..., v_{|V|}\}$) such that for any pair of distinct actors $v_i$ and $v_j$, $v_i$ is to be given higher scheduling priority than $v_j$ if and only if $i < j$. Each actor is mapped to an available processor as soon as it becomes the highest-priority actor — according to $L$ — among all actors that are *ready*. An actor is **ready** if it has not yet been mapped, but its predecessors have all been mapped, and all satisfy $y_T \leq t$, where $t$ is the current value of $c_G$. For self-timed implementation, actors on each processor are ordered according to the order of their associated time intervals.

An important generalization of list scheduling, which we call **ready-list scheduling**, has been formalized by Printz [Pri91]. Ready-list scheduling maintains the list-scheduling convention that a schedule is constructed by repeatedly selecting and scheduling ready actors, but eliminates the notion of a static priority list and a global time clock. Thus, the only list that is fundamental to the scheduling process is the list of actors that are ready at a given scheduling step. The development of effective ready-list algorithms for the scheduling problem was pioneered by Hwang, Chow and Angers [HCA89] (in the *ETF Algorithm*), and by Sih and Lee [SL90, SL93a] (in the *DLS Algorithm*). In this section, we present an overview of list-scheduling and ready-list techniques, including ETF and DLS.

To be effective when IPC costs are not negligible, a list-scheduling or ready-list algorithm must incorporate the latencies associated with IPC operations. This involves either explicitly scheduling IPC operations onto the communication resources of the target architecture as the scheduling process progresses, or incorporating estimates of the time that it takes for data that is produced by an actor on one processor to be available for consumption by an actor that has been assigned to another processor. In either case, an additional constraint is imposed on the earliest possible starting times of actors that depend on the arrival of IPC data.

### 6.3.1    Graham's Bounds

A number of useful results and interesting properties have been shown to hold for list-scheduling algorithms. Most of these results and properties have been demonstrated in the context of an idealized special-case of the scheduling problem, which we call the **ideal scheduling problem**. In ideal scheduling, the target multiprocessor architecture consists of a set of processors that is homogeneous with respect to the task graph actors (the execution time of an actor is independent of the processor it is assigned to), and IPC is performed in zero time. Although, issues of heterogeneous processing times and IPC cost are avoided, the ideal scheduling problem is intractable [GJ79].

When list scheduling is applied to an instance of the ideal scheduling problem and a given priority list for the problem instance, the resulting schedule is not necessarily unique, and generally depends on the details of the particular list scheduling algorithm that is used. Specifically, the schedule depends on the processor selection scheme that is used when more than one processor is available at a given scheduling step. For example, consider the simple task graph in Figure 6.3(a); and suppose that $t(A) = t(B) = t(C) = 1$, and the target multiprocessor architecture consists of two processors $P_1$ and $P_2$. If list scheduling is applied to this example with priority list $L = (A, B, C)$, then any one of the four schedules illustrated in Figure 6.3(b) may result, depending on the processor selection scheme.

Now let $ISP(G, t, n)$ denote the instance of the ideal scheduling problem that consists of task graph $G = (V, E)$; actor execution times (on each processor in the target architecture) $\{t(v) | (v \in V)\}$; and a target, zero-IPC architecture that consists of $n$ identical processors. Then given a list-scheduling algorithm $A$ and a priority list $(v_1, v_2, ..., v_{|V|})$ for $G$, we define $S_A(G, t, n, L)$ to be the schedule produced by $A$ when it is applied to $ISP(G, t, n)$ with priority list $L$, and $\mu_A(G, t, n, L)$ to be the makespan of $S_A(G, t, n, L)$. We also define

$$\Sigma(G, t, n, L) \equiv \{S_A(G, t, n, L) | A \text{ is a list scheduling algorithm}\}. \qquad (6\text{-}7)$$

Thus, $\Sigma(G, t, n, L)$ is the set of schedules that can be produced when a list scheduling is applied to $ISP(G, t, n)$ with priority list $L$. For the example of Figure 6.3, we have

$$\Sigma(G, t, n) = \{S_1, S_2, S_3, S_4\}. \qquad (6\text{-}8)$$

It is easily shown that schedules produced by list-scheduling algorithms on a given instance ISP($G$, $t$, $n$) all have the same makespan. That is,

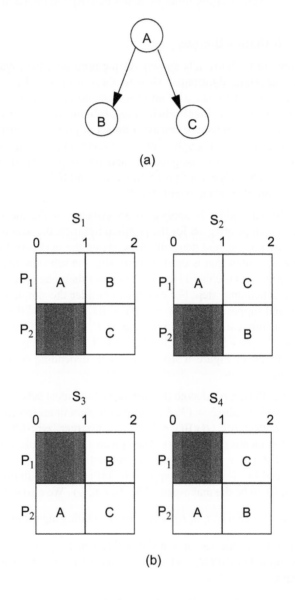

Figure 6.3. An example that illustrates the dependence of list scheduling on processor selection (for a given priority list).

$$|\{\mu_A(G, t, n, L)|(A \in \Sigma(G, t, n, L))\}| = 1. \qquad (6\text{-}9)$$

This property of uniform makespan does not generally hold, however, if we allow heterogeneous processors in the target architecture or if we incorporate nonzero IPC costs.

Clearly, effective construction of the priority list $L$ is critical to achieving high-quality results with list scheduling. Graham has shown that when arbitrary priority lists are allowed, it is possible for the list scheduling approach to produce unusual results. In particular, it is possible that *increasing* the number of processors, *reducing* the execution time of one or more actors, or *relaxing* the precedence constraints in an SDF graph (removing one or more edges) can all cause a list scheduling algorithm to produce results that are *worse* (longer total execution time) than those obtained when the algorithm is applied with the original number of processors, the original set of SDF edges, or original execution times respectively [Gra69].

Graham, however, has established a tight bound on the anomalous performance degradation that can be encountered with list scheduling [Gra69]. This result is summarized by the following theorem.

**Theorem 6.1:** Suppose that $G = (V, E)$ is task graph; $L = (v_1, v_2, ..., v_{|V|})$ is a priority list for $G$; $n$ and $n'$ are positive integers such $n' \geq n$; $t : V \to \{0, 1, 2, ...\}$ and $t' : V \to \{0, 1, 2, ...\}$ are assignments of nonnegative integers to members of $V$ (sets of actor execution times) such that for each $v \in V$, $t'(v) \leq t(v)$; $E' \subseteq E$; $S \in \Sigma(G, t, n, L)$ and $S' \in \Sigma(G', t', n', L)$, where $G' = (V, E')$. Then

$$\frac{\mu(S')}{\mu(S)} \leq 1 + \frac{n-1}{n'}, \qquad (6\text{-}10)$$

and this is the tightest possible bound.

Graham has also established a tight bound on the variation in list scheduling performance that can be encountered when different priority lists are used for the same instance of the ideal scheduling problem.

**Theorem 6.2:** Suppose that $G = (V, E)$ is task graph; $L$ and $L'$ are a priority lists for $G$; $n$ is a positive integer; $t : V \to \{0, 1, 2, ...\}$ are assignments of execution times to members of $V$; $S \in \Sigma(G, t, n, L)$; and $S' \in \Sigma(G, t, n, L')$. Then

$$\frac{\mu(S')}{\mu(S)} \leq 2 - \frac{1}{n}, \qquad (6\text{-}11)$$

and this is the tightest possible bound.

### 6.3.2        The Basic Algorithms — HLFET and ETF

In SDF list scheduling, the **level** of an actor $A$ in an acyclic SDF graph $G$ is defined to be the length of the longest directed path in $G$ that originates at $A$. Here, the length of a path is taken to be the sum of the execution times of the actors on the path. Intuitively, actors with high level values need to be scheduled early since long sequences of computation depend on their completion. One of the earliest and most widely-used list-scheduling algorithms is the HLFET (highest level first with estimated times) algorithm [ACD74, Hu61]. In this algorithm, the priority list $L$ is created by sorting the actors in decreasing order of their levels. HLFET is guaranteed to produce an optimal result if there are only two processors, both processors are identical, and all tasks have identical execution times [Hu61, ACD74]. For the general ideal scheduling problem (any finite number of homogenous processors is allowed, actor execution times need not be identical, and IPC costs are uniformly zero), HLFET has been proven to frequently produce near-optimal schedules [ACD74, Koh75].

Early strategies for incorporating IPC costs into list scheduling include the algorithm of Yu [Yu84]. Yu's algorithm, a modification of HLFET scheduling, repeatedly selects the ready actor that has the highest level, and schedules it on the processor that can finish its execution at the earliest time. The earliest finishing time of a ready actor $A$ on a processor $P$ depends both on the time intervals that have already been scheduled on $P$, and on the IPC time required for the data required from the predecessors of $A$ to arrive at $P$.

In contrast to these early algorithms, the ETF (earliest task first) algorithm of Hwang, Chow and Angers uses the level metric only as a tie-breaking criterion. At each scheduling step in ETF, the value $t_e(A, P)$ — the earliest time at which actor $A$ can commence execution on processor $P$ — is computed for every ready actor $A$ and every target processor $P$. If an actor-processor pair $(A^*, P^*)$ uniquely minimizes $t_e(A, P)$ then $A^*$ is scheduled to execute on $P^*$ starting at time $t_e(A^*, P^*)$; otherwise, the tie is resolved by selecting the actor-processor pair that has the highest level.

### 6.3.3        The Mapping Heuristic

El-Rewini and Lewis have proposed a list-scheduling algorithm, called the **mapping heuristic** (MH) [ERL90], which attempts to account for IPC latencies within an arbitrary, multi-hop interconnection network. Since maintaining and applying a precise accounting of traffic within such a network can be computationally expensive, the MH algorithm has been devised to maintain an approximate view of network state $\Xi(t)$ from which reasonable estimates of communication delay can be derived for scheduling purposes. At any given scheduling time step $t$, $\Xi(t)$ incorporates three $P \times P$ matrices $H$, $L$, and $D$, where $P$ is the set of processors in the target multiprocessor architecture. Given $p_1, p_2 \in P$, $H(p_1, p_2)$ gives the number of hops between $p_1$ and $p_2$ in the interconnection network; $L(p_1, p_2)$ gives the preferred outgoing communication channel of $p_1$ that should be used when communicating data to $p_2$; and $D(p_1, p_2)$ gives the communication delay between $p_1$ and $p_2$ that arises due to contention with other IPC operations in the system.

In the MH algorithm, actors are first prioritized by a static, modified level metric $l_{mh}$ that incorporates the communication costs that are assigned to the task graph edges. For an actor $x$, $l_{mh}(x)$ is the longest path length in the task graph that originates at $x$, where the length of a path is taken as the sum of the actor execution times and edge communication costs along the path. A list-scheduling loop is then carried out in which at any given scheduling time step $t$, an actor $x^*$ that maximizes $l_{mh}(\bullet)$ is selected from among the actors that are ready at $t$. Processor selection is then achieved by assigning $x^*$ to the processor that allows the earliest estimated completion time. This estimated completion time is derived from the network state approximation $\Xi(t)$.

El-Rewini and Lewis observe that there is a significant trade-off between the frequency with which the network state approximation $\Xi(t)$ is updated (which affects the accuracy of the approximation), and the time complexity of the resulting scheduling algorithm. The MH algorithm addresses this trade-off by updating $\Xi(t)$ only when a scheduled actor begins sending IPC data to a successor actor that is scheduled on another processor, or when the IPC data associated with a task graph edge $(x, y)$ arrives at the processor that $y$ is assigned to.

Loosely speaking, the priority mechanism in the MH algorithm is the converse of that employed in ETF. In MH, the "earliest actor-processor mapping" is used as a tie-breaking criterion, while the modified level $l_{mh}$ is used as the primary priority function. Note also that when the target processor set $P$ is homogeneous, selecting an actor-processor pair that minimizes the starting time is equivalent to selecting a pair that minimizes completion time, while this equivalence does not necessarily hold for a heterogeneous architecture. Thus, the concept of "earliest actor-processor mapping" that is employed by ETF is different from that in MH only in the heterogeneous processor case.

### 6.3.4    Dynamic Level Scheduling

In the DLS (dynamic level scheduling) algorithm of Sih and Lee, the use of levels in traditional HLFET scheduling is replaced by a measure of scheduling priority that is to be continually re-evaluated as the schedule is constructed [SL93a]. Sih and Lee demonstrated that such a **dynamic level** concept is preferable because the "scheduling affinity" between actor-processor pairs depends not only on longest paths in the task graph, but also on the current *scheduling state*, which includes the actor/time-interval pairs that have already been scheduled on processing resources, and the IPC operations that have already been scheduled on the communication resources.

As with ETF, the DLS algorithm also abandons the use of the global scheduling clock $c_G$, and allows all target processors to be considered as candidates in every scheduling step (instead of just those processors that are idle at the current value of $c_G$). With the elimination of $c_G$, Sih's metric for prioritizing actors — the *dynamic level* — can be formulated as

$$L(A) - max\{D(A, P, \alpha), F(P, \alpha)\}, \tag{6-12}$$

where $\alpha$ represents the scheduling state at the current scheduling step; $L(A)$ denotes the conventional (static) level of actor $A$; $D(A, P, \alpha)$ denotes the earliest time at which all data required by actor $A$ can arrive at processor $P$; and $F(P, \alpha)$ gives the completion time of the last actor that is presently assigned to $P$.

While the incorporation of scheduling state by the DLS algorithm represents an important advancement in IPC-conscious scheduling, the formulation of the dynamic level in (6-12) contains a subtle limitation, which was observed by Kwok and Ahmad [KA96]. This limitation arises because the relative contributions of the two components in (6-12) (the static level and the data arrival time) to the dynamic level metric vary as the scheduling process progresses. Early in the scheduling process, the static levels are usually high, since the actors considered generally have relatively many topological descendants, and similarly, data arrival times are low, since the scheduling of actors on each processor begins at the origin of the time axis and progresses towards increasing values of time. As more and more scheduling steps are carried out, the static level parameters of ready actors will decrease steadily (implying lower influence on the dynamic level), and the data arrival times will increase (implying higher influence). Thus, the relative weightings of the static level and the data arrival time are not constant, but rather can vary strongly between different scheduling steps. This variation is not taken to account in the DLS algorithm.

### 6.3.5    Dynamic Critical Path Scheduling

Motivated partly by their observation on the limitations of dynamic level scheduling, Kwok and Ahmad have developed an alternative variation of list scheduling, called the DCP (dynamic critical path) algorithm, that also dynamically re-evaluates actor priorities [KA96]. The DCP algorithm is motivated by the observation that the set of *critical paths* in a task graph can change from one scheduling step to the next, where the critical path is defined to be a directed path along which the sum of computation and communication times is maximized.

For example, consider the task graph depicted in Figure 6.4. Here, the number beside each actor gives the execution time of the actor, and each numeric edge weight gives the IPC cost associated with the edge. Initially, in this graph, the critical path is $A \to B \to C \to F$, and the length of this path is 16 time units. If the first two scheduling steps map both actors $A$ and $B$ to the same processor (e.g., to minimize the starting time of $B$), then the weight of the edge $(A, B)$ in Figure 6.4 effectively changes to zero. The critical path in the new "partially scheduled" graph thus becomes the path $A \to D \to E \to F$, which has a length of 14 time units. Because critical paths can change in this manner as the scheduling process progresses, the critical path of the partially scheduled graph is called the **dynamic critical path**.

The DCP algorithm operates by repeatedly selecting and scheduling actors on the dynamic critical path, and updating the partially scheduled graph as scheduling decisions are made. An elaborate processor selection scheme is also incorporated to map the actor selected at each scheduling step. This scheme not only considers the arrival of required data on each candidate processor, but also takes into account the possible starting times of the task graph successors of the selected actor [KA96].

## 6.4    Clustering Algorithms

**Clustering algorithms** for multiprocessor scheduling operate by incrementally constructing groupings, called **clusters**, of actors that are to be executed on the same processor. Clustering and list scheduling can be used in a complementary fashion. Typically, clustering is applied to focus the efforts of a list-scheduler on effective processor assignments. When used efficiently, clustering can significantly enhance the results produced by a list scheduler (and a variety of other scheduling techniques).

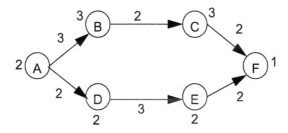

Figure 6.4. An illustration of "dynamic" critical paths in multiprocessor scheduling.

A scheduling algorithm (such as a list scheduler) processes a clustered HSDFG by constraining the vertices of $V$ that are encompassed by each cluster to be assigned to the same processor. More than one cluster may be mapped by the scheduling algorithm to execute on the same processor; thus, a sequence of clustering operations does not necessarily specify a complete processor assignment, even when the target processors are all homogeneous.

The net result of a clustering algorithm is to identify a family of disjoint subsets $M_1, M_2, ..., M_k \subseteq V$ such that the underlying scheduling algorithm is forced to avoid IPC costs between any pair of actors that are members of the same $M_i$. In the remainder of this section, we examine a variety of algorithms for computing such a family of subsets.

### 6.4.1        Linear Clustering

In the *Linear Clustering Algorithm* of Kim and Browne, longest paths in the input task graph are iteratively identified and clustered until every edge in the graph is either encompassed by a cluster or is incident to a cluster at both its source and sink. The path length metric is based on a function of the computation and communication along a given path. If $p = (e_1, e_2, ..., e_n)$ is a task graph path, then the value of Kim and Browne's path length metric for $p$ is given by

$$\alpha \sum_{v \in T_p} t(v) + (1 - \alpha)\left(\beta \sum_{i=1}^{n} c(e_i) + (1 - \beta) \sum_{v \in T_p} c_{\text{adj}}(v)\right), \qquad (6\text{-}13)$$

where

$$T_p = \{src(e_1)\} \cup \left(\bigcup_{i=1}^{n} snk(e_i)\right) \qquad (6\text{-}14)$$

is the set of actors traversed by $p$; $c(e)$ is the IPC cost associated with an edge $e$;

$$c_{\text{adj}}(v) = \sum_{((v,x) \in V) \text{ and } x \notin T_p} c((v, x)) + \sum_{((x,v) \in V) \text{ and } x \notin T_p} c((x, v)) \qquad (6\text{-}15)$$

is the total IPC cost between an actor $v \in T_p$ and actors that are not contained in $T_p$; and the "normalization factors" $\alpha$ and $\beta$ are parameters of the algorithm.

Kim and Browne do not give a systematic technique for determining the normalization factors that should be used with the Linear Clustering Algorithm. Indeed, the derivation of the most appropriate normalization factors — based on characteristics of the input task graph and the target multiprocessor architecture — appears to be an interesting direction for further study. When $\alpha = 0.5$ and

$\beta = 1$, linear clustering reduces to clustering of critical paths.

## 6.4.2    Internalization

Sarkar's *Internalization* algorithm [Sar89] for graph-clustering is based on determining a set of clustering operations that do not degrade the performance of a task graph on a machine with boundless processing resources (i.e., an IRMA). In internalization, the task graph edges are first sorted in decreasing order of their associated IPC costs. The edges in this list are then traversed according to this ordering. When each edge $e$ is visited in this traversal, an estimate $T_e$ of the parallel execution time is computed with the source and sink vertices of $e$ constrained to execute on the same processor. This estimate is derived for an unbounded number of processors, and a fully-connected communication network. If $T_e$ does not exceed the parallel execution time estimate of the current clustered graph, then the current clustered graph is modified by merging the source and sink of $e$ into the same cluster.

An important strength of internalization is its simplicity, which makes it easily adaptable to accommodate additional scheduling objectives beyond minimizing execution time. For example, a hierarchical scheduling framework for multirate DSP systems has been developed using Sarkar's clustering technique as a substrate [PBL95]. This hierarchical framework provides a systematic method for combining multiprocessor scheduling algorithms that minimize execution time with uniprocessor scheduling techniques that optimize a target program's code and data memory requirements.

## 6.4.3    Dominant Sequence Clustering

The DSC (dominant sequence clustering) algorithm of Yang and Gerasoulis incorporates principles similar to those used in the DCP algorithm, but applies these principles under the methodology of clustering rather than list scheduling [YG94]. As with DCP, a "partially scheduled graph" (**PSG**) is repeatedly examined and updated as scheduling steps are carried out. The IPC costs of intracluster edges in the PSG are all zero; all other IPC costs are the same as the corresponding costs in the task graph. Additionally, the DSC algorithm inserts new intracluster edges into the PSG so that a linear (total) ordering of actors is always maintained within each cluster.

Initially, each actor in the task graph is assigned to its own cluster. Each clustering step selects a task graph actor that has not been selected in any previous clustering step, and determines whether or not to merge the selected actor with one of its predecessors in the PSG. The selection process is based on a priority function $f(A)$, which is defined to be the length of the longest path (computation and communication time) in the PSG that traverses actor $A$. This priority function fully captures the concept of dynamic critical paths: an actor maximizes $f(\bullet)$ if, and only if, it lies on a critical path of the PSG.

At a given clustering step, an actor is selected if it is "free" — which means that all of its PSG predecessors have been selected in previous clustering steps — and it maximizes $f(\bullet)$ over all free actors. Thus, if a free actor exists that is on a dynamic critical path, then an actor on the dynamic critical path will be selected. However, it is possible that none of the free actors are on the PSG critical path. In such cases, the selected actor is not on the critical path (in contrast, the DCP algorithm always selects actors that are on the dynamic critical path).

Once an actor $A$ is "selected," its predecessors are sorted in decreasing order of the sum of $t(x) + c(x) + \lambda(x)$, where $t(x)$ is the execution time of predecessor $x$, $c(x)$ is the IPC cost of edge $(x, A)$, and $\lambda(x)$ is the length of the longest direct path in the PSG that terminates at $x$. A set of one or more predecessors $x_1, x_2, ..., x_n$ is then chosen from the head of this sorted list such that "zeroing" (setting the IPC cost to zero) the associated output edges $(x_1, A), (x_2, A), ..., (x_n, A)$ minimizes the value of $\lambda(A)$, and hence $f(A)$, in the new PSG that results from clustering the subset of PSG vertices $\{x_1, x_2, ..., x_n, A\}$.

The DSC algorithm was designed with low computational complexity as the primary objective. The algorithm achieves a time complexity of $O((N + E)\log N)$, where $N$ is the number of task graph actors, and $E$ is the number of edges. In contrast, linear clustering is $O(N(N + E))$ [GY92]; linearization is an $O(E(N + E))$ algorithm; ETF is $O(PN^2)$, where $P$ is the number of target processors; MH is $O(P^3 N^2)$; DLS is $O(N^3 Pg(P))$, where $g$ is the complexity of the data routing algorithm that is used to compute $D(A, P, \alpha)$; DCP is $O(N^3)$; and the *Declustering Algorithm,* discussed in Section 6.4.4 below, has complexity $O(N^3(N + P))$.

As with internalization, DSC is designed for a fully connected network containing an unbounded number of processors, and for practical, processor-constrained systems it can be used as a preprocessing or intermediate compilation phase. As discussed in Section 6.1, optimal scheduling in the presence of IPC costs is intractable even for fully connected, infinite processor systems, and thus, given the polynomial complexity of DSC and internalization, we cannot expect guaranteed optimality from these algorithms. However, DSC is shown to be optimal for a number of nontrivial subclasses of task graphs [YG94].

### 6.4.4      Declustering

Sih and Lee have developed a clustering approach called **declustering** that is based on examining pairs of paths in the task graph to systematically determine which instances of parallelism should be preserved during the clustering process

[SL93b]. Rather than exhaustively examining all pairs of paths (in general, a task that is hopelessly time-consuming) the declustering technique focuses on paths that originate at **branch actors**, which are actors that have multiple successors.

Branch actors are examined in increasing order of their static levels. Examination of a branch actor $B$ begins by sorting its successors in decreasing order of their static levels. The two successors $C_1$ and $C_2$ at the head of this list (highest static levels) are then categorized as being either an *Nbranch* ("non-intersecting branch") or an *Ibranch* ("intersecting branch") pair of path origins. To perform this categorization, it is necessary to compute the **transitive closure** of $C_1$ and $C_2$. The transitive closure of an actor $X$, denoted $TC(X)$, in a task graph $G$ is simply the set of actors $Y$ such that there is a delayless path in $G$ directed from $X$ to $Y$. Given the transitive closures $TC(C_1)$ and $TC(C_2)$, the successor pair $(C_1, C_2)$ is an **Nbranch** instance if

$$TC(C_1) \cap TC(C_2) = \varnothing, \tag{6-16}$$

and otherwise (if the transitive closures have non-empty intersection), $(C_1, C_2)$ is an **Ibranch** instance. Intuitively, the transitive closure is relevant to the derivation of parallel schedules since two actors can execute in parallel (execute over overlapping segments of time) if, and only if, neither actor is in the transitive closure of the other.

Once the branch-actor successor pair $(C_1, C_2)$ is categorized as being an Ibranch or Nbranch instance, a **two-path parallelism instance (TPPI)** is derived from it to determine an effective means for capturing the parallelism associated with $(C_1, C_2)$ within a clustering framework. If $(C_1, C_2)$ is an Nbranch instance, then the TPPI associated with $(C_1, C_2)$ is the subgraph formed by combining a longest path (cumulative execution time) from $C_1$ to any task graph sink actor (an actor that has no output edges), a longest path from $C_2$ to any task graph sink actor, the associated branch actor $B$, and the connecting edges $(B, C_1)$ and $(B, C_2)$.

For example, consider the task graph shown in Figure 6.5(a); for simplicity, assume that the execution time of each actor is unity; observe that the set of branch actors in this graph is $\{Q, T, U\}$; and consider the TPPI computation associated with branch actor $T$. The successors of this branch actor, $U$ and $V$, satisfy $TC(U) = \{W, X, Z\}$ and $TC(V) = \{Y\}$. Thus, we have $TC(U) \cap TC(V) = \varnothing$, which indicates that for branch actor $T$, the successor pair $(U, V)$ is an Nbranch instance. The TPPI associated with this Nbranch instance is shown in Figure 6.5(b).

If $(C_1, C_2)$ is an Ibranch instance then the TPPI associated with $(C_1, C_2)$ is derived by first selecting an actor $M$, called a **merge actor**, from the intersection $TC(U) \cap TC(V)$ that has maximal static level. The TPPI is then formed by combining a longest path from $C_1$ to $M$, a longest path from $C_2$ to $M$, and the connecting edges $(B, C_1)$ and $(B, C_2)$.

Among the branch actors in Figure 6.5(a), only actor $Q$ has an Ibranch instance associated with it. The corresponding TPPI, derived from merge actor $U$, is shown in Figure 6.5(c).

After a TPPI is identified, an optimal schedule of the TPPI onto a two processor architecture is derived. Because of the restricted structure of TPPI topologies, such an optimal schedule can be computed efficiently. Furthermore, depending on whether the TPPI corresponds to an Ibranch or an Nbranch instance, and on whether the optimal two-processor schedule utilizes both target processors, the optimal schedule can be represented by removing zero, one, or two edges, call **cut arcs**, from the TPPI: after removing the cut arcs from the TPPI, the (one or two) connected components in the resulting subgraph give the processor assignment associated with the optimal two-processor schedule.

The declustering algorithm repeatedly applies the branch actor analysis discussed above for all branch actors in the task graph, and keeps track of all cut arcs that are found during this traversal of branch actors. After the traversal is complete, all cut arcs are temporarily removed from the task graph, and the connected components of the resulting graph are clustered. These clusters are then combined (*hierarchical cluster grouping*) in a pairwise fashion — two clusters at a time — to produce a hierarchy of two-actor clusters. Careful graph analysis is used to guide this hierarchy formation to preserve the most useful instances of parallelism for as large a depth as possible within the cluster hierarchy.

Then, during the *cluster hierarchy decomposition* and *cluster breakdown* phases of the declustering algorithm, the cluster hierarchy is systematically broken down and scheduled to match the characteristics of the target multiprocessor architecture. For full details on *hierarchical cluster grouping*, *cluster hierarchy decomposition*, and *cluster breakdown*, the reader is encouraged to consult [Sih91, SL93b].

## 6.5     Integrated Scheduling Algorithms

Due in part to the high complexity of the assignment and ordering problems in the presence of IPC costs, independent comparisons on subsets of algorithms developed for the scheduling problem consistently reveal that no single algorithm dominates as a clear "best-choice" that handles most applications better than all of the other algorithms (for example, see [LAAG94, MKTM94]). Thus, an important challenge facing tool designers for application-specific multiprocessor implementation is the development of efficient methods for integrating the variety of algorithm innovations in IPC-conscious scheduling so that their advantages can be combined in a systematic manner. One example of an initial effort in this direction is the DS (dynamic selection) strategy [LAAG94]. DS is a

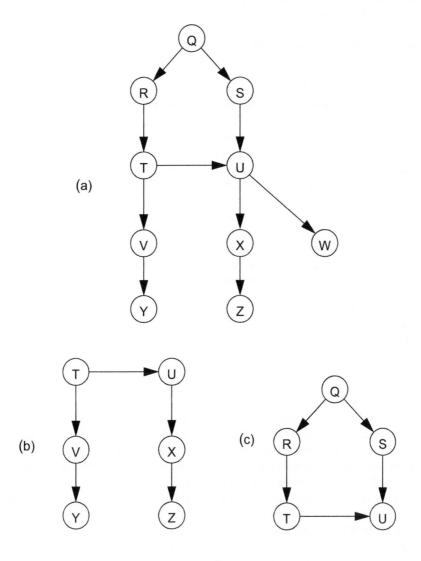

Figure 6.5. An illustration of TPPIs in the Declustering Algorithm.

list scheduling algorithm that compares the number of available processors $n_p$ to the number of executable (ready) actors $n_a$ at each scheduling step. If $(n_p \leq n_a)$, then one step of the DLS algorithm is invoked to complete the current scheduling step; otherwise, a minor variation of HLFET is applied to complete the step. This algorithm was motivated by experiments that revealed certain "regions of operation" in which scheduling algorithms exhibit particularly strong or weak performance compared to others. The performance of DS is shown to be significantly better than that of DLS or HLFET alone.

## 6.6    Pipelined Scheduling

Pipelined scheduling algorithms attempt to efficiently partition a task graph into stages, assign groups of processors to stages, and construct schedules for each pipeline stage. Under such a scheduling model, the *slowest pipeline stage* determines the throughput of the multiprocessor implementation. In general, pipelining can significantly improve the throughput beyond what is achievable by the classical (minimum-makespan) scheduling problem; however, this improvement in throughput may come at the expense of a significant increase in latency (e.g., over the latency that is achievable by employing a minimum makespan schedule). Research on pipelined scheduling is at a significantly less mature state than on the classical *scheduling problem* defined in Section 6.1. Due to its high relevance to DSP and multimedia applications, we expect that in the coming years, there will be increasing activity in the area of pipelined scheduling.

Bokhari developed fundamental results on the mapping of task graphs into pipelined schedules. Bokhari demonstrated an efficient, optimal algorithm for mapping a chain-structured task graph onto a linear chain of processors (the **chain pipelining** problem). Bokhari's algorithm is based on an innovative data structure, called the **layered assignment graph**, for modeling the chain pipelining problem.

Figure 6.6 illustrates an instance of the chain pipelining problem, and the corresponding layered assignment graph. The task graph to be scheduled; the linearly-connected target multiprocessor architecture; and the layered assignment graph associated with the given task graph and target architecture are shown in Figures 6.6(a), 6.6(b) and 6.6(c), respectively. The number above each task graph actor $A_i$ in Figure 6.6(a) gives the execution time $t(A_i)$ of $A_i$, and the number above each task graph edge gives the IPC cost from the associated source and sink actors if the source and sink are mapped to successive stages in the linear chain of target processors. If the source and sink actors of an edge are mapped to the same stage (processor), then the communication cost is taken to be zero.

Given an arbitrary chain-structured task graph $G = (V, E)$ consisting of actors $\{X_1, X_2, ..., X_n\}$ such that

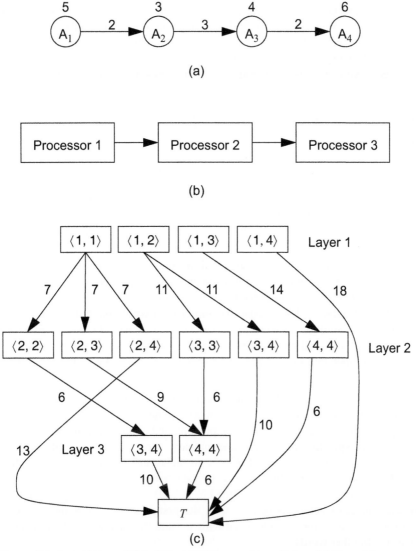

Figure 6.6. An instance of the chain pipelining problem ((a) and (b)), and the associated layered assignment problem.

$$E = \{(X_i, X_{i+1}) | 1 \leq i < n\},$$

and given a linearly connected multiprocessor architecture consisting of processors $P_1, P_2, ..., P_m$ such that each $P_i$ is connected to $P_{i+1}$ by a single unidirectional communication link, the associated layered assignment graph is constructed by first associating a layer $L_i$ of candidate "actor subchains" with each processor $P_i$ in the target architecture. Each layer consists of a vertex for each subsequence $(X_j, X_{j+1}, ..., X_{j+k})$, of task graph actors that may be assigned to processor $i$. When constructing this association of candidate subchains to layers, it is assumed, without loss of generality, that a processor $P_i$ may be assigned a nonempty subset of actors only if $P_1, P_2, ..., P_{i-1}$ are all assigned nonempty subsets as well. It is also assumed that $m < n$. Thus, the set of vertices assigned to a layer $L_i$ can be specified as

$$S_i \equiv \{\langle i, x, y \rangle | (i \leq x \leq y \leq n)\}, \text{ for } i = 1, 2, ..., m-1, \qquad (6\text{-}17)$$

and

$$S_m \equiv \{\langle m, x, n \rangle | (m \leq x \leq n)\}, \qquad (6\text{-}18)$$

where each triple $\langle i, b, c \rangle$ corresponds to the assignment of actors $\{X_b, X_{b+1}, ..., X_c\}$ to processor $P_i$. The set $S_i$ represents the set of all valid assignments of actor subsets to processor $P_i$ under the chain pipelining scheduling model.

A *terminal vertex* $T$ is also added to the layered assignment graph. This vertex models termination of the assignment, and is connected by an edge with each layer vertex that includes the last actor $X_n$ in the associated assignment. We refer to these edges, which model the computation within the last pipeline stage of a candidate solution, as *terminal edges*. The weight $w(v, T)$ associated with a terminal edge directed from vertex $v$ to the terminal vertex $T$ is taken to be the the total computation time of the subset of actors associated with the assignment $v$.

Edges other than the terminal edges in the layered assignment graph model compatibility relationships between elements of successive $S_i$'s, and weights on these edges model the computation and communication costs associated with specific processor assignments. For each pair of vertices $v_1 = \{a, b, c\}$ and $v_2 = \{a+1, b', c'\}$ in "adjacent" layers ($a$ and $a+1$) that satisfy $b' = c+1$, an edge $(v_1, v_2)$ is instantiated in the layered assignment graph. The weight $w((v_1, v_2))$ assigned to $(v_1, v_2)$ is the total computation time of the subset of actors associated with the assignment $v_1$ plus the IPC cost between the last actor, $c$, associated with $v_1$, and the first actor, $b'$, associated with $v_2$. In other words,

$$w((v_1, v_2)) = \sum_{i=b}^{c} t(A_i) + c((A_c, A_{c+1})), \tag{6-19}$$

where $c(e)$ denotes the IPC cost associated with edge $e$ in the input task graph.

From the above formulations for the construction of the layered assignment graph, the graph illustrated in Figure 6.6(c) is easily seen to be the layered assignment graph associated with Figures 6.6(a-b). For clarity, the layer identifier $a$ is omitted from the label of each vertex $\langle a, b, c \rangle$, and instead, the grouping of vertices into layers is designated by the annotations "Layer 1," "Layer 2," "Layer 3" on the right side of Figure 6.6(c).

Each path in Figure 6.6(c) that originates at a vertex in Layer 1 and terminates at $T$ represents a possible solution to the chain pipelining problem for Figure 6.6(a-b). We refer to such a path as a **complete path**. For example, the complete path

$$((\langle 1, 1, 1 \rangle, \langle 2, 2, 3 \rangle), (\langle 2, 2, 3 \rangle, \langle 4, 4, 4 \rangle), (\langle 4, 4, 4 \rangle, T)) \tag{6-20}$$

corresponds to the processor assignment illustrated in Figure 6.7(a), and the complete path

$$((\langle 1, 1, 2 \rangle, \langle 2, 3, 4 \rangle), (\langle 2, 3, 4 \rangle, T)) \tag{6-21}$$

corresponds to the assignment showing in Figure 6.7(b).

In general, the processing rate, or throughput, of the pipelined implementation associated with a given complete path $p = (e_1, e_2, ..., e_n)$, $src(e_1) \in S_1$, $snk(e_n) = T$ in a layered assignment graph is given by

$$T(p) = \frac{1}{max(w(e_i) | (1 \leq i \leq n))}; \tag{6-22}$$

that is, the throughput is simply the reciprocal of the maximum weight of an edge in the path $p$. An edge in $p$ that achieves this maximum weight is called a **bottleneck edge**. Thus, the chain pipelining problem reduces to the problem of computing a *minimum bottleneck, complete path* in the layered assignment graph.

Referring back to the example of Figure 6.6, the throughputs of the assignments corresponding to Figures 6.7(a) and 6.7(b) are easily seen to be $1/9$ and $1/11$, respectively, and thus, Figure 6.7(a) leads to a more efficient implementation. In fact, it can be verified that the path in Figure 6.6(c) that corresponds to the assignment of Figure 6.7(a) is the unique minimum bottleneck, complete path, and therefore, the assignment of Figure 6.7(a) is uniquely optimal for this example.

Bokhari observed that computing minimum bottleneck paths in layered assignment graphs can be performed in polynomial time by applying an adaptation by Edmonds and Karp [EK72] of Dijkstra's shortest path algorithm [Dij59]. Using the Edmonds-Karp adaptation allows an optimal chain pipelining to be computed in $O(m^2n^4)$ time, where $m$ is the number of processors in the target multiprocessor architecture, and $n$ is the number of actors in the chain-structured task graph. However, Bokhari has devised a significantly more efficient algorithm that exploits the layered structure of his assignment graph model. Using this technique, optimal solutions to the chain pipelining problem can be computed in $O(mn^3)$ time.

Bokhari developed a number of extensions to his algorithm for chain pipelining. These compute optimal solutions for *host-satellite* pipeline systems under

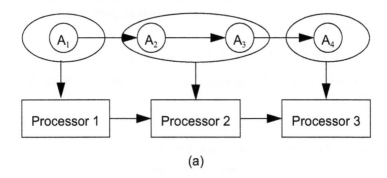

(a)

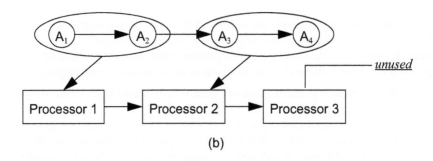

(b)

Figure 6.7. Two possible chain pipelining implementations for the system shown in Figure 6.6. Both of these processor assignments are suboptimal.

various restriction on the application structure [Bok88]. A host-satellite systems consists of an arbitrary number of independent chain pipelining systems that have access to a single, shared host processor. Heuristics for more general forms of the pipelined scheduling problem — for example, for pipelined scheduling that considers arbitrary task graph topologies, and more general classes of target multiprocessor architectures — have been developed by Hoang and Rabaey [HR92]; Banerjee et al. [BHCF95]]; and Liu and Prasanna [LP98]. A more detailed discussion of pipelined scheduling techniques is beyond the scope of this book. For further elaboration on this topic, the reader is encouraged to consult the aforementioned references.

## 6.7    Summary

This chapter has surveyed IPC-conscious scheduling techniques for application-specific multiprocessors, and has emphasized the broad range of fundamental, graph-theoretic insights that have been established in the context of IPC-conscious scheduling. More specifically, we have reviewed key algorithmic developments in four key areas relevant to the scheduling of HSDFGs: static assignment, list scheduling, clustering, and pipeline scheduling. For minimum-makespan scheduling, techniques that jointly address the assignment and ordering sub-problems are typically much more effective than techniques that are based on static assignment algorithms, and thus, list scheduling and clustering have received significantly more attention than static assignment techniques in this problem domain. However, no particular list scheduling or clustering algorithm has emerged as a clear, widely-accepted *best* algorithm so far that outperforms all other algorithms on most applications. Moreover, there is recent evidence that techniques to systematically integrate different scheduling strategies can lead to algorithms that significantly outperform the individual algorithms on which such integrated algorithms are based. The development of such integrated algorithms appears to be a promising direction for further work on scheduling. Another recent trend that is relevant to application-specific multiprocessor implementation is the investigation of pipelined scheduling strategies, which focus on throughput as the key performance metric. In this chapter, we have outlined fundamental results that apply to restricted versions of the pipelined scheduling problem. The development of algorithms that address more general forms of pipelined scheduling, which are commonly encountered in the design of application-specific multiprocessors, is currently an active research area.

# 7

# THE ORDERED-TRANSACTIONS
# STRATEGY

The self-timed scheduling strategy described in Chapter 5 introduces synchronization checks when processors communicate; such checks permit variations in actor execution times, but they also imply run-time synchronization and arbitration costs. In this chapter we present a scheduling model called ordered-transactions that alleviates some of these costs, and in doing so, trades off some of the run-time flexibility afforded by the self-timed approach. The **ordered-transactions** strategy was first proposed by Bier, Lee, and Sriram [LB90][BSL90]. In this chapter, we describe the idea behind the ordered-transactions approach and then we discuss the design and hardware implementation of a shared-bus multiprocessor that makes use of this strategy to achieve a low-cost interprocessor communication using simple hardware. The software environment for this board, for application specification, scheduling, and object code generation for the DSP processors is provided by the Ptolemy system developed at the University of California at Berkeley [BHLM94][Pto98].

## 7.1    The Ordered-Transactions Strategy

In the ordered-transactions strategy, we first obtain a fully-static schedule using the execution time estimates, but we discard the precise timing information specified in the fully-static schedule; as in the self-timed schedule we retain the processor assignment ($\sigma_p$) and actor-ordering on each processor as specified by $\sigma_t$; in addition, we also retain the order in which processors communicate with one another and we enforce this order at run-time. We formalize the concept of transaction order in the following.

Suppose there are $k$ inter-processor communication points $(s_1, r_1), (s_2, r_2), ..., (s_k, r_k)$ — where each $(s_i, r_i)$ is a *send-receive* pair — in the fully-static schedule that we obtain as a first step in the construction of a self-timed schedule. Let $R$ be the set of *receive* actors, and $S$ be the set of *send* actors (i.e., $R \equiv \{r_1, r_2, ..., r_k\}$ and $S \equiv \{s_1, s_2, ..., s_k\}$ ). We define a **transaction order** to be a sequence

$$O = (v_1, v_2, v_3, ..., v_{2k-1}, v_{2k}),$$

where

$$\{v_1, v_2, ..., v_{2k-1}, v_{2k}\} \equiv S \cup R$$

(each communication actor is present in the sequence $O$). We say a transaction order $O$ (as defined above) is **imposed** on a multiprocessor if at run-time the *send* and *receive* actors are forced to execute in the sequence specified by $O$. That is, if $O = (v_1, v_2, v_3, ..., v_{2k-1}, v_{2k})$, then imposing $O$ means ensuring the constraints:

$$end(v_1, k) \leq start(v_2, k), \; end(v_2, k) \leq start(v_3, k), \; ...,$$
$$end(v_{k-1}, k) \leq start(v_k, k); \; \forall k \geq 0.$$

Thus, the ordered-transactions schedule is essentially a self-timed schedule with the added transaction order constraints specified by $O$. Note that the transaction order constraints must satisfy the data precedence constraints of the HSDFG being scheduled; we call such a transaction order an **admissible transaction order**.

One simple mechanism to obtain an admissible transaction order is as follows. After a fully-static schedule is obtained using the execution time estimates, an admissible transaction order is obtained from the $\sigma_t$ function by setting the transaction order to $O = (v_1, v_2, v_3, ..., v_{2k-1}, v_{2k})$, where

$$\sigma_t(v_1) \leq \sigma_t(v_2) \leq ... \leq \sigma_t(v_{2k-1}) \leq \sigma_t(v_{2k}). \tag{7-1}$$

An admissible transaction order can therefore be determined by sorting the set of communication actors $(S \cup R)$ according to their start times $\sigma_t$. Figure 7.1 shows an example of how such an order could be derived from a given fully-static schedule. This fully-static schedule corresponds to the HSDFG and schedule illustrated in Chapter 5 (Figure 5.3). Such an order is clearly not the only admissible transaction order; an order

$$(s_1, r_1, s_2, r_2, s_3, r_3, s_4, r_4, s_6, r_6, s_5, r_5)$$

also satisfies all precedence constraints, and hence is admissible. In the next chapter we will discuss how to choose a good transaction order, which turns out to be close to optimal under certain reasonable assumptions. For the purposes of this chapter, we will assume a given admissible transaction order, and defer the details of how to choose a good transaction order to the next chapter.

The transaction order is enforced at run-time by a controller implemented in hardware. The main advantage of ordering interprocessor transactions is that it allows us to restrict access to communication resources statically, based on the communication pattern determined at compile time. Since communication resources are typically shared between processors, the need for run-time arbitration of these resources, as well as the need for sender-receiver synchronization is eliminated by ordering processor accesses to them; this results in an efficient IPC mechanism at low hardware cost. We have built a prototype four-processor DSP board, called the Ordered Memory Access (OMA) architecture, that demonstrates the ordered-transactions concept. The OMA prototype board utilizes shared memory and a single shared bus for IPC — the sender writes data to a particular shared memory location that is allocated at compile time, and the receiver reads that location. In this multiprocessor, a very simple controller on the board enforces the predetermined transaction order at run-time, thus eliminating the need for run-time bus arbitration or semaphore synchronization. This results in efficient IPC (comparable to the fully-static strategy) at relatively low hardware cost. As in the self-timed scenario, the ordered-transactions strategy is tolerant of variations in execution times of actors, because the transaction order enforces correct sender-receiver synchronization; however, this strategy is more constrained than self-timed scheduling, which allows the order in which communication actors fire to vary at run-time. The ordered-transactions strategy, therefore, falls in between fully-static and self-timed strategies in that, like the self-timed strategy, it is tolerant of variations in execution times and, like the fully-static strategy, has low communication and synchronization costs. These performance issues will be discussed quantitatively in the following chapter; the remainder of this chapter describes the hardware and software implementation of the OMA prototype.

## 7.2    Shared Bus Architecture

The OMA architecture uses a single shared bus and shared memory for interprocessor communication. This kind of shared memory architecture is

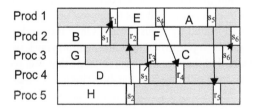

Transaction order: $(s_1, r_1, s_2, r_2, s_3, r_3, s_4, r_4, s_5, r_5, s_6, r_6)$

Figure 7.1. One possible transaction order derived from a fully-static schedule.

attractive for embedded multiprocessor implementations owing to its relative simplicity and low hardware cost and to the fact that it is moderately scalable — a fully interconnected processor topology, for example, would not only be much more expensive than a shared bus topology, but would also suffer from its limited scalability. Bus bandwidth limits scalability in shared bus multiprocessors, but for medium throughput applications (digital audio, music, etc.), a single shared bus provides sufficient bandwidth (of the order of 100MBytes/s). One solution to the scalability problem is the use of multiple busses and hierarchies of busses, for which the ideas behind the OMA architecture directly apply. The reader is referred to Lee and Bier [LB90] for how the OMA concept is extended to such hierarchical bus structures. Although in this book we apply the ordered-transactions strategy to a single shared bus architecture, the synchronization optimization techniques described in Chapters 10 through 12 are applicable to more general platforms and are not restricted to medium throughput applications.

From Figure 5.4 we recall that the self-timed scheduling strategy falls naturally into a message-passing paradigm that is implemented by the send and receive primitives inserted in the HSDFG. Accordingly, the shared memory in an architecture implementing such a scheduling strategy is used solely for message passing: the send primitive corresponds to writes to shared memory locations, and the receive primitive corresponds to reads from shared memory. Thus, the shared memory is not used for storing shared data structures or for storing shared program code. In a self-timed strategy we can further ensure, at compile time, that each shared memory location is written to by only one processor. One way of doing this is to simply assign distinct shared buffers to each of the send primitives; this is the scheme implemented in the multiprocessor DSP code generation domain in the Ptolemy environment [Pto98].

## 7.3    Interprocessor Communication Mechanisms

Let us now consider the implementation of IPC in self-timed schedules on such a shared bus multiprocessor. The sender has to write into shared memory, which involves arbitration costs — it has to request access to the shared bus, and the access must be arbitrated by a bus arbiter. Once the sender obtains access to shared memory, it needs to perform a synchronization check on the shared memory location to ensure that the receiver has read data that was written in the previous iteration, to avoid overwriting previously written data. Such synchronization is typically implemented using a semaphore mechanism; the sender waits until a semaphore is reset before writing to a shared memory location, and upon writing that shared memory location, it sets that semaphore (the semaphore could be a bit in shared memory, one bit for each send operation in the parallel schedule). The receiver, on the other hand, busy-waits until the semaphore is set before reading the shared memory location, and resets the semaphore after completing the read operation. It can easily be verified that this simple protocol guarantees correct sender-receiver synchronization, and, even though the semaphore bits have multiple writers, no atomic test-and-set operation is required of the hardware.

In summary, the operations of the sender are: request bus, wait for arbitration, busy-wait until semaphore is in the correct state, write the shared memory location if semaphore is in the correct state, and then release the bus. The corresponding operations for the receiver are: request bus, wait for arbitration, busy wait on semaphore, read the shared memory location if semaphore is in the correct state, and release the bus. The IPC costs are therefore due to bus arbitration time and due to semaphore checks. If no special hardware support is employed for IPC, such overhead consumes on the order of tens of instruction cycles, and also expends power — an important concern for portable applications. In addition, semaphore checks consume shared bus bandwidth.

An example of this is a four-processor Motorola DSP56000-based shared bus system designed by Dolby Labs for digital audio processing applications. In this machine, processors communicate through shared memory, and a central bus arbiter resolves bus request conflicts between processors. When a processor gets the bus it performs a semaphore check, and continues with the shared memory transaction if the semaphore is in the correct state. It explicitly releases the bus after completing the shared memory transaction. A receive and a send together consume 30 instruction cycles, even if the semaphores are in their correct state and the processor gets the bus immediately upon request. Such a high cost of communication forces the scheduler to insert as few interprocessor communication nodes as possible, which in turn limits the amount of parallelism that can be extracted from the algorithm.

One solution to this problem is to send more than one data sample when a processor gets access to the bus; the arbitration and synchronization costs are then amortized over several data samples. A scheme to "vectorize" data in this manner has been proposed by Zivojinovic, Ritz, and Meyr [ZRM94], where retiming [LS91] is used to move delays in the HSDFG such that data can be transferred in blocks, instead of one sample at a time. Several issues need to be taken care of before the vectorization strategy can be employed. First, retiming HSDFGs has to be done very carefully: moving delays across actors can change the initial state of the HSDFG causing undesirable transients in the algorithm implementation. This can potentially be solved by including preamble code to compute the value of the sample corresponding to the delay when that delay is moved across actors. This, however, results in increased code size, and other associated code generation complications. Second, the work of Zivojinovic et al. does not apply uniformly to all HSDFGs: if there are tight cycles in the graph that need to be partitioned among processors, the samples simply cannot be "vectorized" [Mes88]. Thus, presence of a tight cycle precludes arbitrary blocking of data. Third, vectorizing samples leads to increased latency in the implementation; some signal processing tasks such as interactive speech are sensitive to delay, and hence the delay introduced due to blocking of data may be unacceptable. Finally, the problem of vectorizing data in HSDFGs into blocks, even with all the above limitations, appear to be fundamentally hard; the algorithms proposed by Zivojinovic et al. have exponential worst case run-times. Code generated currently by the Ptolemy system does not support blocking (or vectorizing) of data for many of the above reasons.

Another possible solution is to use special hardware. One could provide a full interconnection network, thus obviating the need to go through shared memory. Semaphores could be implemented in hardware. One could use multiported memories. Needless to say, this solution is not favorable because of cost and potentially higher power consumption, especially when targeting embedded applications.

A general-purpose shared bus machine, the Sequent Balance [PH96] for example, will typically use caches between the processor and the shared bus. Caches lead to increased shared memory bandwidth due to the averaging effect provided by block fetches and due to probabilistic memory access speedup due to cache hits. In signal processing and other real time applications, however, there are stringent requirements for deterministic performance guarantees as opposed to probabilistic speedup. In fact, the unpredictability in task execution times introduced due to the use of caches may be a disadvantage for static scheduling techniques that utilize compile time estimates of task execution times to make scheduling decisions (we recall the discussion in Section 5.9 on techniques for estimating task execution times). In addition, due to the deterministic nature of most signal processing problems (and also many scientific computation problems), shared data can be deterministically prefetched because information about when particular blocks of data are required by a particular processor can often be predicted by a compiler. This feature has been studied in [MM92], where the authors propose memory allocation schemes that exploit predictability in the memory access pattern in DSP algorithms; such a "smart allocation" scheme alleviates some of the memory bandwidth problems associated with high throughput applications.

Processors with caches can cache semaphores locally, so that busy waiting can be done local to the processor without having to access the shared bus, hence saving the bus bandwidth normally expended on semaphore checks. Such a procedure, however, requires special hardware (a snooping cache controller, for example) to maintain cache coherence; cost of such hardware usually makes it prohibitive in embedded scenarios.

Thus, for the embedded signal, image, and video signal processing applications that are the primary focus of this book, we argue that caches do not often have a significant role to play, and we claim that the ordered-transactions approach discussed previously provides a cost-effective solution for minimizing IPC overhead in implementing self-timed schedules.

## 7.4    Using the Ordered-Transactions Approach

The ordered-transactions strategy, we recall, operates on the principle of determining (at compile time) the order in which processor communications occur, and enforcing that order at run-time. For a shared bus implementation, this translates into determining the sequence of shared memory (or, equivalently, shared bus) accesses at compile time and enforcing this predetermined order at run-time. This strategy, therefore, involves no run-time arbitration; processors are simply granted the bus according to the predetermined access order. When a processor obtains access to the bus, it performs the necessary shared memory transaction, and releases the bus; the bus is then granted to the next processor in the ordered list.

The task of maintaining ordered access to shared memory is done by a central **transaction controller**. When the processors are downloaded with code, the controller too is loaded with the predetermined access order list. At run-time the controller simply grants bus access to processors according to this list, granting access to the next processor in the list when the current bus owner releases the bus. Such a mechanism is robust with respect to variations in execution times of the actors; the functionality of the system is unaffected by poor estimates of these execution times, although the real-time performance obviously suffers as in any scheduling strategy that involves static ordering and assignment.

We will show that if we are able to perform accurate compile time analysis, then the new transaction ordering constraints do not significantly impact performance. Also, no arbitration needs to be done since the transaction controller grants exclusive access to the bus to each processor. In addition, no semaphore synchronization needs to be performed, because the transaction ordering constraints respect data precedences in the algorithm; when a processor accesses a shared memory location and is correspondingly allowed access to it, the data accessed by that processor is certain to be valid. As a result, under an ordered-transactions scenario, a send (receive) operation always occupies the shared bus for only one shared memory write (read) cycle. This reduces contention for the bus and reduces the number of shared memory accesses required for each IPC operation by at least a factor of two and possibly much more, depending on the amount of polling required in a conventional arbitration-based shared bus implementation.

The performance of this scheme depends on how accurately the execution times of the actors are known at compile time. If these compile time estimates are reasonably accurate, then an access order can be obtained such that a processor gains access to shared memory whenever necessary. Otherwise, a processor may have to idle until it gets a bus grant, or, even worse, a processor when granted the bus may not complete its transaction immediately, thus blocking all other processors from accessing the bus. This problem would not arise in normal arbitration schemes, because dynamic reordering of independent shared memory accesses is possible.

We will quantify these performance issues in the next chapter, where we show that when reasonably good estimates of actor execution times are available, forcing a run-time access order does not in fact sacrifice performance significantly.

## 7.5     Design of an Ordered Memory Access Multiprocessor

### 7.5.1     High Level Design Description

We chose Motorola DSP96002 processors for the OMA prototype. Although the OMA architecture can be built around any programmable DSP that has built-in bus arbitration logic, the DSP96002 is well-suited for incorporation into an OMA platform because of its dual bus architecture and bus arbitration mechanism. In addition, these processors are powerful DSPs with floating point capability [Mot89].

As an illustration of the OMA concept, a high-level block diagram of a DSP96002-based OMA system is shown in Figure 7.2. Each DSP96002 is provided with a private memory that contains its program code; this local memory resides on one of the processor busses (the "A" bus). The alternate "B" bus of all processors are connected to the shared bus, and shared memory resides on the shared bus. The transaction controller grants access to processors using the bus grant (BG) lines on the processor. A processor attempts to perform a shared memory access when it executes a communication actor (either send or receive). If its BG line is asserted it performs the access, otherwise it stalls and waits for the assertion.

After a processor obtains access to the shared bus, it performs a single shared memory operation (send or receive) and releases the bus. The transaction controller detects the release of the bus and steps through its ordered list, granting the bus to the next processor in its list.

The cost of transfer of one word of data between processors is 3 instruction cycles in the ideal case where the sender and the receiver obtain access to the shared bus immediately upon request; two of these correspond to a shared memory write (by the sender) and a shared memory read (by the receiver), and an extra instruction cycle is expended in bus release by the sender and bus acquisition by the receiver. Such low-overhead interprocessor communication is obtained with the transaction controller providing the only additional hardware support. As described in a subsequent section, this controller can be implemented with very simple hardware.

## 7.5.2    A Modified Design

In the design discussed above, processor-to-processor communication occurs through a central shared memory; two transactions — one write and one read — must occur over the shared bus for each sender-receiver pair. This situation can be improved by distributing the shared memory among processors, as shown in Figure 7.3, where each processor is assigned shared memory in the form of hardware FIFO buffers. Writes to each FIFO are accomplished through the shared bus; the sender simply writes to the FIFO of the processor to which it wants to send data by using the appropriate shared memory address.

Use of a FIFO implies that the receiver must know the exact order in which data is written into its input queue. This, however, is guaranteed by the ordered-transactions strategy. Thus replacing a RAM (random access memory)-based shared memory with distributed FIFOs does not alter the functionality of the design. The sender need only block when the receiving queue is full, which can be accomplished in hardware by using the "Transfer Acknowledge (TA)" sig-

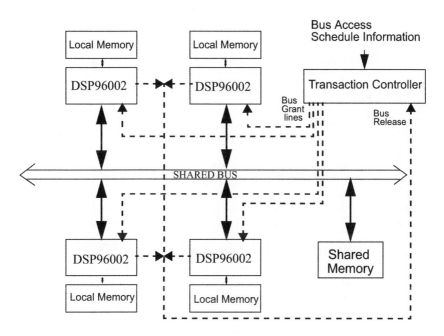

Figure 7.2. Block diagram of the OMA prototype.

nal on the DSP96002; a device can insert an arbitrary number of wait states in the processor memory cycle by de-asserting the TA line. Whenever a particular FIFO is accessed, its "Buffer Full" line is enabled onto the TA line of the processors (Figure 7.4). Thus a full FIFO automatically blocks the processor trying to write into it, and no polling needs to be done by the sender. At the receiving end, reads are local to a processor, and do not consume shared bus bandwidth. The receiver can be made to either poll the FIFO empty line to check for an empty queue, or one can use the same TA signal mechanism to block processor reads from an empty queue. The TA mechanism will then use the local ("A") bus control signals ("A" bus TA signal, "A" address bus, etc.). This is illustrated in Figure 7.4.

Use of such a distributed shared memory mechanism has several advantages. First, the shared bus traffic is effectively halved, because only writes need to go through the shared bus. Second, in the design of Figure 7.2, a processor that is granted the bus is delayed in completing its shared memory access, all other processors waiting for the bus get stalled; this does not happen for half the transactions in the modified design of Figure 7.3 because receiver reads are local. Thus there is more tolerance to variations in the time at which a receiver reads data sent to it. Last, a processor can broadcast data to all (or any subset) of pro-

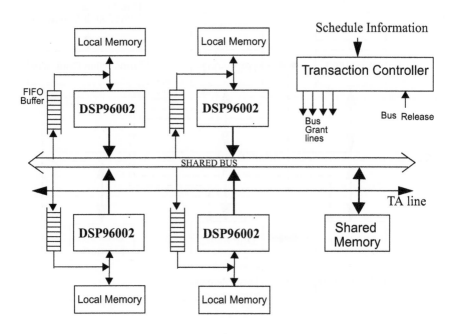

Figure 7.3. Modified design.

cessors in the system by simultaneously writing to more than one FIFO buffer. Such broadcast is not possible with a central shared memory.

The modified design, however, involves a significantly higher hardware cost than the design proposed in Section 7.5.1. As a result, the OMA prototype discussed in the following sections (Sections 7.6 to 7.9) was built around the central shared memory design and not the FIFO based design. In addition, the DSP96002 processor has an on-chip host interface unit that can be used as a 2-deep FIFO; therefore, the potential advantage of using distributed FIFOs can still be evaluated to some degree by using the chip host interface even in the absence of external FIFO hardware.

Simulation models were written for both the above designs using the Thor hardware simulator [Tho86] under the Frigg multiprocessor simulator system [BL89]. Frigg allows the Thor simulator to communicate with a timing-driven functional simulator for the DSP96002 processor provided by Motorola Inc. The Motorola simulator also simulates Input/Output (I/O) operations of the pins of the processor, and Frigg interfaces the signals on the pins to the rest of the Thor

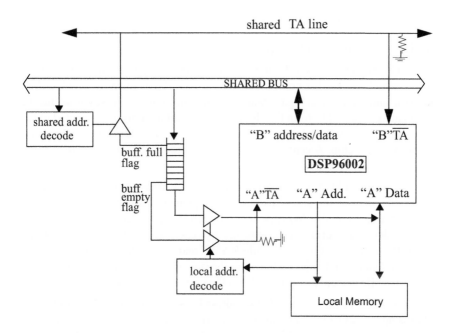

Figure 7.4. Details of the "TA" line mechanism (only one processor is shown).

simulation; as a result, hardware associated with each processor (memories, address-decoding logic, etc.) and interaction between processors can be simulated using Frigg. This allows functionality of the entire system to be verified by running actual programs on the processor simulators. This model was not used for performance evaluation of the OMA prototype, however, because with just a four-processor system the cycle-by-cycle Frigg simulation was far too slow, even for very simple programs. A higher-level (behavioral) simulation would be more useful than a cycle-by-cycle simulation for the purposes of performance evaluation, although such high-level simulation was not carried out on the OMA prototype.

The remainder of this chapter describes hardware and software design details of the OMA board prototype.

## 7.6        Design Details of a Prototype

A proof-of-concept prototype of the OMA architecture has been designed and implemented. The single printed circuit board design is comprised of four DSP96002 processors; the transaction controller is implemented on a Xilinx FPGA (Field Programmable Gate Array). The Xilinx chip also handles the host interface functions, and implements a simple I/O mechanism. A hierarchical description of the hardware design follows.

### 7.6.1        Top Level Design

This section refers to Figure 7.5. At the top level, there are four "processing element" blocks that consist of the processor, local memory, local address decoder, and some glue logic. Address, data, and control busses from the PE blocks are connected to form the shared bus. Shared memory is connected to this bus; address decoding is done by the "shared address decoder" PAL (programmable array logic) chip. A central clock generator provides a common clock signal to all processing elements.

A Xilinx FPGA (XC3090) implements the transaction controller and a simple I/O mechanism, and is also used to implement latches and buffers during bootup, thus saving glue logic. A fast static RAM (up to 32K x 8) stores the bus access order in the form of processor identifications (IDs). The sequence of processor IDs is stored in this "schedule RAM," and this determines the bus access order. An external latch is used to store the processor ID read from the schedule RAM. This ID is then decoded to obtain the processor bus grants.

A subset of the 32 shared bus address lines connect to the Xilinx chip, for addressing the I/O registers and other internal registers. All 32 lines from the shared data bus are connected to the Xilinx. The shared data bus can be accessed from the external connector (the "right side" connector in Figure 7.5) only through the Xilinx chip. This feature can be made use of when connecting multiple OMA boards: shared busses from different boards can be made into one con-

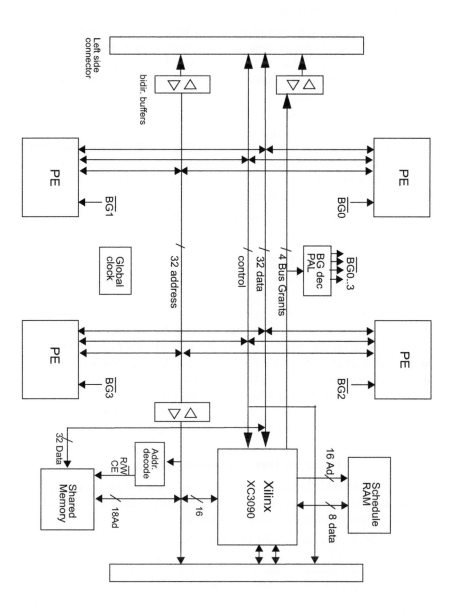

Figure 7.5. Top-level schematic of the OMA prototype.

tiguous bus, or they can be left disconnected, with communication between busses occurring via asynchronous "bridges" implemented on the Xilinx FPGAs. We discuss this further in Section 7.6.7.

Connectors on both ends of the board bring out the shared bus in its entirety. Both left and right side connectors follow the same format, so that multiple boards can be easily connected together. Shared control and address busses are buffered before they go off board via the connectors, and the shared data bus is buffered within the Xilinx.

The DSP96000 processors have on-chip emulation ("OnCE" in Motorola terminology) circuitry for debugging purposes, whereby a serial interface to the OnCE port of a processor can be used for in-circuit debugging. On the OMA board, the OnCE ports of the four processors are multiplexed and brought out as a single serial port; a host may select any one of the four OnCE ports and communicate to it through a serial interface.

We discuss the design details of the individual components of the prototype system next.

### 7.6.2        Transaction Order Controller

The task of the transaction order controller is to enforce the predetermined bus access order at run-time. A given transaction order determines the sequence of processor bus accesses that must be enforced at run-time. We refer to this sequence of bus accesses by the term **bus access order list**. Since the bus access order list is program-dependent, the controller must possess memory into which this list is downloaded after the scheduling and code generation steps are completed, and when the transaction order that needs to be enforced is determined. The controller must step through the access order list, and must loop back to the first processor ID in the list when it reaches the end. In addition, the controller must be designed to effectively use bus arbitration logic present on-chip, to conserve hardware.

### 7.6.2.1        Processor Bus Arbitration Signals

The bus grant ($\overline{BG}$) signal on the DSP chip is used to allow the processor to perform a shared bus access, and the bus request ($\overline{BR}$) signal is used to tell the controller when a processor completes its shared bus access.

Each of the two ports on the DSP96002 has its own set of arbitration signals; the $\overline{BG}$ and $\overline{BR}$ signals are the most relevant signals for the OMA design, and these signals are relevant only for the processor port connected to the shared bus. As the name suggests, the $\overline{BG}$ line (which is an input to the processor) must be asserted before a processor can begin a bus cycle: the processor is forced to wait for $\overline{BG}$ to be asserted before it can proceed with the instruction that requires access to the bus. Whenever an external bus cycle needs to be performed, a processor asserts its $\overline{BR}$ signal, and this signal remains asserted until an instruction that does not access the shared bus is executed. We can therefore use the $\overline{BR}$ signal to determine when a shared bus owner has completed its usage of the shared bus (Figure 7.6 (a)).

The rising edge of the $\overline{BR}$ line is used to detect when a processor releases the bus. To reduce the number of signals going from the processors to the controller, we multiplexed the $\overline{BR}$ signals from all processors onto a common $\overline{BR}$ signal. The current bus owner has its $\overline{BR}$ output enabled onto this common

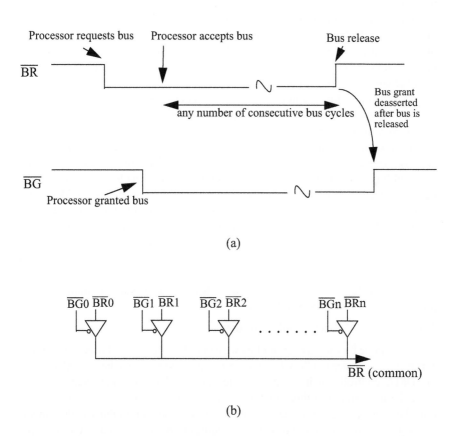

(a)

(b)

Figure 7.6. Using processor bus arbitration signals for controlling bus access.

reverse signal; this provides sufficient information to the controller because the controller only needs to observe the $\overline{\text{BR}}$ line from the current bus owner. This arrangement is shown in Figure 7.6 (b); the controller grants access to a processor by asserting the corresponding $\overline{\text{BG}}$ line, and then it waits for an upper edge on the reverse $\overline{\text{BR}}$ line. On receiving a positive going edge on this line it grants the bus to the next processor in its list.

### 7.6.2.2          A Simple Implementation

One straightforward implementation of the above functionality is to use a counter addressing a RAM that stores the access order list in the form of processor IDs. We call this counter the **schedule counter** and the memory that stores the processor IDs is called the **schedule RAM**. Decoding the output of the RAM provides the required $\overline{\text{BG}}$ lines. The counter is incremented at the beginning of a processor transaction by the negative going edge of the common $\overline{\text{BR}}$ signal and the output of the RAM is latched at the positive going edge of $\overline{\text{BR}}$, thus granting the bus to the next processor as soon as the current processor completes its shared memory transaction. The counter is reset to zero after it reaches the end of the list (i.e., the counter counts modulo the bus access list size). This is shown in Figure 7.7. Incrementing the counter as soon as $\overline{\text{BR}}$ goes low ensures enough time for the counter outputs and the RAM outputs to stabilize. For a 33MHz processor with zero wait states, $\overline{\text{BR}}$ width is a minimum of 60 nanoseconds. Thus, the counter incrementing and the RAM access must both finish before this time. Consequently, we need a fast counter and fast static RAM for the schedule memory. The width of the counter determines the maximum allowable size of the access list (a counter width of size $n$ implies a maximum list size of $2^n$ ); a wider counter, however, implies a slower counter. If, for a certain width, the counter (implemented on the Xilinx part in our case) turns out to be too slow — i.e., the output of the schedule memory will not stabilize at least one latch set up period before the positive going edge of $\overline{\text{BR}}$ arrives — wait states may have to be inserted in the processor bus cycle to delay the positive edge of $\overline{\text{BR}}$. We found that a 10-bit-wide counter does not require any wait states, and allows a maximum of 1024 processor IDs in the access order list.

### 7.6.2.3          Presettable Counter

A single bus access list implies we can only enforce one bus access pattern at run-time. In order to allow for some run-time flexibility, we have implemented the OMA controller using a presettable counter. The processor that currently owns the bus can preset this counter by writing to a certain shared memory location. This causes the controller to jump to another location in the schedule memory, allowing the multiple bus access schedules to be maintained in the schedule RAM and switching between them at run-time depending on the outcome of computations in the program. The counter appears as an address in the shared memory map of the processors. The presettable counter mechanism is shown in

Figure 7.8.

An arbitrary number of lists may, in principle, be maintained in the schedule memory. This feature can be used to support algorithms that display data

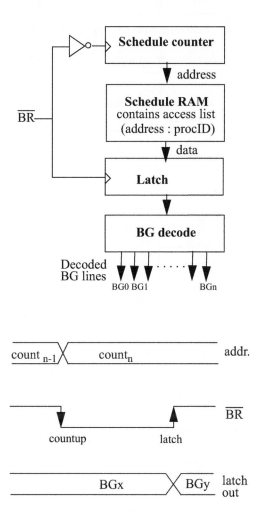

Figure 7.7.  Transaction Controller implementation.

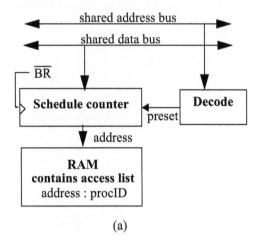

(a)

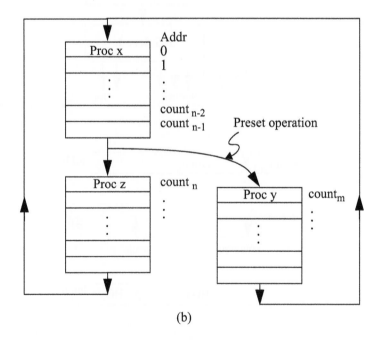

(b)

Figure 7.8. Presettable counter implementation.

dependency in their execution. For example, a dataflow graph with a conditional construct will, in general, require a different access schedule for each outcome of the conditional. One of two different SDF subgraphs are executed in this case, depending on the branch outcome, and the processor that determines the branch outcome can also be assigned the task of presetting the counter, making it branch to the access list of the appropriate SDF subgraph. The access controller behaves as in Figure 7.8 (b).

We discuss the use of this presettable feature in detail later in the book.

### 7.6.3     Host Interface

The function of the host interface is to allow downloading programs onto the OMA board, controlling the board, setting parameters of the application being run, and debugging from a host workstation. The host for the OMA board connects to the shared bus through the Xilinx chip, via one of the shared bus connectors. Since part of the host interface is configured inside the Xilinx, different hosts (32 bit, 16 bit) with different handshake mechanisms can be used with the board.

The host that is being used for the prototype is a Motorola DSP56000-based DSP board called the **S-56X card**, manufactured by Ariel Corp [Ari91]. The S-56X card is designed to fit into one of the Sbus slots in a Sun Sparc workstation; a user level process can communicate with the S-56X card via a unix device driver. Thus the OMA board too can be controlled (via the S-56X card) by a user process running on the workstation. The host interface configuration is depicted in Figure 7.9.

Unlike the DSP56000 processors, the DSP96002 processors do not have built-in serial ports, so the S-56X board is also used as a serial I/O processor for the OMA board. It essentially performs serial-to-parallel conversion of data, buffering of data, and interrupt management. The Xilinx on the OMA board implements the necessary transmit & receive registers, and synchronization flags — we discuss the details of the Xilinx circuitry in Section 7.6.5.

The S-56X card communicates with the Sparc Sbus using DMA (direct memory access). A part of the DSP56000 bus and control signals are brought out of the S-56X card through another Xilinx FPGA (XC3040) on the S-56X. For the purpose of interfacing the S-56X board with the OMA board, the Xilinx on the S-56X card is configured to bring out 16 bits of data and 5 bits of address from the DSP56000 processor onto the cable connected to the OMA (see Figure 7.9). In addition, the serial I/O port (the SSI port) is also brought out, for interface with I/O devices such as A/D and D/A convertors. By making the DSP56000 write to appropriate memory locations, the 5 bits of address and 16 bits of data going into the OMA may be set and strobed for a read or a write, to or from the OMA board. In other words, the OMA board occupies certain locations in the DSP56000 memory map; host communication is done by reading and writing to these memory locations.

### 7.6.4    Processing Element

Each processing element (PE) consists of a DSP96002 processor, local memory, address buffers, local address decoder, and some address decoding logic. The circuitry of each processing element is very similar to the design of the Motorola 96000 ADS (Application Development System) board [Mot90]. The local address, control, and data busses are brought out into a 96 pin euro-connector, following the format of the 96ADS. This connector can be used for local memory expansion; we have used it for providing local I/O interface to the processing element (as an alternative to using the shared bus for I/O). Port A of the processor forms the local bus, connecting to local memory and address decoding PAL. Each PE also contains address buffers, and logic to set up the bootup mode upon reset and power-up. Port B of the processor is connected to the shared bus.

The DSP96002 processor has a Host Interface (HI) on each of its ports. The port B HI is memory-mapped to the shared bus, so that HI registers may be

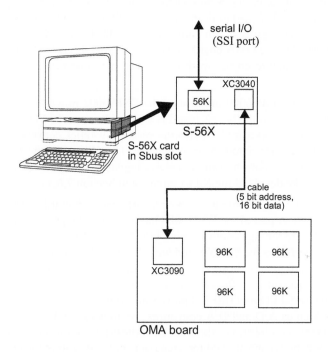

Figure 7.9. Host interface.

read and written from the shared bus. This feature allows a host to download code and control information into each processor through the shared bus. Furthermore, a processor, when granted the shared bus, may also access the port B HI of other processors. This allows processors to bypass the shared memory while communicating with one another and to broadcast data to all processors. In effect, the HI on each processor can be used as a two-deep local FIFO, similar to the scheme in Section 7.5.2, except that the FIFO is internal to each processor.

### 7.6.5    FPGA Circuitry

As mentioned previously, the XC3090 Xilinx FPGA is used to implement the transaction controller as well as a simple I/O interface. It is also configured to provide latches and buffers for addressing the Host Interface (HI) ports on the DSP96002 during bootup and downloading of code onto the processors. For this to work, the Xilinx is first configured to implement the bootup- and download-related circuitry, which consists of latches to drive the shared address bus and to access the schedule memory. After downloading code onto the processors, and downloading the bus access order into the schedule RAM, the Xilinx chip is reconfigured to implement the transaction controller and the I/O interface. Thus the process of downloading and running a program requires configuring the Xilinx chip twice.

There are several possible ways in which a Xilinx part may be pro-

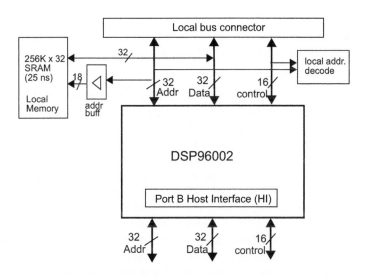

Figure 7.10. Processing element.

grammed. For the OMA board, the configuration bitmap is downloaded byte-wise by the host (Sun workstation through the S-56X card). The bitmap file, generated and stored as a binary file on a workstation, is read in by a function implemented in the *qdm* software (discussed in Section 7.7, which describes the OMA software interface) and the bytes thus read are written into the appropriate memory location on the S-56X card. The DSP56K processor on the S-56X then strobes these bytes into the Xilinx configuration port on the OMA board. The user can reset and reconfigure the Xilinx chip from a Sun Sparc workstation by manipulating the Xilinx control pins by writing to a "Xilinx configuration latch" on the OMA board. Various configuration pins of the Xilinx chip are manipulated by writing different values into this latch.

We use two different Xilinx circuits, one during bootup and the other during run-time. The Xilinx configuration during bootup helps eliminate some glue logic that would otherwise be required to latch and decode address and data from the S-56X host. This configuration allows the host to read and write from any of the HI ports of the processors, and also to access the schedule memory and the shared memory on board.

Run-time configuration on the Xilinx consists of the transaction controller implemented as a presettable counter. The counter can be preset through the shared bus. It addresses an external fast RAM (8 nanosecond access time) that contains processor IDs corresponding to the bus access schedule. Output from the schedule memory is externally latched and decoded to yield bus grant ($\overline{\text{BG}}$) lines (Figure 7.7).

A schematic of the Xilinx configuration at run-time is given in Figure 7.11. This configuration is for I/O with an S-56X (16 bit data) host, although it can easily be modified to work with a 32-bit host.

### 7.6.5.1          I/O Interface

The S-56X board reads data from the Transmit (Tx) register and writes into the receive (Rx) register on the Xilinx. These registers are memory-mapped to the shared bus, such that any processor that possesses the bus may write to the Tx register or read from the Rx register. For a 16-bit host, two transactions are required to perform a read or write with the 32-bit Tx and Rx registers. The processors themselves need only one bus access to load or unload data from the I/O interface. Synchronization on the S-56X (host) side is done by polling status bits that indicate an Rx empty flag (if true, the host performs a write, otherwise it busy-waits) and a Tx full flag (if true, the host performs a read, otherwise it busy-waits). On the OMA side, synchronization is done by the use of the TA (transfer acknowledge) pin on the processors. When a processor attempts to read Rx or write Tx, the appropriate status flags are enabled onto the TA line, and wait states are automatically inserted in the processor bus cycle whenever the TA line is not asserted, which in our implementation translates to wait states whenever the sta-

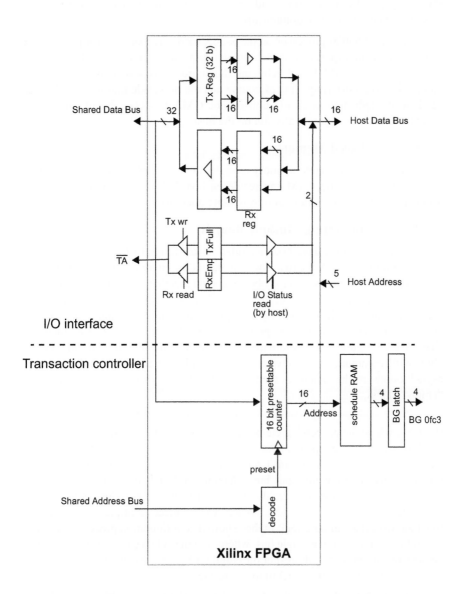

Figure 7.11. Xilinx configuration at run-time.

tus flags are false. Thus, processors do not have the overhead of polling the I/O status flags; an I/O transaction is identical to a normal bus access, with zero or more wait states inserted automatically.

The DSP56000 processor on the S-56X card is responsible for performing I/O with the actual (possibly asynchronous) data source and acts as the interrupt processor for the OMA board, relieving the board of tasks such as interrupt servicing and data buffering. This of course has the downside that the S-56X host needs to be dedicated as an I/O unit for the OMA processor board, and limits other tasks that could potentially run on the host.

### 7.6.6     Shared Memory

Space for two shared memory modules are provided, so that up to 512K x 32 bits of shared static RAM can reside on board. The memory must have an access time of 25ns to achieve zero wait state operation.

### 7.6.7     Connecting Multiple Boards

Several features have been included in the design to facilitate connecting together multiple OMA boards. The connectors on either end of the shared bus are compatible, so that boards may be connected together in a linear fashion (Figure 7.12). As mentioned before, the shared data bus goes to the "right side connector" through the Xilinx chip. By configuring the Xilinx to "short" the external and internal shared data busses, processors on different boards can be made to share one contiguous bus. Alternatively, busses can be "cleaved" on the Xilinx chip, with communication between busses implemented on the Xilinx via an asynchronous mechanism (e.g., read and write latches synchronized by "full" and "empty" flags).

This concept is similar to the idea used in the SMART processor array [Koh90], where the processing elements are connected to a switchable bus: when the bus switches are open, processors are connected only to their neighbors (forming a linear processor array), and when the switches are closed, processors are connected onto a contiguous bus. Thus, the SMART array allows formation of clusters of processors that reside on a common bus; these clusters then communicate with adjacent clusters. When we connect multiple OMA boards together, we get a similar effect: in the "shorted" configuration processors on different boards connect to a single bus, whereas in the "cleaved" configuration processors on different boards reside on common busses, and neighboring boards communicate through an asynchronous interface.

Figure 7.12 illustrates the above scheme. The highest 3 bits of the shared address bus are used as the "board ID" field. Memory, processor Host Interface ports, configuration latches, etc., decode the board ID field to determine if a shared memory or host access is meant for them. Thus, a total of 8 boards can be hooked onto a common bus in this scheme.

## 7.7      Hardware and Software Implementation

### 7.7.1      Board Design

We used single-sided through-hole printed circuit board technology for the OMA prototype [Sri92]. The printed circuit board design was done using the "SIERA" system developed at the University of California at Berkeley. Under this system, a design is entered hierarchically using a netlist language called SDL (Structure Description Language). Geometric placement of components can be easily specified in the SDL netlist itself. A "tiling" feature is also provided to ease compact fitting of components. The SDL files were written in a modular fashion; the schematics hierarchy is shown in Figure 7.12. The SIERA design

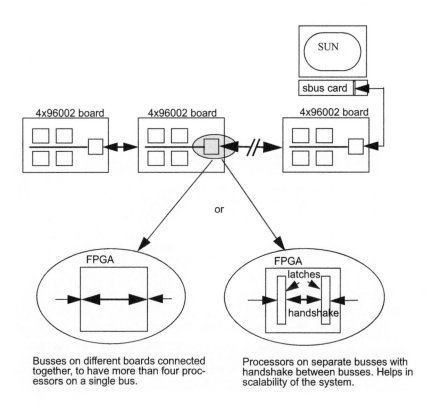

Busses on different boards connected together, to have more than four processors on a single bus.

Processors on separate busses with handshake between busses. Helps in scalability of the system.

Figure 7.12. Connecting multiple boards.

manager (DMoct) was then used to translate the netlists into an input file accept-able by Racal, a commercial PCB layout tool, which was then used to auto-route the board. Figure 7.13 shows a photograph of the board.

### 7.7.2    Software Interface

As discussed earlier, we use an S-56X card attached to a Sparc as a host for the OMA board. The Xilinx chip on the S-56X card is configured to provide 16 bits of data and 5 bits of address. We use the *qdm* [Lap91] software as an interface for the S-56X board; *qdm* is a debugger/monitor that has several useful built-in routines for controlling the S-56X board, for example data can be written and read from any location in the DSP56000 address space through function calls in *qdm*. Another useful feature of *qdm* is that it uses '*Tcl*', an embeddable, exten-sible, shell-like interpreted command language [Ous94]. Tcl provides a set of built-in functions (such as an expression evaluator, variables, control-flow state-ments, etc.) that can be executed via user commands typed at its textual interface, or from a specified command file. Tcl can be extended with application-specific commands; in our case, these commands correspond to the debugging/monitor commands implemented in *qdm* as well as commands specific to the OMA. Another useful feature of Tcl is the scripting facility it provides; sequences of commands can be conveniently integrated into scripts, which are in turn executed

Figure 7.13. OMA prototype board photograph.

by issuing a single command.

Some functions specific to the OMA hardware that have been compiled into *qdm* are the following:

**omaxiload** *fileName.bit* :

load OMA Xilinx with configuration specified by file.bit
**omapboot** *fileName.lod proc#* :

load bootstrap monitor code into the specified processor
**omapload** *fileName.lod proc#* :

load DSP96002 .lod file into the specified processor
**schedload** *accessOrder* :

load OMA bus access schedule memory

These functions use existing *qdm* functions for reading and writing values to the DSP56000 memory locations that are mapped to the OMA board host interface.

Each processor is programmed through its Host Interface via the shared bus. First, a monitor program (omaMon.lod) consisting of interrupt routines is loaded and run on the selected processor. Code is then loaded into processor memory by writing address and data values into the HI port and interrupting the processor. The interrupt routine on the processor is responsible for inserting data into the specified memory location. The S-56X host forces different interrupt routines, for specifying which of the three (X, Y, or P) memories the address refers to and for specifying a read or a write to or from that location. This scheme is similar to that employed in downloading code onto the S-56X card [Ari91].

Status and control registers on the OMA board are memory mapped to the S-56X address space and can be accessed to reset, reboot, monitor, and debug the board. Tcl scripts were written to simplify commands that used are most often (e.g., "change y:fff0 0x0" was aliased to "omareset").

A Ptolemy multiprocessor hardware target [Pto98] was written for the OMA board, for automatic partitioning, code generation, and execution of an algorithm from a block diagram specification. A simple heterogeneous multiprocessor target was also written in Ptolemy for the OMA and S-56X combination; this target generates DSP56000 code for the S-56X card, and generates DSP96000 multiprocessor code for the OMA.

## 7.8    Ordered I/O and Parameter Control

A mechanism has been implemented for the OMA whereby I/O can be done over the shared bus. This mechanism makes use of the fact that I/O for DSP applications is periodic; samples (or blocks of samples) typically arrive at constant, periodic intervals, and the processed output is again required (by, say, a digital-to-analog convertor) at periodic intervals. With this observation, it is in fact possible to schedule the I/O operations within the multiprocessor schedule, and consequently determine when, relative to the other shared bus accesses due to IPC, the shared bus is required for I/O. This allows us to include bus accesses for I/O in the bus access order list. In our particular implementation, I/O is implemented as shared address locations that address the Tx and Rx registers in the Xilinx chip (Section 7.6.5), which in turn communicate with the S-56X board; a processor accesses these registers as if they were a part of shared memory. It obtains access to these registers when the transaction controller grants access to the shared bus; bus grants for the purpose of I/O are taken into account when constructing the access order list. Thus, accesses to shared I/O resources can be ordered much as accesses to shared bus and memory.

The ordered memory access strategy can also be applied to run-time parameter control. By run-time parameter control we mean controlling parameters in the DSP algorithm (gain of some component, bit-rate of a coder, pitch of synthesized music sounds, etc.) while the algorithm is running in real time on the hardware. Such a feature is obviously very useful and sometimes indispensable. Usually, one associates such parameter control with an asynchronous user input: the user changes a parameter (ideally by means of a suitable GUI on his or her computer) and this change causes an interrupt to occur on a processor, and the interrupt handler then performs the appropriate operations that cause the parameter change that the user requested.

For the OMA architecture, however, unpredictable interrupts are not desirable, as was noted earlier in this chapter; on the other hand, shared I/O and IPC are relatively inexpensive owing to the ordered-transactions mechanism. To exploit this trade-off, parameter control is implemented in the following fashion: The S-56X host handles the task of accepting user interrupts; whenever a parameter is altered, the DSP56000 on the S-56X card receives an interrupt and it modifies a particular location in its memory (call it $M$). The OMA board, on the other hand, receives the contents of $M$ *on every schedule period,* whether $M$ was actually modified or not. Thus, the OMA processors never "see" a user-created interrupt; they in essence update the parameter corresponding to the value stored in $M$ in *every* iteration of the dataflow graph. Since reading in the value of $M$ costs two instruction cycles, the overhead involved in this scheme is minimal.

An added practical advantage of the above scheme is that the tcl/tk-based [Ous94] GUI primitives that have been implemented in Ptolemy for the S-56X (see "CG56 Domain" in Volume 1 of [Pto98]) can be directly used with the OMA board for parameter control purposes.

## 7.9 Application Examples

In this section we discuss several applications that are implemented using the OMA prototype.

### 7.9.1 Music Synthesis

The Karplus-Strong algorithm [KS83] is a well-known approach for synthesizing the sound of a plucked string. The basic idea is to pass a noise source in a feedback loop containing a delay, a low pass filter, and a multiplier with a gain of less than one. The delay determines the pitch of the generated sound, and the multiplier gain determines the rate of decay. Multiple voices can be generated and combined by implementing one feedback loop for each voice and then adding the outputs from all the loops. If we want to generate sound at a sampling rate of 44.1 KHz (compact disc sampling rate), we can implement 7 voices on a single processor in real time using the blocks from the Ptolemy DSP96000 code generation library (CG96). These 7 voices consume 370 instruction cycles out of the 380 instruction cycles available per sample period.

Using four processors on the OMA board, we implemented 28 voices in real time. The hierarchical block diagram for this is shown in Figure 7.14. At the top-level the block diagram consists of a noise source whose output is multiplied by a pulse; this excitation is fed into four hierarchical blocks consisting of 7 copies of the basic feedback loop for each voice. The outputs are added together, and this sum fed to an analog to digital convertor after being converted into a fixed-point representation from a floating point representation. A schedule for this application is shown in Figure 7.15. The makespan for this schedule is 377 instruction cycles, which is just within the maximum allowable limit of 380. This schedule uses 15 pairs of sends and receives, and is therefore not communication-intensive. Even so, a higher IPC cost than the three instruction cycles the OMA architecture affords us would not allow this schedule to execute in real time at a 44.1 KHz sampling rate, because there is only a three-instruction-cycle margin between the makespan of this schedule and the maximum allowable makespan. To schedule this application, we employed Hu-level scheduling along with manual assignment of some of the blocks.

### 7.9.2 QMF Filter Bank

A Quadrature Mirror Filter (QMF) bank consists of a set of *analysis* filters used to decompose a signal (usually audio) into frequency bands, and a bank of *synthesis* filters is used to reconstruct the decomposed signal [Vai93]. In the analysis bank, a filter pair is used to decompose the signal into high pass and low pass components, which are then decimated by a factor of two. The low pass component is then decomposed again into low pass and high pass components, and this process proceeds recursively. The synthesis bank performs the complementary

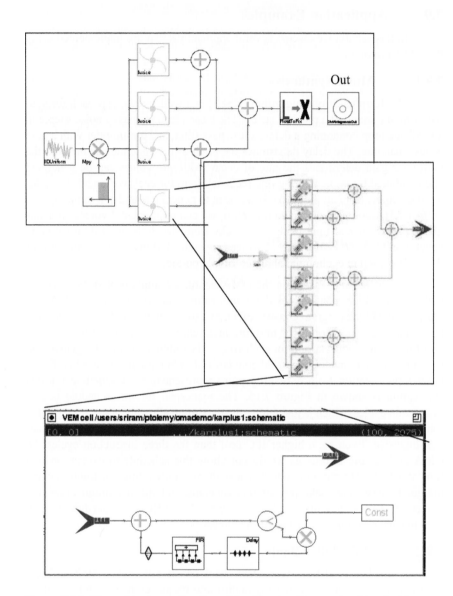

Figure 7.14. Hierarchical specification of the Karplus-Strong algorithm in 28 voices.

operation of upsampling, filtering, and combining the high pass and low pass components; this process is again performed recursively to reconstruct the input signal. Figure 7.16(a) shows a block diagram of a synthesis filter bank followed by an analysis bank.

QMF filter banks are designed such that the analysis bank cascaded with the synthesis bank yields a transfer function that is a pure delay (i.e., has unity response except for a delay between the input and the output). Such filter banks are also called **perfect reconstruction** filter banks, and they find applications in high quality audio compression; each frequency band is quantized according to its energy content and its perceptual importance. Such a coding scheme is employed in the audio portion of the MPEG standard.

We implemented a perfect-reconstruction QMF filter bank to decompose audio from a compact disc player into 15 bands. The synthesis bank was implemented together with the analysis part. There are a total of 36 multirate filters of 18 taps each. This is shown hierarchically in Figure 7.16(a). Note that delay blocks are required in the first 13 output paths of the analysis bank to compensate for the delay through successive stages of the analysis filter bank.

There are 1010 instruction cycles of computation per sample period in this example. Using Sih's Dynamic Level (DL) scheduling heuristic [SL93a], we were able to achieve an average iteration period of 366 instruction cycles, making use of 40 IPCs. The schedule that is actually constructed (Gantt chart of Fig-

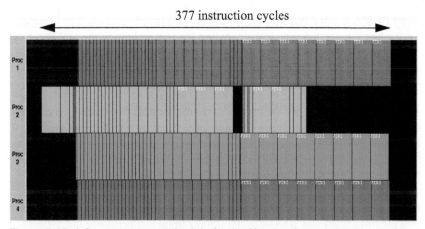

Figure 7.15. A four-processor schedule for the Karplus-Strong algorithm in 28 voices. Three processors are assigned 8 voices each, and the fourth (Proc 1) is assigned 4 voices along with the noise source.

ure 7.16(b)) operates on a block of 512 samples because this number of samples is needed before all the actors in the graph fire at least once; this makes manual scheduling very difficult. We found that the DL heuristic performs close to 20% better than the classic Hu-level heuristic [Hu61] in this example, although the DL heuristic takes more than twice the time to compute the schedule compared to the Hu-level technique.

### 7.9.3          1024 Point Complex Fast Fourier Transform (FFT)

For this example, input data (1024 complex numbers) is assumed to be present in shared memory, and the transform coefficients are written back to shared memory. A single 96002 processor on the OMA board performs a 1024 point complex FFT in 3.0 milliseconds (ms). For implementing the transform on all four processors, we used the first stage of a radix four, decimation in frequency FFT computation, after which each processor independently performs a 256-point FFT. In this scheme, each processor reads all 1024 complex inputs at the beginning of the computation, combines them into 256 complex numbers on which it performs a 256 point FFT, and then writes back its result to shared memory using bit-reversed addressing. The entire operation takes 1.0 ms. Thus, we achieve a speedup of 3 over a single processor. This example is communication intensive; the throughput is limited by the available bus bandwidth. Indeed, if all processors had independent access to the shared memory (if the shared memory were 4-ported, for example), we could achieve an ideal speedup of four, because each 256 point FFT is independent of the others except for data input and output.

For this example, data partitioning, shared memory allocation, scheduling, and tuning the assembly program was done by hand, using the 256-point complex FFT block in the Ptolemy CG96 domain as a building block. The Gantt chart for the hand-generated schedule, including IPC costs through the OMA controller, is shown in Figure 7.17.

## 7.10     Summary

In this chapter, we discussed the ideas behind the ordered-transactions scheduling strategy. This strategy combines compile time analysis of the IPC pattern with simple hardware support to minimize interprocessor communication overhead. We discussed the hardware design and implementation details of a prototype shared bus multiprocessor — the Ordered Memory Access architecture — that uses the ordered-transactions scheduling model to statically assign the sequence of processor accesses to shared memory. External I/O and user-specified control inputs can also be taken into account when scheduling accesses to the shared bus. We also discussed the software interface details of the prototype and illustrated some applications that were implemented on the OMA prototype.

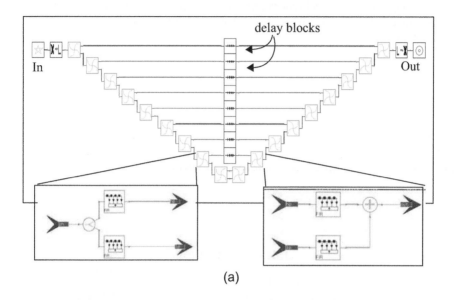

(a)

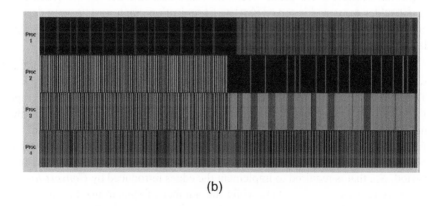

(b)

Figure 7.16. (a) Hierarchical block diagram for a 15 band analysis and synthesis filter bank. (b) Schedule on four processors (using Sih's DL heuristic [SL93a]).

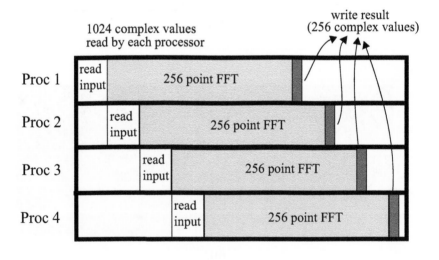

Figure 7.17. Schedule for the FFT example.

# 8

## ANALYSIS OF THE ORDERED-TRANSACTIONS STRATEGY

In this chapter the limits of the ordered-transactions scheduling strategy are systematically analyzed. Recall that the self-timed schedule is obtained by first generating a fully-static schedule $\{\sigma_p(v), \sigma_t(v), T_{FS}\}$, and then ignoring the exact firing times specified by the fully-static schedule; the fully-static schedule itself is derived using compile time estimates of actor execution times of actors. As defined in the previous chapter, the ordered-transactions strategy is essentially the self-timed strategy with added ordering constraints $O$ that force processors to communicate in an order predetermined at compile time. The questions addressed in this chapter are: What exactly are we sacrificing by imposing such a restriction? Is it possible to choose a transaction such that this penalty is minimized? What is the effect of variations of task (actor) execution times on the throughput achieved by a self-timed strategy and by an ordered transactions strategy?

The effect of imposing a transaction order on a self-timed schedule is best illustrated by the following example. Let us assume that we use the dataflow graph and its schedule that was introduced in Chapter 5 (Figure 5.3), and that we enforce the transaction order (obtained by sorting the $\sigma_t$ values) of Figure 7.1; we reproduce these for convenience in Figure 8.1 (a) and (b).

If we observe how the schedule "evolves" as it is executed in a self-timed manner (essentially a simulation in time of when each processor executes actors assigned to it), we get the "unfolded" schedule of Figure 8.2; successive iterations of the HSDFG overlap in a natural manner. This is of course an idealized scenario where IPC costs are ignored; we do so to avoid unnecessary detail in the diagram, since IPC costs can be included in our analysis in a straightforward manner. Note that the self-timed schedule in Figure 8.2 eventually settles to a periodic pattern consisting of two iterations of the HSDFG; the average iteration period under the self-timed schedule is 9 units. The average iteration period (which we will refer to as $T_{ST}$) for such an idealized (zero IPC cost) self-timed schedule represents a **lower bound** on the iteration period achievable by *any* schedule that maintains the same processor assignment and actor-ordering. This is because the only run-time constraint on processors that the self-timed schedule imposes is due to data dependencies: each processor executes actors assigned to it (including the communication actors) according to the compile-time-determined order. An actor at the head of this ordered list is executed as soon as data is available for it. Any other schedule that maintains the same processor assignment and actor ordering, and respects data precedences in $G$, cannot result in an execution where actors fire earlier than they do in the idealized self-timed schedule. In particular, the overlap of successive iterations of the HSDFG in the idealized self-timed schedule ensures that $T_{ST} \leq T_{FS}$ in general.

The self-timed schedule allows reordering among IPCs at run-time. In fact, we observe from Figure 8.2 that once the self-timed schedule settles into a

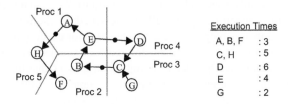

(a) HSDFG "G"

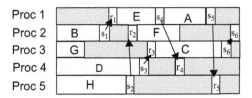

Transaction order:  $(s_1, r_1, s_2, r_2, s_3, r_3, s_4, r_4, s_5, r_5, s_6, r_6)$

(b) Schedule and Transaction order

Figure 8.1. Fully-static schedule on five processors.

periodic pattern, IPCs in successive iterations are ordered differently: in the first iteration, the order in which IPCs occur is indeed the unique $\sigma_t$-sorted transaction order shown in Figure 8.1(b):

$$(s_1, r_1, s_2, r_2, s_3, r_3, s_4, r_4, s_5, r_5, s_6, r_6) . \tag{8-1}$$

However, once the schedule settles into a periodic pattern, the order alternates between:

$$(s_3, r_3, s_1, r_1, s_2, r_2, s_4, r_4, s_6, r_6, s_5, r_5) \tag{8-2}$$

and

$$(s_1, r_1, s_3, r_3, s_4, r_4, s_2, r_2, s_5, r_5, s_6, r_6) . \tag{8-3}$$

In contrast, if we impose the transaction order in Figure 8.1(b) that enforces the order

$$(s_1, r_1, s_2, r_2, s_3, r_3, s_4, r_4, s_5, r_5, s_6, r_6) ,$$

the resulting ordered transactions schedule evolves as shown in Figure 8.3. Notice that enforcing this schedule introduces idle time (hatched rectangles); as a result, $T_{OT}$, the average iteration period for the ordered transactions schedule, is 10 units, which is (as expected) larger than the iteration period of the ideal self-timed schedule with zero arbitration and synchronization overhead $T_{ST}$ (9 units) but is smaller than $T_{FS}$ (11 units). In general $T_{FS} \geq T_{OT} \geq T_{ST}$: the self-timed schedule only has assignment and ordering constraints, the ordered transactions schedule has the transaction ordering constraints in addition to the constraints in the self-timed schedule, whereas the fully-static schedule has exact timing constraints that subsume the constraints in the self-timed and ordered transactions schedules. The question we would like to answer is: is it possible to choose the transaction ordering more intelligently than the straightforward $\sigma_t$-sorted order

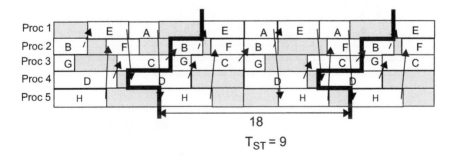

Figure 8.2. Self-timed schedule.

chosen in Figure 8.1(b)?

As a first step towards determining how such a "best" possible access order might be obtained, we attempt to model the self-timed execution itself and try to determine the precise effect (e.g., increase in the iteration period) of adding transaction ordering constraints. Note again that as the schedule evolves in a self-timed manner in Figure 8.2, it eventually settles into a periodically repeating pattern that spans two iterations of the dataflow graph, and the average iteration period, $T_{ST}$, is 9. We would like to determine these properties of self-timed schedules analytically without having to resort to simulation.

## 8.1    Inter-processor Communication Graph ($G_{ipc}$)

In a self-timed strategy a schedule $S$ specifies the actors assigned to each processor, including the IPC actors *send* and *receive*, and specifies the order in which these actors must be executed. At run-time each processor executes the actors assigned to it in the prescribed order. When a processor executes a send it writes into a certain buffer of finite size, and when it executes a receive, it reads from a corresponding buffer, and it checks for buffer overflow (on a send) and buffer underflow (on a receive) before it performs communication operations; it blocks, or suspends execution, when it detects one of these conditions.

We model a self-timed schedule using an HSDFG $G_{ipc} = (V, E_{ipc})$ derived from the application graph $G = (V, E)$ and the given self-timed schedule. The graph $G_{ipc}$, which we will refer to as the **inter-processor communication modeling graph**, or **IPC graph** for short, models the fact that actors of $G$ assigned to the same processor execute sequentially, and it models constraints due to interprocessor communication. For example, the self-timed schedule in Figure 8.1 (b)

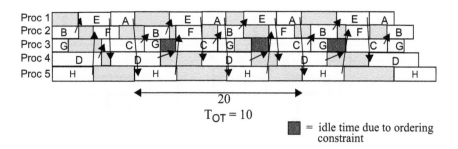

Figure 8.3. Schedule evolution when the transaction order of Figure 7.1 is enforced.

can be modeled by the IPC graph in Figure 8.4.

The IPC graph has the same vertex set $V$ as $G$, corresponding to the set of actors in $G$. The self-timed schedule specifies the actors assigned to each processor, and the order in which they execute. For example in Figure 8.1, processor 1 executes $A$ and then $E$ repeatedly. We model this in $G_{ipc}$ by drawing a cycle around the vertices corresponding to $A$ and $E$, and placing a delay on the edge from $E$ to $A$. The delay-free edge from $A$ to $E$ represents the fact that the $k$th execution of $A$ precedes the $k$th execution of $E$, and the edge from $E$ to $A$ with a delay represents the fact that the $k$th execution of $A$ can occur only after the $(k-1)$th execution of $E$ has completed. Thus, if actors $v_1, v_2, ..., v_n$ are assigned to the same processor in that order, then $G_{ipc}$ would have a cycle $((v_1, v_2), (v_2, v_3), ..., (v_{n-1}, v_n), (v_n, v_1))$, with $delay((v_n, v_1)) = 1$ (because $v_1$ is executed first). If there are $P$ processors in the schedule, then we have $P$ such cycles corresponding to each processor. The additional edges due to these constraints are shown as dashed arrows in Figure 8.4.

As mentioned before, edges in $G$ that cross processor boundaries after scheduling represent interprocessor communication. Communication actors (*send* and *receive*) are inserted for each such edge; these are shown in Figure 8.1.

The IPC graph has the same semantics as an HSDFG, and its execution models the execution of the corresponding self-timed schedule. The following definitions are useful to formally state the constraints represented by the IPC

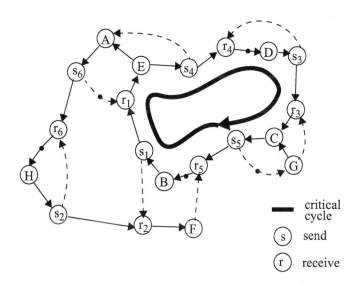

Figure 8.4. The IPC graph for the schedule in Figure 8.1.

graph. Time is modeled as an integer that can be viewed as a multiple of a base clock.

Recall that the function $start(v, k) \in Z^+$ represents the time at which the $k$th execution of actor $v$ starts in the self-timed schedule. The function $end(v, k) \in Z^+$ represents the time at which the $k$th execution of the actor $v$ ends and $v$ produces data tokens at its output edges, and we set $start(v, k) = 0$ and $end(v, k) = 0$ for $k < 0$ as the "initial conditions." The $start(v, 0)$ values are specified by the schedule: $start(v, 0) = \sigma_t(v)$.

Recall from Definition 5.2, as per the semantics of an HSDFG, each edge $(v_j, v_i)$ of $G_{ipc}$ represents the following data dependence constraint:

$$start(v_i, k) \geq end(v_j, k - delay((v_j, v_i))),$$

$$\text{for all } (v_j, v_i) \in E_{ipc}, \text{ for all } k \geq delay(v_j, v_i). \tag{8-4}$$

The constraints in (8-4) are due both to communication edges (representing synchronization between processors) and to edges that represent sequential execution of actors assigned to the same processor.

Also, to model execution times of actors we associate execution time $t(v)$ with each vertex of the IPC graph; $t(v)$ assigns a positive integer execution time to each actor $v$ (which can be interpreted as $t(v)$ cycles of a base clock). Interprocessor communication costs can be represented by assigning execution times to the *send* and *receive* actors. Now, we may substitute

$$end(v_j, k) = start(v_j, k) + t(v_j)$$

in (8-4) to obtain

$$start(v_i, k) \geq start(v_j, k - delay((v_j, v_i))) + t(v_j), \text{ for all } (v_j, v_i) \in E_{ipc} \tag{8-5}$$

In the self-timed schedule, actors fire as soon as data is available at all their input edges. Such an "as soon as possible" (ASAP) firing pattern implies:

$$start(v_i, k) = max(\{start(v_j, k - delay((v_j, v_i))) + t(v_j) | (v_j, v_i) \in E_{ipc}\}) \tag{8-6}$$

In contrast, recall that in the fully-static schedule we would force actors to fire periodically according to

$$start(v, k) = \sigma_t(v) + kT_{FS}. \tag{8-7}$$

The IPC graph has the same semantics as a Timed Marked graph in Petri net theory [Pet81] — the *transitions* of a marked graph correspond to the nodes of the IPC graph, the *places* of a marked graph correspond to edges, and the *initial marking* of a marked graph corresponds to initial tokens on the edges. The

IPC graph is also similar to Reiter's computation graph [Rei68]. The same properties hold for it, and we state some of the relevant properties here. The proofs listed here are similar to the proofs for the corresponding properties in marked graphs and computation graphs in the references above.

**Lemma 8.1:** The number of tokens in any cycle of the IPC graph is always conserved over all possible valid firings of actors in the graph, and is equal to the path delay of that cycle.

*Proof:* For each cycle $C$ in the IPC graph, the number of tokens on $C$ can only change when actors that are on it fire because actors not on $C$ remove and place tokens only on edges that are not part of $C$. If

$$C = ((v_1, v_2), (v_2, v_3), ..., (v_{n-1}, v_n), (v_n, v_1)), \qquad (8\text{-}8)$$

and any actor $v_k$ ($1 \le k \le n$) fires, then exactly one token is moved from the edge $(v_{k-1}, v_k)$ to the edge $(v_k, v_{k+1})$, where $v_0 \equiv v_n$ and $v_{n+1} \equiv v_1$. This conserves the total number of tokens on $C$. *QED.*

**Definition 8.1:** An HSDFG $G$ is said to be **deadlocked** if at least one of its actors cannot fire an infinite number of times in any valid sequence of firings of actors in $G$. Thus, when executing a valid schedule for a deadlocked HSDFG, some actor $v$ fires $k < \infty$ number of times, and is never enabled to fire subsequently.

**Lemma 8.2:** An HSDFG $G$ (in particular, an IPC graph) is free of deadlock if and only if it does not contain delay free cycles.

*Proof:* Suppose there is a delay free cycle

$$C = ((v_1, v_2), (v_2, v_3), ..., (v_{n-1}, v_n), (v_n, v_1))$$

in $G$ (i.e., $Delay(C) = 0$). By Lemma 8.1 none of the edges $(v_1, v_2), (v_2, v_3), ..., (v_{n-1}, v_n), (v_n, v_1)$, can contain tokens during any valid execution of $G$. Then each of the actors $v_1, ..., v_n$ has at least one input that never contains any data. Thus, none of the actors on $C$ are ever enabled to fire, and hence $G$ is deadlocked.

Conversely, suppose $G$ is deadlocked, i.e., there is one actor $v_1$ that never fires after a certain sequence of firings of actors in $G$. Thus, after this sequence of firings, there must be an input edge $(v_2, v_1)$ that never contains data. This implies that the actor $v_2$ in turn never gets enabled to fire, which in turn implies that there must be an edge $(v_3, v_2)$ that never contains data. In this manner we can trace a path $p = ((v_n, v_{n-1}), ..., (v_3, v_2), (v_2, v_1))$ for $n = |V|$ back from $v_1$ to $v_n$ that never contains data on its edges after a certain sequence of firing of actors in $G$. Since $G$ contains only $|V|$ actors, $p$ must visit some actor twice, and hence must contain a cycle $C$. Since the edges of $p$ do not contain data, $C$ is a delay-free cycle. *QED.*

**Definition 8.2:** A schedule $S$ is said to be **deadlocked** if after a certain finite time at least one processor blocks (on a buffer full or buffer empty condition) and stays blocked.

If the specified schedule is deadlock-free then the corresponding IPC graph is deadlock-free. This is because a deadlocked IPC graph would imply that a set of processors depend on data from one another in a cyclic manner, which in turn implies a schedule that displays deadlock.

**Lemma 8.3:** The iteration period for a *strongly connected* IPC graph $G$ when actors execute as soon as data is available at all inputs is given by:

$$T = \frac{max}{cycle\ C\ in\ G} \left\{ \frac{\sum\limits_{v\ is\ on\ C} t(v)}{Delay(C)} \right\} \tag{8-9}$$

Note that $Delay(C) > 0$ for an IPC graph constructed from an admissible schedule. This result has been proved in so many different contexts ([KLL87][Pet81][RH80][Rei68][RN81]) that we do not present another proof of this fact here.

The quotient in (8-9),

$$\frac{\sum\limits_{v\ is\ on\ C} t(v)}{Delay(C)} \tag{8-10}$$

is called the **cycle mean** of the cycle $C$. The entire quantity on the right hand side of (8-9) is called the **maximum cycle mean** of the strongly connected IPC graph $G$. If the IPC graph contains more than one SCC, then different SCCs may have different asymptotic iteration periods, depending on their individual maximum cycle means. In such a case, the iteration period of the overall graph (and hence the self-timed schedule) is the *maximum* over the maximum cycle means of all the SCCs of $G$, because the execution of the schedule is constrained by the slowest component in the system. Henceforth, we will define the maximum cycle mean as follows.

**Definition 8.3:** The **maximum cycle mean** of an IPC graph $G_{ipc}$, denoted by $MCM(G_{ipc})$, is the maximal cycle mean over all strongly connected components of $G_{ipc}$: That is,

$$MCM(G_{ipc}) = \max_{\text{cycle } C \text{ in } G_{ipc}} \left\{ \frac{\sum_{v \text{ is on } C} t(v)}{Delay(C)} \right\}. \tag{8-11}$$

Note that $MCM(G)$ may be a non-integer rational quantity. We will use the term $MCM$ instead of $MCM(G)$ when the graph being referred to is clear from the context. A fundamental cycle in $G_{ipc}$ whose cycle mean is equal to $MCM$ is called a **critical cycle** of $G_{ipc}$. Thus the throughput of the system of processors executing a particular self-timed schedule is equal to the corresponding $\frac{1}{MCM}$ value.

For example, in Figure 8.4, $G_{ipc}$ has one SCC, and its maximal cycle mean is 7 time units. This corresponds to the critical cycle

$$((B, E), (E, I), (I, G), (G, B)).$$

We have not included IPC costs in this calculation, but these can be included in a straightforward manner by appropriately setting the execution times of the *send* and *receive* actors.

As explained in Section 3.15 the maximum cycle mean can be calculated in time $O(|E|\Gamma)$, where $\Gamma$ is the sum of $t(v)$ over all actors in the HSDFG.

## 8.2    Execution Time Estimates

If we only have execution time estimates available instead of exact values, and we set $t(v)$ in the previous section to be these estimated values, then we obtain the *estimated* iteration period by calculating $MCM$. Henceforth, we will assume that we know the *estimated throughput*

$$\frac{1}{MCM}$$

calculated by setting the $t(v)$ values to the available timing estimates. As discussed in Chapter 1, for most practical scenarios, we can only assume such compile time estimates, rather than clock-cycle accurate execution time estimates. In fact, this is the reason we had to rely on self-timed scheduling, and we proposed the ordered transaction strategy as a means of achieving efficient IPC despite the fact that we do not assume knowledge of exact actor execution times. Section 5.9 discusses estimation techniques for actor execution times.

## 8.3      Ordering Constraints Viewed as Added Edges

The ordering constraints can be viewed as edges added to the IPC graph: an edge $(v_j, v_i)$ with zero delays represents the constraint $start(v_i, k) \geq end(v_j, k)$. The ordering constraints can therefore be expressed as a set of edges between communication actors. For example, the constraints $O = (s_1, r_1, s_2, r_2, s_3, r_3, s_4, r_4, s_5, r_5, s_6, r_6)$ applied to the IPC graph of Figure 8.4 is represented by the graph in Figure 8.5. If we call these additional ordering constraint edges $E_{OT}$ (solid arrows in Figure 8.5), then the graph $(V, E_{ipc} \cup E_{OT})$ represents constraints in the ordered transactions schedule, as it evolves in Figure 8.3. Thus, the maximum cycle mean of $(V, E_{ipc} \cup E_{OT})$ represents the effect of adding the ordering constraints. The critical cycle $C$ of this graph is drawn in

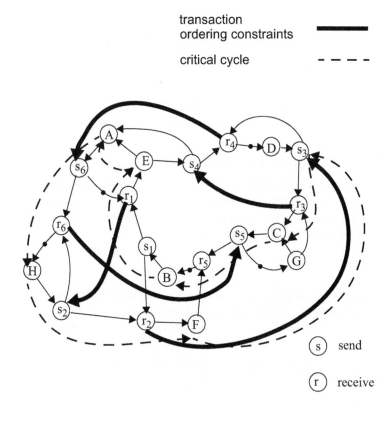

Figure 8.5. Transaction ordering constraints.

Figure 8.5; it is different from the critical cycle in Figure 8.4 because of the added transaction ordering constraints. Ignoring communication costs, the *MCM* is 9 units, which was also observed from the evolution of the transaction constrained schedule in Figure 8.3.

The problem of finding an "optimal" transaction order can therefore be stated as: Determine a transaction order $O$ such that the resultant constraint edges $E_{OT}$ do not increase the *MCM*, i.e.,

$$MCM((V, E_{ipc})) = MCM((V, E_{ipc} \cup E_{OT})). \qquad (8\text{-}12)$$

## 8.4     Periodicity

We noted earlier that as the self-timed schedule in Figure 8.2 evolves, it eventually settles into a periodic repeating pattern that spans two iterations of the dataflow graph. It can be shown that a self-timed schedule always settles down into a periodic execution pattern; in [BCOQ92] the authors show that the firing times of transitions in a marked graph are periodic asymptotically. Interpreted in our notation, for any strongly connected HSDFG there exist $K$ and $N$ such that:

$$start(v_i, k + N) = start(v_i, k) + MCM(G_{ipc}) \times N, \qquad (8\text{-}13)$$

for all $v_i \in V$, and for all $k > K$.

Thus, after a "transient" that lasts K iterations, the self-timed schedule evolves into a periodic pattern. The periodic pattern itself spans N iterations; we call N the **periodicity**. The periodicity depends on the number of delays in the critical cycles of $G_{ipc}$; it can be as high as the least common multiple of the number of delays in the critical cycles of $G_{ipc}$ [BCOQ92]. For example, the IPC graph of Figure 8.4 has one critical cycle with two delays on it, and thus we see a periodicity of two for the schedule in Figure 8.2. The "transient" region defined by $K$ (which is 1 in Figure 8.2) can also be exponential.

The effect of transients followed by a periodic regime is essentially due to properties of longest paths in weighted directed graphs. These effects have been studied in the context of instruction scheduling for VLIW processors [AN88][ZS89], as-soon-as-possible firing of transitions in Petri nets [Chr83], and determining clock schedules for sequential logic circuits [SBSV92]. In [AN88] the authors note that if instructions in an iterative program for a VLIW processor (represented as a dependency graph) are scheduled in an as-soon-as-possible fashion, a pattern of parallel instructions "emerges" after an initial transient, and the authors show how determining this pattern (essentially by simulation) leads to efficient loop parallelization. In [ZS89], the authors propose a max-algebra [CG79] based technique for determining the "steady state" pattern in the VLIW program. In [Chr85] the author studies periodic firing patterns of transitions in timed Petri nets. The iterative algorithms for determining clock schedules in

[SBSV92] have convergence properties similar to the transients in self-timed schedules (their algorithm converges when an equivalent self-timed schedule reaches a periodic regime).

Returning to the problem of determining the optimal transaction order, one possible scheme is to derive the transaction order from the repeating pattern that the self-timed schedule settles into. That is, instead of using the transaction order of Figure 7.1, if we enforce the transaction order that repeats over two iterations in the evolution of the self-timed schedule of Figure 8.2, the ordered transactions schedule would "mimic" the self-timed schedule exactly, and we would obtain an ordered transactions schedule that performs as well as the ideal self-timed schedule, and yet involves low IPC costs in practice. However, as pointed out above, the number of iterations that the repeating pattern spans depends on the critical cycles of $G_{ipc}$, and it can be exponential in the size of the HSDFG [BCOQ92]. In addition the "transient" region before the schedule settles into a repeating pattern can also be exponential. Consequently, the memory requirements for the controller that enforces the transaction order can be prohibitively large in certain cases; in fact, even for the example of Figure 8.2, the doubling of the controller memory that such a strategy entails may be unacceptable. We therefore restrict ourselves to determining and enforcing a transaction order that spans only one iteration of the HSDFG; in the following section we show that there is no sacrifice in imposing such a restriction and we discuss how such an "optimal" transaction order is obtained.

## 8.5    Optimal Order

In this section we show how to determine an order $O^*$ on the IPCs in the schedule such that imposing $O^*$ yields an ordered transactions schedule that has iteration period within one unit of the ideal self-timed schedule ($T_{ST} \leq T_{OT} \leq \lceil T_{ST} \rceil$). Thus, imposing the order $O^*$ results in essentially no loss in performance over an unrestrained schedule, and at the same time we get the benefit of cheaper IPC.

Our approach to determining the transaction order $O^*$ is to modify a given fully-static schedule so that the resulting fully-static schedule has $T_{FS}$ equal to $\lceil T_{ST} \rceil$, and then to derive the transaction order from that modified schedule. Intuitively it appears that, for a given processor assignment and ordering of actors on processors, the self-timed approach *always* performs better than the fully-static or ordered transactions approaches ($T_{FS} > T_{OT} > T_{ST}$) simply because it allows successive iterations to overlap. The following result, however, tells us that it is always possible to modify any given fully-static schedule so that it performs nearly as well as its self-timed counterpart. Stated more precisely:

**Theorem 8.1:**  Given a fully-static schedule $S \equiv \{\sigma_p(v), \sigma_t(v), T_{FS}\}$, let $T_{ST}$ be the average iteration period for the corresponding self-timed schedule (as mentioned before, $T_{FS} \geq T_{ST}$). Suppose $T_{FS} > T_{ST}$; then, there exists a valid fully-static schedule $S'$ that has the same processor assignment as $S$, the same order of execution of actors on each processor, but an iteration period of $\lceil T_{ST} \rceil$. That is, $S' \equiv \{\sigma_p(v), \sigma'_t(v), \lceil T_{ST} \rceil\}$ where, if actors $v_i$, $v_j$ are on the same processor (i.e., $\sigma_p(v_i) = \sigma_p(v_j)$) then $\sigma_t(v_i) > \sigma_t(v_j) \Rightarrow \sigma'_t(v_i) > \sigma'_t(v_j)$. Furthermore, $S'$ is obtained by solving the following set of linear inequalities for $\sigma'_t$:

$$\sigma'_t(v_j) - \sigma'_t(v_i) \leq \lceil T_{ST} \rceil \times d(v_j, v_i) - t(v_j) \quad \text{for each edge } (v_j, v_i) \text{ in } G_{\text{ipc}}. \quad (8\text{-}14)$$

*Proof:* Let $S'$ have a period equal to $T$. Then, under the schedule $S'$, the $k$ th starting time of actor $v_i$ is given by

$$start(v_i, k) = \sigma'_t(v_i) + kT. \quad (8\text{-}15)$$

Also, data precedence constraints imply (as in (8-5))

$$start(v_i, k) \geq start(v_j, k-delay(v_j, v_i)) + t(v_j), \text{ for all } (v_j, v_i) \in E_{\text{ipc}}. \quad (8\text{-}16)$$

Substituting (8-15) in (8-16), we have

$$\sigma'_t(v_i) + kT \geq \sigma'_t(v_j) + (k-delay(v_j, v_i))T + t(v_j),$$

for all $(v_j, v_i) \in E_{\text{ipc}}$. That is,

$$\sigma'_t(v_j) - \sigma'_t(v_i) \leq T \times d(v_j, v_i) - t(v_j), \text{ for all } (v_j, v_i) \in E_{\text{ipc}}. \quad (8\text{-}17)$$

Note that the construction of $G_{\text{ipc}}$ ensures that processor assignment constraints are automatically met: if $\sigma_p(v_i) = \sigma_p(v_j)$ and $v_i$ is to be executed immediately after $v_j$ then there is an edge $(v_j, v_i)$ in $G_{\text{ipc}}$. The relations in (8-17) represent a system of $|E_{\text{ipc}}|$ inequalities in $|V|$ unknowns (the quantities $\sigma'_t(v_i)$).

The system of inequalities in (8-17) is a difference constraint problem that can be solved in polynomial time ($O(|E_{\text{ipc}}||V|)$) using the Bellman-Ford shortest-path algorithm, as described in Section 3.14. Recall that a feasible solution to a given set of difference equations exists if and only if the corresponding constraint graph does not contain a negative weight cycle; this is equivalent to condition

$$T \geq \max_{\text{cycle } C \text{ in } G_{\text{ipc}}} \left\{ \frac{\displaystyle\sum_{v \in C} t(v)}{D(C)} \right\}. \quad (8\text{-}18)$$

and, from (8-9), this is equivalent to $T \geq T_{ST}$.

If we set $T = \lceil T_{ST} \rceil$, then the right-hand sides of the system of inequalities in 8-17 are integers, and the Bellman-Ford algorithm yields integer solutions for $\sigma'_t(v)$. This is because the weights on the edges of the constraint graph, which are equal to the right hand side of the difference constraints, are integers if

$T$ is an integer; consequently, the shortest paths calculated on the constraint graph are integers.

Thus, $S' \equiv \{\sigma_p(v), \sigma'_t(v), \lceil T_{ST} \rceil\}$ is a valid fully-static schedule. *QED*.

**Remark 8.1:** Theorem 8.1 essentially states that a fully-static schedule can be modified by skewing the relative starting times of processors so that the resulting schedule has iteration period less than $(T_{ST} + 1)$; the resulting iteration period lies within one time unit of its lower bound for the specified processor assignment and actor ordering. It is possible to unfold the graph and generate a fully-static schedule with average period exactly $T_{ST}$, but the resulting increase in code size is usually not worth the benefit of (at most) one time unit decrease in the iteration period. Recall that a "time unit" is essentially the clock period; therefore, one time unit can usually be neglected.

For example, the static schedule $S$ corresponding to Figure 8.1 has $T_{FS} = 11 > T_{ST} = 9$ units. Using the procedure outlined in the proof of Theorem 8.1, we can skew the starting times of processors in the schedule $S$ to obtain a schedule $S'$, as shown in (8-16), that has a period equal to 9 units (Figure 8.6). Note that the processor assignment and actor ordering in the schedule of Figure 8.6 is identical to that of the schedule in Figure 8.1. The values $\sigma'_t(v)$ are: $\sigma'_t(A) = 9$, $\sigma'_t(B) = \sigma'_t(G) = 2$, $\sigma'_t(C) = 6$, $\sigma'_t(D) = 0$, $\sigma'_t(E) = 5$, $\sigma'_t(F) = 8$, and $\sigma'_t(H) = 3$.

Theorem 8.1 may not seem useful at first sight: why not obtain a fully-static schedule that has a period $\lceil T_{ST} \rceil$ to begin with, thus eliminating the post-processing step suggested in Theorem 8.1? Recall from Chapters 5 and 6 that a fully-static schedule is usually obtained using heuristic techniques that are either based on blocked nonoverlapped scheduling (which use critical path based heuristics) [Sih91] or are based on overlapped scheduling techniques that employ list scheduling heuristics [dGH92][Lam88]. None of these techniques guarantee that the generated fully-static schedule will have an iteration period within one unit of

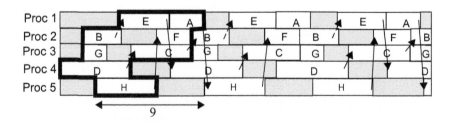

Figure 8.6. Modified schedule $S$.

the period achieved if the same schedule were run in a self-timed manner. Thus, for a schedule generated using any of these techniques, we might be able to obtain a gain in performance, essentially for free, by performing the post-processing step suggested in Theorem 8.1. What we propose can therefore be added as an efficient postprocessing step in existing schedulers. Of course, an exhaustive search procedure like the one proposed in [SI85] will certainly find the schedule $S'$ directly.

We set the transaction order $O^*$ to be the transaction order suggested by the modified schedule $S'$ (as opposed to the transaction order from $S$ used in Figure 7.1). Thus for the example of Figure 8.1(a),

$$O^* = (s_1, r_1, s_3, r_3, s_2, r_2, s_4, r_4, s_6, r_6, s_5, r_5).$$

Imposing the transaction order $O^*$ as in Figure 8.6 results in $T_{OT}$ of 9 units instead of 10 that we get if the transaction order of Figure 8.1(b) is used. Under the transaction order specified by $S'$, $T_{ST} \leq T_{OT} \leq \lceil T_{ST} \rceil$; thus imposing the order $O^*$ ensures that the average period is within one unit of the unconstrained self-timed strategy. Again, unfolding may be required to obtain a transaction-ordered schedule that has period exactly equal to $T_{ST}$, but the extra cost of a larger controller (to enforce the transaction ordering) outweighs the small gain of at most one unit reduction in the iteration period. Thus for all practical purposes $O^*$ is the *optimal* transaction order. The "optimality" is in the sense that the transaction order $O^*$ we determine statically is the best possible one, given the timing information available at compile time.

## 8.6    Effects of Changes in Execution Times

We recall that the execution times we use to determine the actor assignment and ordering in a self-timed schedule are compile time estimates, and we have been stating that static scheduling is advantageous when we have "reasonably good" compile time estimates of execution time of actors. Also, intuitively we expect an ordered transaction schedule to be more sensitive to changes in execution times than an unconstrained self-timed schedule. In this section we attempt to formalize these notions by exploring the effect of changes in execution times of actors on the throughput achieved by a static schedule.

Compile time estimates of actor execution times may be different from their actual values at run-time due to errors in estimating execution times of actors that otherwise have fixed execution times, and due to actors that display run-time variations in their execution times, because of conditionals or data-dependent loops within them, for example. The first case is simple to model, and we will show in Section 8.6.1 how the throughput of a given self-timed schedule changes as a function of actor execution times. The second case is inherently difficult; how do we model run-time changes in execution times due to data-dependencies, or due to events such as error-handling, cache misses, and pipeline effects? In Section 8.6.2 below we briefly discuss a very simple model for such run-time variations; we assume actors have random execution times according to some known probability distribution. We conclude that analysis of even such a simple model for the expected value of the throughput is often intractable, and we discuss efficiently computable upper and lower bounds for the expected throughput.

## 8.6.1    Deterministic Case

Consider the IPC graph in Figure 8.7, which is the same IPC graph as in Figure 8.4 except that we have used a different execution time for actor H to make the example more illustrative. The number next to each actor represents execution times of the actors. We let the execution time of actor C be $t(C) = t_C$, and we determine the iteration period as a function of given a particular value of $t_C$ ($T_{ST}(t_C)$). The iteration period is given by $MCM(G_{ipc})$, the maximum cycle mean. The function $T_{ST}(t_C)$ is shown in Figure 8.8. When $0 \le t_C \le 1$, the cycle $((A, s_6)(s_6, r_1)(r_1, E)(E, A))$ is critical, and the $MCM$ is constant at 7, since $C$ is not on this cycle; when $1 \le t_C \le 9$, the cycle

$$((B, s_1)(s_1, r_1)(r_1, E)(E, s_4)(s_4, r_4)(r_4, D)(D, s_3)(s_3, r_3)(r_3, C)(C, s_5)$$
$$(s_5, r_5)(r_5, B))$$

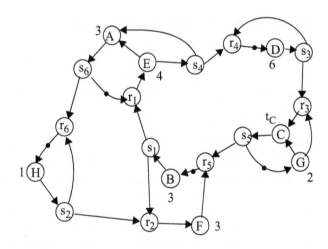

Figure 8.7. $G_{ipc}$, where actor $C$ has execution time $t_c$, constant over all invocations of $C$.

is critical, and since this cycle has two delays, the slope of $T_{ST}(t_C)$ is 0.5 in this region; finally, when $9 \leq t_C$ the cycle $((C, s_5)(s_5, G)(G, C))$ becomes critical, and the slope now is one because there is only one delay on that cycle.

Thus, the iteration period is a *piecewise linear* function of actor execution times. The slope of this function is zero if the actor is not on a critical cycle, otherwise it depends on the number of delays on the critical cycle(s) that the actor lies on. The slope is at most one (when the critical cycle containing the particular actor has a single delay on it). The iteration period is a **convex function** of actor execution times.

**Definition 8.4:** A function $f(x)$ is said to be **convex** over an interval $(a, b)$ if for every $x_1, x_2 \in (a, b)$ and $0 \leq \lambda \leq 1$,

$$f(\lambda x_1 + (1 - \lambda)x_2) \leq \lambda f(x_1) + (1 - \lambda)f(x_2).$$

Geometrically, if we plot a convex function $f(x)$ along $x$, a line drawn between two points on the curve lies above the curve (but it may overlap sections of the curve).

It is easily verified geometrically that $T_{ST}(t_C)$ is convex: since this function is piecewise linear with a slope that is positive and nondecreasing, a line

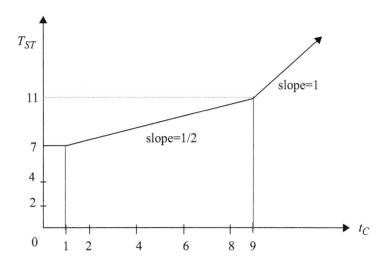

Figure 8.8. $T_{ST}(t_C)$ .

joining two points on it must lie above (but may coincide with) the curve.

We can also plot $T_{ST}$ as a function of execution times of more than one actor (e.g., $T_{ST}(t_A, t_B, \dots)$); this function will be a convex surface consisting of intersecting planes. Slices of this surface along each variable look like Figure 8.8, which is a slice parallel to the $t_C$ axis, with the other execution times held constant ($t_A = 3$, $t_B = 3$, etc.).

The modeling described in this section is useful for determining how "sensitive" the iteration period is to fixed changes in execution times of actors, given a processor assignment and actor ordering. We observe that the iteration period increases linearly (with slope one) at worst, and does not change at all at best, when execution time of an actor is increased beyond its compile time estimate.

### 8.6.2     Modeling Run-time Variations in Execution Times

The effect of variations in execution times of actors on the performance of statically scheduled hardware is inherently difficult to quantify, because these variations could occur due to a large number of factors — conditional branches or data-dependent loops within an actor, error handling, user interrupts, etc. — and because these variations could have a variety of different characteristics, from being periodic, to being dependent on the input statistics, and to being completely random. As a result, thus far we have had to resort to statements like "for a static scheduling strategy to be viable, actors must not show significant variations in execution times." In this section we point out the issues involved in modeling the effects of variations in execution times of actors.

A very simple model for actors with variable execution times is to assign to each actor an execution time that is a random variable (r.v.) with a discrete probability distribution (p.d.f.); successive invocations of each actor are assumed statistically independent, execution times of different actors are also assumed to be independent, and the statistics of the random execution times are assumed to be time-invariant. Thus, for example, an actor $A$ could have execution time $t_1$ with probability (w.p.) $p$ and execution time $t_2$ w.p. $(1-p)$. The model is essentially that $A$ flips a coin each time it is invoked to decide what its execution time should be for that invocation. Such a model could describe a data-dependent conditional branch for example, but it is of course too simple to capture many real scenarios.

Dataflow graphs where actors have such random execution times have been studied by Olsder et al. [Ols89][ORVK90] in the context of modeling data-driven networks (also called **wave-front arrays** [KLL87]) where the multiply operations in the array display data-dependent execution times. The authors show that the behavior of such a system can be described by a discrete-time Markov chain. The idea behind this, briefly, is that such a system is described by a state space consisting of a set of state vectors $s$. Entries in each vector $s$ represent the

$k$ th starting time of each actor normalized with respect to one (any arbitrarily chosen) actor:

$$s = \begin{pmatrix} 0 \\ start(v_2, k) - start(v_1, k) \\ start(v_3, k) - start(v_1, k) \\ \dots \\ start(v_n, k) - start(v_1, k) \end{pmatrix} \tag{8-19}$$

The normalization (with respect to actor $v_1$ in the above case) is done to make the state space finite; the number of distinct values that the vector $s$ (as defined above) can assume is shown to be finite in [ORVK90]. The states of the Markov chain correspond to each of the distinct values of $s$. The average iteration period, which is defined as:

$$T = \lim_{K \to \infty} \frac{start(v_i, K)}{K} \tag{8-20}$$

can then be derived from the stationary distribution of the Markov chain. There are several technical issues involved in this definition of the average iteration period; for example, when does the limit in (8-20) exist, and how do we show that the limit is in fact the same for all actors (assuming that the HSDFG is strongly connected)? These questions are fairly nontrivial because the random process $\{start(v_i, k)\}_k$ may not even be stationary. These questions are answered rigorously in [BCOQ92], where it is shown that:

$$T = \lim_{K \to \infty} \frac{start(v_i, K)}{K} = E[T] \quad \forall v_i \in V. \tag{8-21}$$

Thus, the limit $T$ is in fact a constant *almost surely* [Pap91].

The problem with such exact analysis, however, is the very large state space that can result. We found that for an IPC Graph similar to Figure 8.4, with certain choices of execution times, and assuming that only $t_C$ is random (takes two different values based on a weighted coin flip), we could get several thousand states for the Markov chain. A graph with more vertices leads to an even larger state space. The size of the state space can be exponential in the number of vertices (exponential in $|V|$). Solving the stationary distribution for such Markov chains would require solving a set of linear equations equal in number to the number of states, which is highly compute intensive. Thus, we conclude that this approach has limited use in determining effects of varying execution times; even for unrealistically simple stochastic models, computation of exact solutions is prohibitive.

If we assume that all actors have exponentially distributed execution times, then the system can be analyzed using continuous-time Markov chains [Mol82]. This is done by exploiting the memoryless property of the exponential distribution: when an actor fires, the state of the system at any moment does not depend on how long that actor has spent executing its function; the state changes only when that actor completes execution. The number of states for such a system is equal to the number of different valid token configurations on the edges of the dataflow graph, where by "valid" we imply any token configuration that can be reached by a sequence of firings of enabled actors in the HSDFG. This is also equal to the number of *valid retimings* [LS91] that exist for the HSDFG. This number, unfortunately, can again be exponential in the size of the HSDFG.

Analysis of such graphs with exponentially distributed execution times has been extensively studied in the area of stochastic Petri nets (in [Mur89] Murata provides a large and comprehensive list of references on Petri nets — 315 in all — a number of which focus on stochastic Petri nets). There is a considerable body of work that attempts to cope with the state explosion problem. Some of these works attempt to divide a given Petri net into parts that can be solved separately (e.g., [YW93]), some others propose simplified solutions when the graphs have particular structures (e.g., [CS93]), and others propose approximate solutions for values such as the expected firing rate of transitions (e.g., [Hav91]). None of these methods are general enough to handle even a significant class of IPC graphs. Again, exponentially distributed execution times for *all* actors is clearly a crude approximation to any realistic scenario to make the computations involved in exact calculations worthwhile.

### 8.6.3     Bounds on the Average Iteration Period

As an alternative to determining the exact value of $E[T]$, we discuss how to determine efficiently computable bounds for it.

**Definition 8.5:** Given an HSDFG $G = V, E$ that has actors with random execution times, define $G_{ave} = (V, E)$ to be an equivalent graph with actor execution times equal to the expected value of their execution times in $G$.

**Fact 8.1:** [Dur91] (Jensen's inequality) If $f(x)$ is a convex function of $x$, then: $E[f(x)] \geq f(E[x])$.

In [RS94] the authors use Fact 8.1 to show that $E[T] \geq MCM(G_{ave})$. This follows from the fact that $MCM(G_{ave})$ is a convex function of the execution times of each of its actors. This result is especially interesting because of its generality; it is true no matter what the statistics of the actor execution times are (even the various independence assumptions we made can be relaxed!).

One might wonder what the relationship between $E[T]$ and $E[MCM(G)]$ might be. We can again use Fact 8.1, along with the fact that the maximum cycle mean is a convex function of actor execution times, to show the following:

$$E[MCM(G)] \geq MCM[G_{ave}] . \tag{8-22}$$

However, we cannot say anything about $E[T]$ in relation to $E[MCM(G)]$; there are IPC graphs where $E[T] > E[MCM(G)]$, and others where $E[T] < E[MCM(G)]$.

If the execution times of actors are all bounded $(t_{min}(v) \leq t(v) \leq t_{max}(v)$ $\forall v \in V$, e.g., if all actors have execution times uniformly distributed in some interval $[a, b]$) then we can say the following:

$$MCM(G_{max}) \geq E[T] \geq MCM(G_{ave}) \geq MCM(G_{min}) \tag{8-23}$$

where $G_{max} = (V, E)$ is same as $G$ except the random actor execution times are replaced by their upper bounds $(t_{max}(v))$, and similarly $G_{min} = (V, E)$ is the same as $G$ except the random actor execution times are replaced by their lower bounds $(t_{min}(v))$.

Equation (8-23) summarizes the useful bounds we know for expected value of the iteration period for graphs that contain actors with random execution times. It should be noted that good upper bounds on $E[T]$ are not known. Rajsbaum and Sidi propose upper bounds for exponentially distributed execution times [RS94]; these upper bounds are typically more than twice the exact value of $E[T]$, and hence not very useful in practice. We attempted to simplify the Markov chain model (i.e., reduce the number of states) for the self-timed execution of a stochastic HSDFG by representing such an execution by a set of self-timed schedules of deterministic HSDFGs, between which the system makes transitions randomly. This representation reduces the number of states of the Markov chain to the number of different deterministic graphs that arise from the stochastic HSDFG. We were able to use this idea to determine an upper bound for $E[T]$; however, this bound also proved to be too loose in general (hence, we omit the details of this construction here).

### 8.6.4    Implications for the Ordered Transactions Schedule

Intuitively, an ordered transactions schedule is more sensitive to variations in execution times; even though in a functional sense, the computations performed using the ordered transactions schedule are robust with respect to execution time variations (the transaction order ensures correct sender-receiver synchronization). The ordering restriction makes the iteration period more dependent on execution time variations than the ideal ST schedule. This is apparent from our IPC graph model; the transaction ordering constraints add additional edges $(E_{OT})$ to $G_{ipc}$. For example, an IPC graph with transaction ordering constraints represented as dashed arrows is shown in Figure 8.9 (we use the transaction order $O^* = (s_1, r_1, s_3, r_3, s_2, r_2, s_4, r_4, s_6, r_6, s_5, r_5)$ determined in Section

8.5 and, again, communication times are not included). The graph for $T_{OT}(t_C)$ is now different and is plotted in Figure 8.8. Note that the $T_{OT}(t_C)$ curve for the ordered transactions schedule (solid) is "above" the corresponding curve for the unconstrained ST schedule (dashed): this shows precisely what we mean by an ordered transactions schedule being more sensitive to variations in execution times of actors. The "optimal" transaction order $O^*$ we determined ensures that the transaction constraints do not sacrifice throughput (ensures $T_{OT} = T_{ST}$) when actor execution times are *equal to their compile time estimates;* $O^*$ was calculated using $t_C = 3$ in Section 8.5, and sure enough, $T_{OT}(t_C) = T_{ST}(t_C)$ when $t_C = 3$.

Modeling using random variables for the ordered transactions schedule can again be done as before, and since we have more constraints in this schedule, the expected iteration period will in some cases be larger than that for a self-timed schedule.

## 8.7    Effects of Interprocessor Communication Costs

The techniques developed in Section 8.5 for determining optimal transaction orders assume that the time required to perform interprocessor communica-

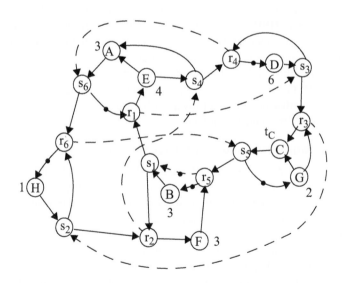

Figure 8.9. IPC graph $G_{ipc}$ with transaction ordering constraints represented as dashed lines.

tion is negligible. When IPC costs are not negligible, as is increasingly the case in practice, the techniques of Section 8.5 are still useful in that they can be used to derive upper bounds on the performance of the ordered transaction approach, and they also provide a low-complexity heuristic for determining a transaction order.

The problem of efficient ordered transaction implementation under non-negligible IPC costs has been studied in [KBB06]. In this work, it is shown that the when IPC costs are nonzero, the problem of determining an optimal transaction order becomes NP-hard. However, it is also shown that when IPC costs are nonzero, the corresponding self-timed schedule does not in general provide a performance bound for an ordered transaction schedule, as it does for the case of negligible IPC costs. In fact, ordered transaction schedules can significantly outperform their self-timed counterparts when IPC costs are nonzero. Thus, in the presence of nonnegligible IPC costs, the problem of determining an optimal transaction order becomes more difficult, but the potential benefits of applying an ordered transaction schedule become greater. In addition to developing these points of contrast between the cases of nonnegligible and negligible IPC costs, the work of [KBB06] develops heuristic techniques for constructing efficient transaction orders in a manner that takes IPC costs into account.

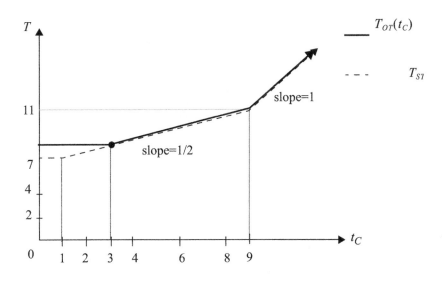

Figure 8.10. $T_{ST}(t_C)$ and $T_{OT}(t_C)$ for the example of Figure 8.9.

## 8.8    Summary

In this chapter we presented a quantitative analysis of self-timed and ordered transactions schedules and showed how to determine the effects of imposing a transaction order on a self-timed schedule. If the actual execution times do not deviate significantly from the estimated values, the difference in performance of the self-timed and ordered transactions strategies is minimal. If the execution times do in fact vary significantly, then even a self-timed strategy is not practical; it then becomes necessary to use a more dynamic strategy such as static assignment or fully dynamic scheduling [LH89] to make the best use of computing resources. Under the assumption that the variations in execution times are small enough so that a self-timed or an ordered transactions strategy is viable may be wiser to use the ordered transactions strategy rather than self-timed because of the more efficient IPC of the ordered transactions strategy. This is because a transaction order $O^*$ can be efficiently determined such that the ordering constraints do not sacrifice performance; if the execution times of actors are close to their estimates, the ordered transactions schedule with $O^*$ as the transaction order has iteration period close to the minimum achievable period $T_{ST}$. Thus, we make the best possible use of compile time information when we determine the transaction order $O^*$.

The complexities involved in modeling run-time variations in execution times of actors were also discussed; even highly simplified stochastic models are difficult to analyze precisely. We pointed out bounds that have been proposed in Petri net literature for the value of the expected iteration period, and concluded that although a lower bound is available for this quantity for rather general stochastic models (using Jensen's inequality), tight upper bounds are not known to date, except for the trivial upper bound using maximum execution times of actors ($MCM(G_{max})$).

We also discussed the incorporation of IPC costs into the problem of constructing transaction orders, and we described how this introduces significant new complexity, but also new opportunity into the ordered transaction approach.

# 9

---

# EXTENDING THE OMA
# ARCHITECTURE

---

The techniques of the previous chapters apply compile time analysis to static schedules for application graphs that have no decision-making at the dataflow graph (inter-task) level. This chapter considers graphs with data-dependent control flow. Recall that atomic actors in an SDF graph are allowed to perform data-dependent decision making within their body, as long as their input/output behavior respects SDF semantics. We show how some of the ideas we explored previously can still be applied to dataflow graphs containing actors that display data-dependent firing patterns, and therefore are not SDF actors. We do this by studying OMA concepts in the context of the more general Boolean dataflow (BDF) model, which was reviewed in Section 4.7.1.

## 9.1    Scheduling BDF Graphs

Buck presents techniques for statically scheduling BDF graphs on a single processor; his methods attempt to generate a sequential program without a dynamic scheduling mechanism, using **if-then-else** and **do-while** control constructs where required. Because of the inherent undecidability of determining deadlock behaviour and bounded memory usage, these techniques are not always guaranteed to generate a static schedule, even if one exists; a dynamically scheduled implementation, where a run-time kernel decides which actors to fire, can be used when a static schedule cannot be found in a reasonable amount of time.

Automatic parallel scheduling of general BDF graphs is still an unsolved problem. A *naive* mechanism for scheduling graphs that contain SWITCH and SELECT actors is to generate an Acyclic Precedence Extension Graph (APEG), similar to the APEG generated for SDF graphs discussed in Section 3.8, for every possible assignment of the Boolean valued control tokens in the BDF graph. For example, the if-then-else graph in Figure 4.7(a) could have two different APEGs, shown in Figure 9.1, and APEGs thus obtained can be scheduled individually using a self-timed strategy; each processor now gets several lists of actors, one list for each possible assignment of the control tokens. The problem with this approach is that for a graph with $n$ different control tokens, there are $2^n$ possible distinct APEGs, each corresponding to each execution path in the graph. Such a set of APEGs can be compactly represented using the so-called Annotated Acyclic Precedence Graph (AAPG) of [Buc93] in which actors and arcs are annotated with conditions under which they exist in the graph. Buck uses the AAPG construct to determine whether a bounded-length uniprocessor schedule exists. In the case of multiprocessor scheduling, it is not clear how such an AAPG could be used to explore scheduling options for the different values that the control tokens could take, without explicitly enumerating all possible execution paths.

A useful body of work in parallel scheduling of dataflow graphs that have dynamic actors is the **quasi-static scheduling** approach, discussed in Section

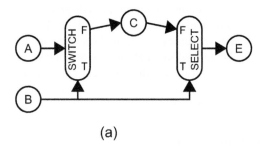

(a)

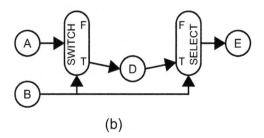

(b)

Figure 9.1. Acyclic precedence extension graphs (APEGs) corresponding to the if-then-else graph of Figure 4.7. (a) corresponds to the TRUE assignment of the control token, (b) to the FALSE assignment.

5.6. In this work, techniques have been developed that statically schedule standard dynamic constructs such as data-dependent conditionals, data-dependent iterations, and recursion. Such a quasi-static scheduling approach clearly does not handle a general BDF graph, although it is a good starting point for doing so.

We will consider only the conditional and the iteration construct here. We will assume that we are given a quasi-static schedule, obtained either manually or using Ha's techniques [HL97] that were described briefly in Section 5.6. We then explore how the techniques proposed in the previous chapters for multiprocessors that utilize a self-timed scheduling strategy apply when we implement a quasi-static schedule on a multiprocessor. First, we propose an implementation of a quasi-static schedule on a shared memory multiprocessor, and then we show how we can implement the same program on the OMA architecture, using the hardware support provided in the OMA architecture prototype.

## 9.2 Parallel Implementation on Shared Memory Machines

### 9.2.1 General Strategy

A quasi-static schedule ensures by means of the execution profile that the pattern of processor availability is identical regardless of how the data-dependent construct executes at run-time; in the case of the conditional construct this means that irrespective of which branch is actually taken, the pattern of processor availability after the construct completes execution is the same. This has to be ensured by inserting idle time on processors when necessary. Figure 9.2 shows a quasi-static schedule for a conditional construct. Maintaining the same pattern of processor availability allows static scheduling to proceed after the execution of the conditional; the data-dependent nature of the control construct can be ignored at that point. In Figure 9.2 for example, the scheduling of subgraph-1 can proceed independent of the conditional construct because the pattern of processor availability after this construct is the same independent of the branch outcome; note that "nops" (idle processor cycles) have been inserted to ensure this.

Multiprocessor implementation of a quasi-static schedule directly, however, implies enforcing global synchronization after each dynamic construct in order to ensure a particular pattern of processor availability. We therefore use a mechanism similar to the self-timed strategy; we first determine a quasi-static schedule using the methods of Lee and Ha, and then discard the timing information and the restrictions of maintaining a processor availability profile. Instead, we only retain the assignment of actors to processors, the order in which they execute, and also under what conditions on the Boolean tokens in the system the actor should execute. Synchronization between processors is done at run-time whenever processors communicate. This scheme is analogous to constructing a self-timed schedule from a fully-static schedule, as discussed in Section 5.4.

Thus, the quasi-static schedule of Figure 9.2 can be implemented by the set of programs in Figure 9.3, for the three processors. Here, $\{r_{c1}, r_{c2}, r_1, r_2\}$ are the receive actors, and $\{s_{c1}, s_1, s_2\}$ are the send actors. The subscript "$c$" refers to actors that communicate control tokens.

The main difference between such an implementation and the self-timed

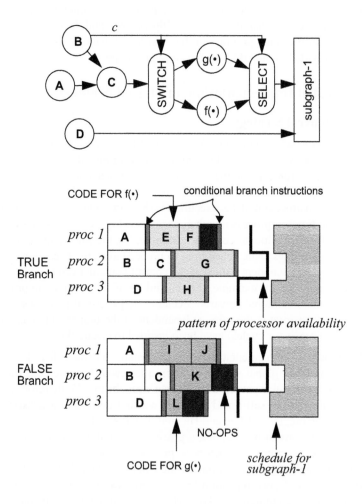

Figure 9.2. Quasi-static schedule for a conditional construct.

implementation we discussed in earlier chapters are the control tokens. Whenever a conditional construct is partitioned across more than one processor, the control token(s) that determine its behavior must be broadcast to all the processors that execute that construct. Thus, in Figure 9.2, the value $c$, which is computed by Processor 2 (since the actor that produces $c$ is assigned to Processor 2), must be broadcast to the other two processors. In a shared memory machine this broadcast can be implemented by allowing the processor that evaluates the control token (Processor 2 in our example) to write its value to a particular shared memory location preassigned at compile time; the processor will then update this location once for each iteration of the graph. Processors that require the value of a particular control token simply read that value from shared memory, and the processor that writes the value of the control token needs to do so only once. In this way, actor executions can be conditioned upon the value of control tokens evaluated at run-time. In the previous chapters, we discussed synchronization associated with data transfer between processors. Synchronization checks must also be performed for the control tokens; the processor that writes the value of a token must not overwrite the shared memory location unless all processors requiring the value of that token have in fact read the shared memory location, and processors reading a control token must ascertain that the value they read corresponds to the current iteration rather than a previous iteration.

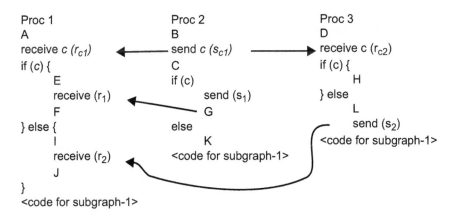

Figure 9.3. Programs on three processors for the quasi-static schedule of Figure 9.2.

The need for broadcast of control tokens creates additional communication overhead that should ideally be taken into account during scheduling. The methods of Lee and Ha, and also prior research related to quasi-static scheduling that they refer to in their work, do not take this cost into account. Static multiprocessor scheduling applied to graphs with dynamic constructs taking costs of distributing control tokens into account is thus an interesting problem for further study.

## 9.2.2    Implementation on the OMA

Recall that the OMA architecture imposes an order in which shared memory is accessed by processors in the machine. This is done to implement the OT strategy, and is feasible because the pattern of processor communications in a self-timed schedule of an HSDFG is in fact predictable. What happens when we want to run a program derived from a quasi-static schedule, such as the parallel program in Figure 9.3, which was derived from the schedule in Figure 9.2? Clearly, the order of processor accesses to shared memory is no longer predictable; it depends on the outcome of run-time evaluation of the control token $c$. The quasi-static schedule of Figure 9.2 specifies the schedules for the TRUE and FALSE branches of the conditional. If the value of $c$ were always TRUE, then we can determine from the quasi-static schedule that the transaction order would be $(s_{c1}, r_{c1}, r_{c2}, s_1, r_1, <\text{access order for subgraph-1}>)$, and if the value of $c$ were always FALSE, the transaction order would be

$$(s_{c1}, r_{c1}, r_{c2}, s_2, r_2, <\text{access order for subgraph-1}>) . \qquad (9\text{-}1)$$

Note that writing the control token $c$ once to shared memory is enough since the same shared location can be read by all processors requiring the value of $c$.

For the OMA architecture, a possible strategy is to switch between these two access orders at run-time. This is enabled by the preset feature of the transaction controller (Section 7.6.2). Recall that the transaction controller is implemented as a presettable schedule counter that addresses memory containing the processor IDs corresponding to the bus access order. To handle conditional constructs, we derive two bus access lists corresponding to each path in the program, and the processor that determines the branch condition (processor 2 in our example) forces the controller to switch between access lists by loading the schedule counter with the appropriate value (address "7" in the bus access schedule of Figure 9.5). Note from Figure 9.5 that there are two points where the schedule counter can be set; one is at the completion of the TRUE branch, and the other is a jump into the FALSE branch. The branch into the FALSE path is best taken care of by processor 2, since it computes the value of the control token $c$, whereas the branch after the TRUE path (which bypasses the access list of the FALSE branch) is best taken care of by processor 1, since processor 1 already possesses the bus at the time when the counter needs to be loaded. The schedule counter load operations are easily incorporated into the sequential programs of processors 1 and 2.

The mechanism of switching between bus access orders works well when the number of control tokens is small. But if the number of such tokens is large, then this mechanisms breaks down, even if we can efficiently compute a quasi-static schedule for the graph. To see why this is so, consider the graph in Figure 9.6, which contains $k$ conditional constructs in parallel paths going from the input to the output. The functions "$f_i$" and "$g_i$" are assumed to be subgraphs that are assigned to more than one processor. In Ha's hierarchical scheduling approach, each conditional is scheduled independently; once scheduled, it is converted into an atomic node in the hierarchy, and a profile is assigned to it. Scheduling of the other conditional constructs can then proceed based on these profiles. Thus, the scheduling complexity in terms of the number of parallel paths is $O(k)$ if there are $k$ parallel paths. If we implement the resulting quasi-static schedule in the manner stated in the previous section, and employ the OMA mechanism above, we would need one bus access list for every combination of the Booleans $b_1,..., b_k$. This is because each $f_i$ and $g_i$ will have its own associated bus access list, which then has to be combined with the bus access lists of all the other branches to yield one list. For example, if all Booleans $b_i$ are true, then all the $f_i$'s are executed, and we get one access list. If $b_1$ is TRUE, and $b_2$ through $b_k$ are FALSE, then $g_1$ is executed, and $f_2$ through $f_k$ are executed. This corresponds to another

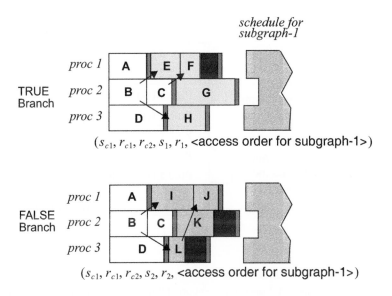

Figure 9.4. Transaction order for the TRUE and FALSE branches.

bus access list. This implies $2^k$ bus access lists for each of the combination of $f_i$ and $g_i$ that execute, i.e., for each possible execution path in the graph.

### 9.2.3        Improved Mechanism

Although the idea of maintaining separate bus access lists is a simple mechanism for handling control constructs, it can sometimes be impractical, as in the example above. We propose an alternative mechanism based on *masking* that handles parallel conditional constructs more effectively.

The main idea behind masking is to store an ID of a Boolean variable along with the processor ID in the bus access list. The Boolean ID determines whether a particular bus grant is "enabled." This allows us to combine the access lists of all the nodes $f_1$ through $f_k$ and $g_1$ through $g_k$. The bus grant corresponding to each $f_i$ is tagged with the boolean ID of the corresponding $b_i$, and an additional bit indicates that the bus grant is to be enabled when $b_i$ is TRUE. Similarly, each bus grant corresponding to the access list of $g_i$ is tagged with the ID of $b_i$, and an additional bit indicates that the bus grant must be enabled only if the correspond-

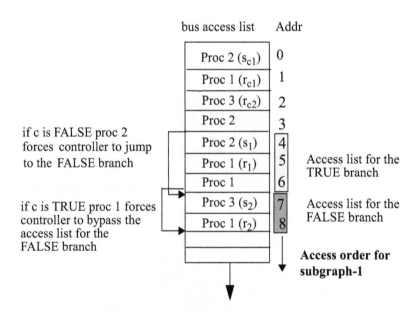

Figure 9.5. Bus access list that is stored in the schedule RAM for the quasi-static schedule of Figure 9.4. Loading operation of the schedule counter conditioned on value of $c$ is also shown.

ing control token has a FALSE value. At run-time, the controller steps through the bus access list as before, but instead of simply granting the bus to the processor at the head of the list, it first checks that the control token corresponding to the Boolean ID field of the list is in its correct state. If it is in the correct state (i.e., it is TRUE for a bus grant corresponding to an $f_i$ and FALSE for a bus grant corresponding to a $g_i$), then the bus grant is performed, otherwise it is masked. Thus, the run-time values of the Booleans must be made available to the transaction controller for it to decide whether to mask a particular bus grant or not.

More generally, a particular bus grant should be enabled by a product (AND) function of the Boolean variables in the dataflow graph, and the complement of these Booleans. Nested conditionals in parallel branches of the graph necessitate bus grants that are enabled by a product function; a similar need arises when bus grants must be reordered based on values of the Boolean variables. Thus, in general we need to implement an *annotated* bus access list of the form $\{(c_1)ProcID_1, (c_2)ProcID_2, \dots \}$ ; each bus access is annotated with a Boolean val-

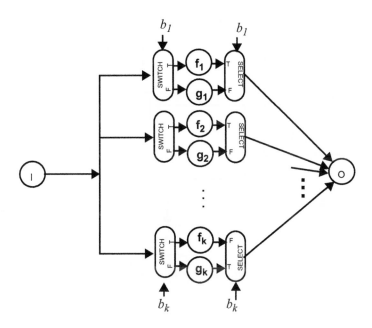

Figure 9.6. Conditional constructs in parallel paths.

ued condition $c_i$, indicating that the bus should be granted to the processor corresponding to $ProcID_i$ when $c_i$ evaluates to TRUE; $c_i$ could be an arbitrary product function of the Booleans $\{b_1, b_2, ..., b_n\}$ in the system, and the complements of these Booleans (e.g., $c_j = b_2 \cdot \overline{b_4}$, where the bar over a variable indicates its complement).

This scheme is implemented as shown in Figure 9.7. The schedule memory now contains two fields corresponding to each bus access: <Condition>:<ProcID> instead of the <ProcID> field alone that we had before. The <Condition> field encodes a unique product $c_i$ associated with that particular bus

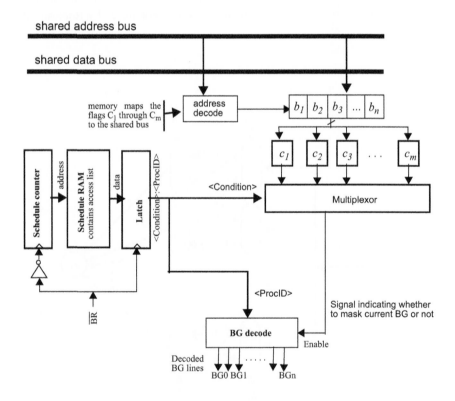

Figure 9.7. A bus access mechanism that selectively "masks" bus grants based on values of control tokens that are evaluated at run-time.

access. In the OMA prototype, we can use 3 bits for <ProcID>, and 5 bits for the <Condition> field. This would allow us to handle 8 processors and 32 product combinations of Booleans. There can be up to $m = 3^n$ product terms in the worst case corresponding to $n$ Booleans in the system, because for each Boolean $b_i$, a product term could contain $b_i$, or it could contain its complement $\overline{b}_i$, or else $b_i$ could be a "don't care". It is unlikely that all $3^n$ possible product terms will be required in practice; we therefore expect such a scheme to be practical. The necessary product terms ($c_j$) can be implemented within the controller at compile time, based on the bus access pattern of the particular dynamic dataflow graph to be executed.

In Figure 9.7, the flags $b_1, b_2, ..., b_n$, are 1-bit memory elements (flip-flops) that are memory mapped to the shared bus, and store the values of the Boolean control tokens in the system. The processor that computes the value of each control token updates the corresponding $b_i$ by writing to the shared memory location that maps to $b_i$. The product combinations $c_1, c_2, ..., c_n$, are just AND functions of the $b_i$s and the complement of the $b_i$s, e.g. $c_j$ could be $b_2 \cdot \overline{b}_4$. As the schedule counter steps through the bus access list, the bus grant is actually granted only if the condition corresponding to that access evaluates to TRUE; thus if the entry <$c_2$><Proc1> appears at the head of the bus access list, and $c_2 = b_2 \cdot \overline{b}_4$, then processor 1 receives a bus grant only if the control token $b_2$ is TRUE and $b_4$ is FALSE, otherwise the bus grant is masked and the schedule counter moves up to the next entry in the list.

This scheme can be incorporated into the transaction controller in our existing OMA architecture prototype, since the controller is implemented on an FPGA. The product terms $c_1, c_2, ..., c_n$ may be programmed into the FPGA at compile time; when we generate programs for the processors, we can also generate the *annotated* bus access list (a sequence of <Condition><Proc ID> entries), and a hardware description for the FPGA (in a hardware description language such as VHDL, say) that implements the required product terms.

### 9.2.4 Generating the Annotated Bus Access List

Consider the problem of obtaining an annotated bus access list $\{(c_1)ProcID_1, (c_2)ProcID_2, ...\}$, from which we can derive the sequence of <Condition><Proc ID> entries for the mask-based transaction controller. A straightforward, even if inefficient, mechanism for obtaining such a list is to use enumeration; we simply enumerate all possible combinations of Booleans in the system ($2^n$ combinations for $n$ Booleans), and determine the bus access sequence (sequence of ProcID's) for each combination. Each combination corresponds to an execution path in the graph, and we can estimate the time of occurrence of bus accesses corresponding to each combination from the quasi-static schedule. For example, bus accesses corresponding to one schedule period of the two execution paths in the quasi-static schedule of Figure 9.4 may be marked along the time axis as shown in Figure 9.8 (we have ignored the bus access sequence corresponding to subgraph-1 to keep the illustration simple).

The bus access schedules for each of the combinations can now be collapsed into one annotated list, as in Figure 9.8; the fact that accesses for each combination are ordered with respect to time allows us to enforce a global order on the accesses in the collapsed bus access list. The bus accesses in the collapsed list are annotated with their respective Boolean condition.

The collapsed list obtained above can be used, as is, in the masked controller scheme; however, there is a potential for optimizing this list. Note, however, that the same transaction may appear in the access list corresponding to different Boolean combinations, because a particular Boolean token may be a "don't care" for that bus access. For example, the first three bus accesses in Figure 9.8 appear in both execution paths, because they are independent of the value of $c$. In the worst case, a bus access that is independent of all Booleans will end up appearing in the bus access lists of all the Boolean combinations. If these bus accesses appear contiguously in the collapsed bus access sequence, we can combine them into one. For example, "$(c)$ Proc2, $(\bar{c})$ Proc2" in the annotated schedule of Fig-

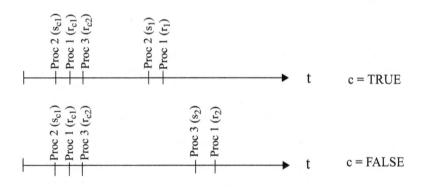

$\{(c)$ Proc2, $(\bar{c})$ Proc2, $(c)$ Proc1, $(\bar{c})$ Proc1, $(c)$ Proc3, $(\bar{c})$ Proc3,

$(c)$ Proc2, $(c)$ Proc1, $(\bar{c})$ Proc3, $(\bar{c})$ Proc 1$\}$

Figure 9.8. Bus access lists and the annotated list corresponding to Figure 9.4.

ure 9.8 can be combined into a single "Proc 2" entry, which is not conditioned on any control token. Consider another example: if we get contiguous entries "$(b_1 \cdot \overline{b_2})$ Proc3" and "$(b_1 \cdot b_2)$ Proc3" in the collapsed list, we can replace the two entries with a single entry "$(b_1)$ Proc3."

More generally, if the collapsed list contains a contiguous segment of the form:

$$\{..., (c_1)ProcID_k, (c_2)ProcID_k, (c_3)ProcID_k, ..., (c_l)ProcID_k, ...\},$$

each of the contiguous segments can be written as:

$$\{..., (c_1 + c_2 + ... + c_l)ProcID_k, ...\},$$

where the bus grant condition is an expression $(c_1 + c_2 + ... + c_l)$, which is a sum of products (SOP) function of the Booleans in the system. Two-level logic minimization can then be applied to determine a minimal representation of each of these expressions. Such 2-level minimization can be done by using a logic minimization tool such as ESPRESSO [BHMSVc84], which simplifies a given SOP expression into an SOP representation with minimal number of product terms. Suppose the expression $(c_1 + c_2 + ... + c_l)$ can be minimized into another SOP expression $(c_1' + c_2' + ... + c_p')$, where $p < l$. The segment

$$\{..., (c_1)ProcID_k, (c_2)ProcID_k, (c_3)ProcID_k, ..., (c_l)ProcID_k, ...\}$$

can then be replaced with an equivalent segment of the form:

$$\{..., (c_1')ProcID_k, (c_2')ProcID_k, (c_3')ProcID_k, ..., (c_p')ProcID_k, ...\}.$$

This procedure results in a minimal set of contiguous appearances of a bus grant to the same processor.

Another optimization that can be performed is to combine annotated bus access lists with the switching mechanism of Section 9.2.1. Suppose we have the following annotated bus access list:

$$\{..., (b_1 \cdot \overline{b_2})ProcID_i, (b_1 \cdot \overline{b_3})ProcID_j, (b_1 \cdot b_4 \cdot b_5)ProcID_k, ...\}.$$

Then, by "factoring" $b_1$ out, the above list may be equivalently written as:

$$\{..., (b_1)\{(\overline{b_2})ProcID_i, (\overline{b_3})ProcID_j, (b_4 \cdot b_5)ProcID_k\}, ...\}.$$

Now, all the three bus accesses may be skipped whenever the Boolean $b_1$ is FALSE by loading the schedule counter and forcing it to increment its count by three, instead of evaluating each access separately, and skipping over each one individually. This strategy reduces overhead, because it costs an extra bus cycle to disable a bus access when a condition corresponding to that bus access evaluates to FALSE; by skipping over three bus accesses that we know are going to be disabled, we save three idle bus cycles. There is an added cost of one cycle for loading the schedule counter; the total savings in this example is therefore two bus cycles.

One of the problems with the above approach is that it involves explicit enumeration of all possible combinations of Booleans, the complexity of which limits the size of problems that can be tackled with this approach. An implicit mechanism for representing all possible execution paths is therefore desirable. One such mechanism is the use of Binary Decision Diagrams (BDDs), which have been used to efficiently represent and manipulate Boolean functions for the purpose of logic minimization [Bry86]. BDDs have been used to compactly represent large state spaces, and to perform operations implicitly over such state spaces when methods based on explicit techniques are infeasible. One difficulty encountered in applying BDDs to the problem of representing execution paths is that it is not obvious how precedence and ordering constraints can be encoded in a BDD representation. The execution paths corresponding to the various Boolean combinations can be represented using a BDD, but it isn't clear how to represent the relative order between bus accesses corresponding to the different execution paths. We leave this as an area for future exploration.

## 9.3　　Data-dependent Iteration

Recall the data-dependent iteration construct shown in Figure 4.7(b). A quasi-static schedule for such a construct may look like the one in Figure 9.9. The actors $A$, $B$, $C$, and $D$ of Figure 4.7(b) are assumed to be subgraphs rather than atomic actors.

Such a quasi-static schedule can also be implemented in a straightforward fashion on the OMA architecture, provided that the data-dependent construct spans all the processors in the system. The bus access schedule corresponding to the iterated subgraph is simply repeated until the iteration construct terminates. The processor responsible for determining when the iteration terminates can be

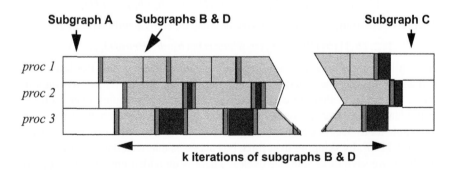

Figure 9.9. Quasi-static schedule for the data-dependent iteration graph of Figure 4.7(b).

made to force the schedule counter to loop back until the termination condition is reached. This is shown in Figure 9.10.

## 9.4 Summary

This chapter has dealt with extensions of the ordered-transactions approach to graphs with data-dependent control flow. The Boolean dataflow model was briefly reviewed, and the quasi-static approach to scheduling conditional and data-dependent iteration constructs. A scheme was then described whereby the Ordered Transactions approach could be used when such control constructs are included in the dataflow graph. In this scheme, bus access schedules are computed for each set of values that the control tokens in the graph evaluate to, and the bus access controller is made to select between these lists at run-time based on which set of values the control tokens actually take at any given time. This was also shown to be applicable to data-dependent iteration constructs. Such a scheme is feasible when the number of execution paths in the graph is small. A mechanism based on masking of bus accesses depending on run-time values of control tokens may be used for handling the case when there are multiple conditional constructs in "parallel."

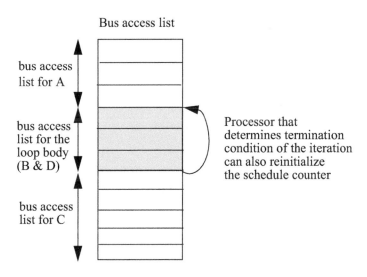

Figure 9.10. A possible access order list corresponding to the quasi-static schedule of Figure 9.9.

# 10

## SYNCHRONIZATION IN SELF-TIMED SYSTEMS

The previous three chapters have been concerned with the Ordered Transactions strategy, which is a hardware approach to reducing IPC and synchronization costs in self-timed schedules. In this chapter and the following two chapters, we discuss software-based strategies for minimizing synchronization costs in the final implementation of a given self-timed schedule. These software-based techniques are widely-applicable to shared-memory multiprocessors that consist of homogeneous or heterogeneous collections of processors, and they do not require the availability of hardware support for employing the OT approach or any other form of specialized hardware support.

Recall that the self-timed scheduling strategy introduces synchronization checks whenever processors communicate. A straightforward implementation of a self-timed schedule would require that for each interprocessor communication (IPC), the sending processor ascertains that the buffer it is writing to is not full, and the receiver ascertains that the buffer it is reading from is not empty. The processors block (suspend execution) when the appropriate condition is not met. Such sender-receiver synchronization can be implemented in many ways depending on the particular hardware platform under consideration: in shared memory machines, such synchronization involves testing and setting semaphores in shared memory; in machines that support synchronization in hardware (such as barriers), special synchronization instructions are used; and in the case of systems that consist of a mix of programmable processors and custom hardware elements, synchronization is achieved by employing interfaces that support blocking reads and writes.

In each type of platform, each IPC that requires a synchronization check costs performance, and sometimes extra hardware complexity. Semaphore checks cost execution time on the processors, synchronization instructions that make use of special synchronization hardware such as barriers also cost execution time, and blocking interfaces between a programmable processor and custom hardware in a combined hardware/software implementation require more hardware than nonblocking interfaces [H+93].

In this chapter, we present algorithms and techniques that reduce the rate at which processors must access shared memory for the purpose of synchronization in multiprocessor implementations of SDF programs. One of the procedures we present, for example, detects when the objective of one synchronization operation is guaranteed as a side effect of other synchronizations in the system, thus enabling us to eliminate such superfluous synchronization operations. The optimization procedure that we propose can be used as a postprocessing step to any static scheduling technique (for example, to any one of the techniques presented in Chapter 6) for reducing synchronization costs in the final implementation. As before, we assume that "good" estimates are available for the execution times of actors and that these execution times rarely display large variations so that self-timed scheduling is viable for the applications under consideration. If additional timing information is available, such as guaranteed upper and lower bounds on the execution times of actors, it is possible to use this information to further optimize synchronizations in the schedule. However, use of such timing bounds will be left as future work; we mention this again in Chapter 14.

## 10.1    The Barrier MIMD Technique

Among the prior work that is most relevant to this chapter is the **barrier MIMD** principle of Dietz, Zaafrani, and O'Keefe, which is a combined hardware and software solution to reducing run-time synchronization overhead [DZO92]. In this approach, a shared-memory MIMD computer is augmented with hardware support that allows arbitrary subsets of processors to synchronize precisely with respect to one another by executing a synchronization operation called a *barrier*. If a subset of processors is involved in a barrier operation, then each processor in this subset will wait at the barrier until all other processors in the subset have reached the barrier. After all processors in the subset have reached the barrier, the corresponding processes resume execution in *exact synchrony*.

In [DZO92], the barrier mechanism is applied to minimize synchronization overhead in a self-timed schedule with hard lower and upper bounds on the task execution times. The execution time ranges are used to detect situations where the earliest possible execution time of a task that requires data from another processor is guaranteed to be later than the latest possible time at which the required data is produced. When such an inference cannot be made, a barrier is instantiated between the sending and receiving processors. In addition to performing the required data synchronization, the barrier resets (to zero) the uncertainty between the relative execution times for the processors that are involved in the barrier, and thus enhances the potential for subsequent timing analysis to eliminate the need for explicit synchronizations.

The techniques of barrier MIMD do not apply to the problem that we address because they assume that a hardware barrier mechanism exists; they assume that tight bounds on task execution times are available; they do not address iterative, self-timed execution, in which the execution of successive iterations of the dataflow graph can overlap; and even for noniterative execution, there is no obvious correspondence between an optimal solution that uses barrier synchronizations and an optimal solution that employs decoupled synchronization checks at the sender and receiver end (**directed synchronization**). This last point is illustrated in Figure 10.1. Here, in the absence of execution time bounds, an optimal application of barrier synchronizations can be obtained by inserting two barriers — one barrier across $A_1$ and $A_3$, and the other barrier across $A_4$ and $A_5$. This is illustrated in Figure 10.1(c). However, the corresponding collection of directed synchronizations ($A_1$ to $A_3$, and $A_5$ to $A_4$) is not sufficient since it does not guarantee that the data required by $A_6$ from $A_1$ is available before $A_6$ begins execution.

## 10.2  Redundant Synchronization Removal in Non-Iterative Dataflow

In [Sha89], Shaffer presents an algorithm that minimizes the number of directed synchronizations in the self-timed execution of a dataflow graph. However, this work, like that of Dietz et al., does not allow the execution of successive iterations of the dataflow graph to overlap. It also avoids having to consider dataflow edges that have delay. The technique that we discuss in this chapter for removing redundant synchronizations can be viewed as a generalization of Shaffer's algorithm to handle delays and overlapped, iterative execution, and we will discuss this further in Section 10.7. The other major software-based techniques for synchronization optimization that we discuss in this book — handling the feedforward edges of the *synchronization graph* (to be defined in Section 10.5.2), discussed in Section 10.8, and "resynchronization," discussed in Chapters 11 and 12 — are fundamentally different from Shaffer's technique since they address issues that are specific to the more general context of overlapped, iterative execution.

As discussed in Chapter 5, a multiprocessor executing a self-timed schedule is one where each processor is assigned a sequential list of actors, some of which are *send* and *receive* actors, which it executes in an infinite loop. When a processor executes a communication actor, it synchronizes with the processor(s) it communicates with. Thus, exactly when a processor executes each actor depends on when, at run time, all input data for that actor is available, unlike the fully-static case where no such run-time check is needed. In this chapter we use

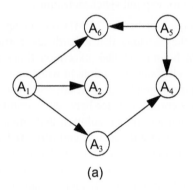

(a)

Proc 1: $A_1, A_2$

Proc 2: $A_3, A_4$

Proc 3: $A_5, A_6$

(b)

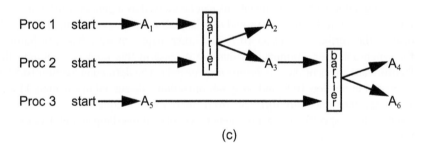

(c)

Figure 10.1. (a) An HSDFG. (b) A three-processor self-timed schedule for (a). (c) An illustration of execution under the placement of barriers.

"processor" in slightly general terms: a processor could be a programmable component, in which case the actors mapped to it execute as software entities, or it could be a hardware component, in which case actors assigned to it are implemented and executed in hardware. See [KL93] for a discussion on combined hardware/software synthesis from a single dataflow specification. Examples of application-specific multiprocessors that use programmable processors and some form of static scheduling are described in [B+88][Koh90], which were also discussed in Chapter 2.

Interprocessor communication between processors is assumed to takes place via shared memory. Thus, the sender writes to a particular shared memory location and the receiver reads from that location. The shared memory itself could be global memory between all processors, or it could be distributed between pairs of processors (as hardware FIFO queues or dual ported memories for example). Each interprocessor communication edge in an HSDFG thus translates into a buffer of a certain size in shared memory.

Sender-receiver synchronization is also assumed to take place by setting flags in shared memory. Special hardware for synchronization (barriers, semaphores implemented in hardware, etc.) would be prohibitive for the embedded multiprocessor machines for applications such as DSP that we are considering. Interfaces between hardware and software are typically implemented using memory-mapped registers in the address space of the programmable processor (again a kind of shared memory), and synchronization is achieved using flags that can be tested and set by the programmable component, and the same can be done by an interface controller on the hardware side [H+93].

Under the model above, the benefits of synchronization optimization become obvious. Each synchronization that is eliminated directly results in one less synchronization check, or, equivalently, one less shared memory access. For example, where a processor would have to check a flag in shared memory before executing a *receive* primitive, eliminating that synchronization implies there is no longer need for such a check. This translates to one less shared memory read. Such a benefit is especially significant for simplifying interfaces between a programmable component and a hardware component: a *send* or a *receive* without the need for synchronization implies that the interface can be implemented in a nonblocking fashion, greatly simplifying the interface controller. As a result, eliminating a synchronization directly results in simpler hardware in this case.

Thus, the metric for the optimizations we present in this chapter is the total number of accesses to shared memory that are needed for the purpose of synchronization in the final multiprocessor implementation of the self-timed schedule. This metric will be defined precisely in Section 10.6.

## 10.3    Analysis of Self-Timed Execution

We model synchronization in a self-timed implementation using the IPC graph model introduced in the previous chapter. As before, an IPC graph $G_{ipc}(V, E_{ipc})$ is extracted from a given HSDFG $G$ and multiprocessor schedule; Figure 10.2 shows one such example, which we use throughout this chapter.

We will find it useful to partition the edges of the IPC graph in the following manner: $E_{ipc} \equiv E_{int} \cup E_{comm}$, where $E_{comm}$ are the **communication edges** (shown dashed in Figure 10.2(d)) that are directed from the send to the receive actors in $G_{ipc}$, and $E_{int}$ are the "internal" edges that represent the fact that actors assigned to a particular processor (actors internal to that processor) are executed sequentially according to the order predetermined by the self-timed schedule. A communication edge $e \in E_{comm}$ in $G_{ipc}$ represents two functions: 1) reading and writing of data values into the buffer represented by that edge; and 2) synchronization between the sender and the receiver. As mentioned before, we assume the use of shared memory for the purpose of synchronization; the synchronization operation itself must be implemented using some kind of software *protocol* between the sender and the receiver. We discuss these synchronization protocols shortly.

### 10.3.1    Estimated Throughput

Recall from Lemma 8.3 that the average iteration period corresponding to a self-timed schedule with an IPC graph $G_{ipc}$ is given by the maximum cycle mean of the graph $MCM(G_{ipc})$. If we only have execution time estimates available instead of exact values, and we set the execution times of actors $t(v)$ to be equal to these estimated values, then we obtain the *estimated* iteration period by computing $MCM(G_{ipc})$. Henceforth, we will assume that we know the *estimated throughput* $MCM^{-1}$ calculated by setting the $t(v)$ values to the available timing estimates.

In all the transformations that we present in the rest of the chapter, we will preserve the estimated throughput by preserving the maximum cycle mean of $G_{ipc}$, with each $t(v)$ set to the estimated execution time of $v$. In the absence of more precise timing information, this is the best we can hope to do.

## 10.4    Strongly Connected Components and Buffer Size Bounds

In dataflow semantics, the edges between actors represent infinite buffers. Accordingly, the edges of the IPC graph are potentially buffers of infinite size. However, from Lemma 8.1, every **feedback edge** (an edge that belongs to a strongly connected component, and hence to some cycle) can only have a finite number of tokens at any time during the execution of the IPC graph. We will call

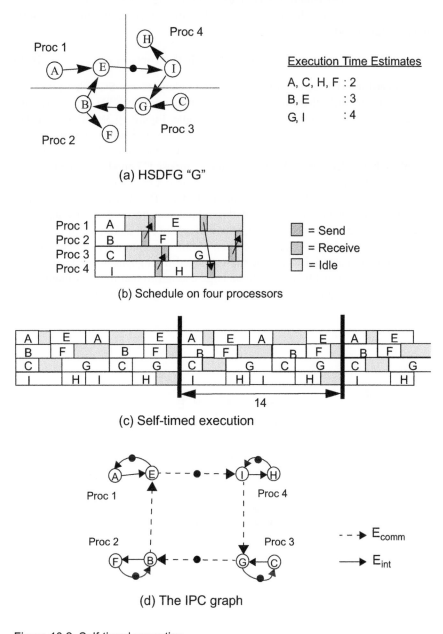

(a) HSDFG "G"

(b) Schedule on four processors

(c) Self-timed execution

(d) The IPC graph

Figure 10.2. Self-timed execution.

this constant the **self-timed buffer bound** of that edge, and for a feedback edge $e$ we will represent this bound by $B_{fb}(e)$. Lemma 8.1 yields the following self-timed buffer bound:

$$B_{fb}(e) = min(\{Delay(C)|C \text{ is a cycle that contains } e\}) \qquad (10\text{-}1)$$

**Feedforward edges** (edges that do not belong to any SCC) have no such bound on buffer size; therefore for practical implementations we need to *impose* a bound on the sizes of these edges. For example, Figure 10.3(a) shows an IPC graph where the communication edge $(s, r)$ could be unbounded when the execution time of $A$ is less than that of $B$. In practice, we need to bound the buffer size of such an edge; we will denote such an "imposed" bound for a feedforward edge $e$ by $B_{ff}(e)$. Since the effect of placing such a restriction includes "artificially" constraining $src(e)$ from getting more than $B_{ff}(e)$ invocations ahead of $snk(e)$, its effect on the estimated throughput can be modeled by adding a reverse edge that has $m$ delays on it, where $m = B_{ff}(e) - delay(e)$, to $G_{ipc}$ (grey edge in Figure 10.3(b)). Since the addition of this edge introduces a new cycle in $G_{ipc}$, it has the potential to reduce the estimated throughput; to prevent such a reduction, $B_{ff}(e)$ must be chosen to be large enough so that the maximum cycle mean remains unchanged upon adding the reverse edge with $m$ delays.

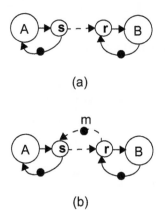

(a)

(b)

Figure 10.3. An IPC graph with a feedforward edge: (a) original graph, (b) imposing bounded buffers.

Sizing buffers optimally such that the maximum cycle mean remains unchanged has been studied by Kung, Lewis and Lo in [KLL87]. In this approach, the authors propose an integer linear programming formulation of the problem, with the number of constraints equal to the number of fundamental cycles in the HSDFG (potentially an exponential number of constraints).

An efficient heuristic procedure to determine $B_{ff}$ is to note that if

$$B_{ff}(e) \geq \left\lceil \left( \sum_{x \in V} t(x) \right) / (MCM(G_{ipc})) \right\rceil \qquad (10\text{-}2)$$

holds for each feedforward edge $e$, then the maximum cycle mean of the resulting graph does not exceed $MCM$.

Then, a binary search on $B_{ff}(e)$ for each feedforward edge, while computing the maximum cycle mean at each search step and ascertaining that it is less than $MCM(G_{ipc})$, results in a buffer assignment for the feedforward edges. Although this procedure is efficient, it is suboptimal because the order that the edges $e$ are chosen is arbitrary and may effect the quality of the final solution.

As we will see in Section 10.8, however, imposing such a bound $B_{ff}$ is a *naive* approach for bounding buffer sizes, because such a bound entails an added synchronization cost. In Section 10.8 we show that there is a better technique for bounding buffer sizes; this technique achieves bounded buffer sizes by transforming the graph into a strongly connected graph by adding a minimal number of additional synchronization edges. Thus, in the final algorithm, it is not in fact necessary to use or compute these bounds $B_{ff}$.

## 10.5    Synchronization Model

### 10.5.1    Synchronization Protocols

We define two basic synchronization protocols for a communication edge based on whether or not the length of the corresponding buffer is guaranteed to be bounded from the analysis presented in the previous section. Given an IPC graph $G$, and a communication edge $e$ in $G$, if the length of the corresponding buffer is not bounded — that is, if $e$ is a feedforward edge of $G$ — then we apply a synchronization protocol called **unbounded buffer synchronization (UBS)**, which guarantees that (a) an invocation of $snk(e)$ never attempts to read data from an empty buffer; *and* (b) an invocation of $src(e)$ never attempts to write data into the buffer unless the number of tokens in the buffer is less than some prespecified limit $B_{ff}(e)$, which is the amount of memory allocated to the buffer as discussed in the previous section.

On the other hand, if the topology of the IPC graph guarantees that the buffer length for $e$ is bounded by some value $B_{fb}(e)$ (the self-timed buffer bound of $e$ ), then we use a simpler protocol, called **bounded buffer synchronization (BBS)**, that only explicitly ensures (a) above. Below, we outline the mechanics of the two synchronization protocols defined so far.

*BBS.* In this mechanism, a buffer of size $B^* = (B_{fb} + 1)$ is allocated; a *write pointer* $wr(e)$ for $e$ is maintained on the processor that executes $src(e)$; a *read pointer* $rd(e)$ for $e$ is maintained on the processor that executes $snk(e)$; and a copy of $wr(e)$ is maintained in some shared memory location $sv(e)$. The addition of one additional unit of memory (as specified by $B^*$) beyond that given by the $B_{fb}$ bound helps to simplify the requirements for manipulating the read and write pointers into the buffer. The pointers $rd(e)$ and $wr(e)$ are initialized to zero and $delay(e)$, respectively. Just after each execution of $src(e)$, the new data value produced onto $e$ is written into the shared memory buffer for $e$ at off-set $wr(e)$; $wr(e)$ is updated by the following operation — $wr(e) \leftarrow (wr(e) + 1) \bmod B^*$; and $sv(e)$ is updated to contain the new value of $wr(e)$. Just before each execution of $snk(e)$, the value contained in $sv(e)$ is repeatedly examined until it is found to be *not equal* to $rd(e)$; then the data value residing at offset $rd(e)$ of the shared memory buffer for $e$ is read; and $rd(e)$ is updated by the operation $rd(e) \leftarrow (rd(e) + 1) \bmod B^*$.

*UBS.* This mechanism also uses the read/write pointers $rd(e)$ and $wr(e)$, and these are initialized the same way; however, rather than maintaining a copy of $wr(e)$ in the shared memory location $sv(e)$, we maintain a count (initialized to $delay(e)$) of the number of unread tokens that currently reside in the buffer. Just after $src(e)$ executes, $sv(e)$ is repeatedly examined until its value is found to be less than $B_{ff}(e)$; then the new data value produced onto $e$ is written into the shared memory buffer for $e$ at offset $wr(e)$; $wr(e)$ is updated as in BBS (except that the new value is not written to shared memory); and the count in $sv(e)$ is incremented. Just before each execution of $snk(e)$, the value contained in $sv(e)$ is repeatedly examined until it is found to be nonzero; then the data value residing at offset $rd(e)$ of the shared memory buffer for $e$ is read; the count in $sv(e)$ is decremented; and $rd(e)$ is updated as in BBS.

Note that we are assuming that there is enough shared memory to hold a separate buffer of size $B_{ff}(e)$ for each feedforward communication edge $e$ of $G_{ipc}$, and a separate buffer of size $(B_{fb}(e) + 1)$ for each feedback communication edge $e$. When this assumption does not hold, smaller bounds on some of the buffers must be imposed, possibly for feedback edges as well as for feedforward edges, and in general, this may require some sacrifice in estimated throughput. Note that whenever a buffer bound smaller than $B_{ff}(e)$ is imposed on a feedback edge $e$, then a protocol identical to UBS must be used. The problem of optimally choosing which edges should be subject to stricter buffer bounds when there is a shortage of shared memory, and the selection of these stricter bounds is an interesting area for further investigation.

An important parameter in an implementation of FFS or FBS is the **back-off time** $T_b$. If a receiving processor finds that the corresponding IPC buffer is full, then the processor releases the shared memory bus, and waits $T_b$ time units before requesting the bus again to recheck the shared memory synchronization variable. Similarly, a sending processor waits $T_b$ time units between successive accesses of the same synchronization variable. The back-off time can be selected experimentally by simulating the execution of the given synchronization graph (with the available execution time estimates) over a wide range of candidate back-off times, and selecting the back-off time that yields the highest simulated throughput.

## 10.5.2      The Synchronization Graph

As we discussed in the beginning of this chapter, some of the communication edges in $G_{ipc}$ need not have explicit synchronization, whereas others require synchronization, which need to be implemented either using the UBS protocol or the BBS protocol. All communication edges also represent buffers in shared memory. Thus, we divide the set of communication edges as follows: $E_{comm} \equiv E_s \cup E_r$, where the edges $E_s$ need explicit synchronization operations to be implemented, and the edges $E_r$ need no explicit synchronization. We call the edges $E_s$ **synchronization edges**.

Recall that a communication edge $(v_j, v_i)$ of $G_{ipc}$ represents the **synchronization constraint**:

$$start(v_i, k) \geq end(v_j, k - delay((v_j, v_i))) \; \forall k > delay((v_j, v_i)) \qquad (10\text{-}3)$$

Thus, before we perform any optimization on synchronizations, $E_{comm} \equiv E_s$ and $E_r \equiv \phi$, because every communication edge represents a synchronization point. However, in the following sections we describe how we can move certain edges from $E_s$ to $E_r$, thus reducing synchronization operations in the final implementation. After all synchronization optimizations have been applied, the communication edges of the IPC graph fall into either $E_s$ or $E_r$. At this point the edges $E_s \cup E_r$ in $G_{ipc}$ represent buffer activity, and must be implemented as buffers in shared memory, whereas the edges $E_s$ represent synchronization constraints, and are implemented using the UBS and BBS protocols introduced in the previous section. For the edges in $E_s$, the synchronization protocol is executed before the buffers corresponding to the communication edge are accessed so as to ensure sender-receiver synchronization. For edges in $E_r$, however, no synchronization needs to be done before accessing the shared buffer. Sometimes we will also find it useful to introduce synchronization edges without actually communicating data between the sender and the receiver (for the purpose of ensuring finite buffers for example), so that no shared buffers need to be assigned to these edges, but the corresponding synchronization protocol is invoked for these edges.

All optimizations that move edges from $E_s$ to $E_r$ must respect the synchronization constraints implied by $G_{ipc}$. If we ensure this, then we only need to implement the synchronization protocols for the edges in $E_s$. We call the graph $G_s = (V, E_{int} \cup E_s)$ the **synchronization graph**. The graph $G_s$ represents the synchronization constraints in $G_{ipc}$ that need to be explicitly ensured, and the algorithms we present for minimizing synchronization costs operate on $G_s$. Before any synchronization-related optimizations are performed $G_s \equiv G_{ipc}$, because $E_{comm} \equiv E_s$ at this stage, but as we move communication edges from $E_s$ to $E_r$, $G_s$ has fewer and fewer edges. Thus, moving edges from $E_s$ to $E_r$ can be viewed as removal of edges from $G_s$. Whenever we remove edges from $G_s$ we have to ensure, of course, that the synchronization graph $G_s$ at that step respects all the synchronization constraints of $G_{ipc}$, because we only implement synchronizations represented by the edges $E_s$ in $G_s$. The following theorem is useful to formalize the concept of when the synchronization constraints represented by one synchronization graph $G_s^1$ imply the synchronization constraints of another graph $G_s^2$. This theorem provides a useful constraint for synchronization optimization, and it underlies the validity of the main techniques that we will present in this chapter.

**Theorem 10.1:** The synchronization constraints in a synchronization graph $G_s^1 = (V, E_{int} \cup E_s^1)$ imply the synchronization constraints of the synchronization graph $G_s^2 = (V, E_{int} \cup E_s^2)$ if the following condition holds: $\forall \varepsilon$ s.t. $\varepsilon \in E_s^2, \varepsilon \notin E_s^1$, $\rho_{G_s^1}(src(\varepsilon), snk(\varepsilon)) \leq delay(\varepsilon)$; that is, if for each edge $\varepsilon$ that is present in $G_s^2$ but not in $G_s^1$ there is a minimum delay path from $src(\varepsilon)$ to $snk(\varepsilon)$ in $G_s^1$ that has total delay of at most $delay(\varepsilon)$.

(Note that since the vertex sets for the two graphs are identical, it is meaningful to refer to $src(\varepsilon)$ and $snk(\varepsilon)$ as being vertices of $G_s^1$ even though there are edges $\varepsilon$ s.t. $\varepsilon \in E_s^2, \varepsilon \notin E_s^1$.)

First we prove the following lemma.

**Lemma 10.1:** If there is a path $p = (e_1, e_2, e_3, ..., e_n)$ in $G_s^1$, then

$$start(snk(e_n), k) \geq end(src(e_1), k - Delay(p)).$$

*Proof of Lemma 10.1:*

The following constraints hold along such a path $p$ (as per (5-1))

$$start(snk(e_1), k) \geq end(src(e_1), k - delay(e_1)). \tag{10-4}$$

Similarly,

$$start(snk(e_2), k) \geq end(src(e_2), k - delay(e_2)).$$

Noting that $src(e_2)$ is the same as $snk(e_1)$, we get

$$start(snk(e_2), k) \geq end(snk(e_1), k - delay(e_2)).$$

Causality implies $end(v, k) \geq start(v, k)$, so we get

$$start(snk(e_2), k) \geq start(snk(e_1), k - delay(e_2)). \tag{10-5}$$

Substituting (10-4) in (10-5),

$$start(snk(e_2), k) \geq end(src(e_1), k - delay(e_2) - delay(e_1)).$$

Continuing along $p$ in this manner, it can easily be verified that

$$start(snk(e_n), k) \geq end(src(e_1), k - delay(e_n) - delay(e_{n-1}) -$$
$$... - delay(e_1))$$

that is,

$$start((snk(e_n), k) \geq end(src(e_1), k - Delay(p))). \; QED.$$

*Proof of Theorem 10.1:* If $\varepsilon \in E_s^2, \varepsilon \in E_s^1$, then the synchronization constraint due to the edge $\varepsilon$ holds in both graphs. But for each $\varepsilon$ s.t. $\varepsilon \in E_s^2, \varepsilon \notin E_s^1$ we need to show that the constraint due to $\varepsilon$ :

$$start(snk(\varepsilon), k) > end(src(\varepsilon), k - delay(\varepsilon)) \tag{10-6}$$

holds in $G_s^1$ provided $\rho_{G_s^1}(src(\varepsilon), snk(\varepsilon)) \leq delay(\varepsilon)$, which implies there is at least one path $p = (e_1, e_2, e_3, ..., e_n)$ from $src(\varepsilon)$ to $snk(\varepsilon)$ in $G_s^1$ ($src(e_1) = src(\varepsilon)$ and $snk(e_n) = snk(\varepsilon)$) such that $Delay(p) \leq delay(\varepsilon)$.

From Lemma 10.1, existence of such a path $p$ implies

$$start((snk(e_n), k) \geq end(src(e_1), k - Delay(p))).$$

that is,

$$start((snk(\varepsilon), k) \geq end(src(\varepsilon), k - Delay(p))). \tag{10-7}$$

If

$$Delay(p) \leq delay(\varepsilon),$$

then $end(src(\varepsilon), k - Delay(p)) \geq end(src(\varepsilon), k - delay(\varepsilon))$. Substituting this in (10-7) we get

$$start((snk(\varepsilon), k) \geq end(src(\varepsilon), k - delay(\varepsilon))).$$

The above relation is identical to (10-6), and this proves the Theorem. *QED.*

The above theorem motivates the following definition.

**Definition 10.1:** If $G_s^1 = (V, E_{int} \cup E_s^1)$ and $G_s^2 = (V, E_{int} \cup E_s^2)$ are synchronization graphs with the same vertex-set, we say that $G_s^1$ **preserves** $G_s^2$ if $\forall \varepsilon$ s.t. $\varepsilon \in E_2, \varepsilon \notin E_1$, we have $\rho_{G_s^1}(src(\varepsilon), snk(\varepsilon)) \leq delay(\varepsilon)$.

Thus, Theorem 10.1 states that the synchronization constraints of $(V, E_{int} \cup E_s^1)$ imply the synchronization constraints of $(V, E_{int} \cup E_s^2)$ if $(V, E_{int} \cup E_s^1)$ preserves $(V, E_{int} \cup E_s^2)$.

Given an IPC graph $G_{ipc}$, and a synchronization graph $G_s$ such that $G_s$ preserves $G_{ipc}$, suppose we implement the synchronizations corresponding to the synchronization edges of $G_s$. Then, the iteration period of the resulting system is determined by the maximum cycle mean of $G_s$ ($MCM(G_s)$). This is because the synchronization edges alone determine the interaction between processors; a communication edge without synchronization does not constrain the execution of the corresponding processors in any way.

## 10.6    A Synchronization Cost Metric

We refer to each access of the shared memory "synchronization variable" $sv(e)$ by $src(e)$ and $snk(e)$ as a **synchronization access** to shared memory. If synchronization for $e$ is implemented using UBS, then we see that on average, 4 synchronization accesses are required for $e$ in each iteration period, while BBS implies 2 synchronization accesses per iteration period. We define the **synchronization cost** of a synchronization graph $G_s$ to be the average number of synchronization accesses required per iteration period. Thus, if $n_{ff}$ denotes the number of synchronization edges in $G_s$ that are feedforward edges, and $n_{fb}$ denotes the number of synchronization edges that are feedback edges, then the synchronization cost of $G_s$ can be expressed as $(4n_{ff} + 2n_{fb})$. In the remainder of this chapter, we develop techniques that apply the results and the analysis framework developed in the previous sections to minimize the synchronization cost of a self-timed implementation of an HSDFG without sacrificing the integrity of any interprocessor data transfer or reducing the estimated throughput.

Note that in the measure defined above of the number of shared memory accesses required for synchronization, some accesses to shared memory are not taken into account. In particular, the "synchronization cost" metric does not consider accesses to shared memory that are performed while the sink actor is waiting for the required data to become available, or the source actor is waiting for an "empty slot" in the buffer. The number of accesses required to perform these "busy-wait" or "spin-lock" operations is dependent on the exact relative execution times of the actor invocations. Since in the problem context under consideration, this information is not generally available to us, the *best case* number of accesses — the number of shared memory accesses required for synchronization assuming that IPC data on an edge is always produced before the corresponding sink invocation attempts to execute — is used as an approximation.

In the remainder of this chapter, we discuss two mechanisms for reducing synchronization accesses. The first (presented in Section 10.7) is the detection and removal of *redundant* synchronization edges, which are synchronization edges whose respective synchronization functions are subsumed by other synchronization edges, and thus need not be implemented explicitly. This technique essentially detects the set of edges that can be moved from the $E_s$ to the set $E_r$. In Section 10.8, we examine the utility of adding additional synchronization edges to convert a synchronization graph that is not strongly connected into a strongly connected graph. Such a conversion allows us to implement all synchronization edges with BBS. We address optimization criteria in performing such a conversion, and we will show that the extra synchronization accesses required for such a conversion are always (at least) compensated by the number of synchronization accesses that are saved by the more expensive UBS synchronizations that are converted to BBS synchronizations.

Chapters 11 and 12 discuss a mechanism, called **resynchronization**, for inserting synchronization edges in a way that the number of original synchronization edges that become redundant exceeds the number of new edges added.

## 10.7    Removing Redundant Synchronizations

The first technique that we explore for reducing synchronization overhead is removal of *redundant synchronization edges* from the synchronization graph, i.e., finding a minimal set of edges $E_s$ that need explicit synchronization.

**Definition 10.2:** A synchronization edge is **redundant** in a synchronization graph $G$ if its removal yields a synchronization graph that preserves $G$. Equivalently, from definition 10.1, a synchronization edge $e$ is redundant in the synchronization graph $G$ if there is a path $p \neq (e)$ in $G$ directed from $src(e)$ to $snk(e)$ such that $Delay(p) \leq delay(e)$. The synchronization graph $G$ is **reduced** if $G$ contains no redundant synchronization edges.

Thus, the synchronization function associated with a redundant synchronization edge "comes for free" as a by-product of other synchronizations. Figure 10.4 shows an example of a redundant synchronization edge. Here, before executing actor $D$, the processor that executes $\{A, B, C, D\}$ does not need to synchronize with the processor that executes $\{E, F, G, H\}$ because, due to the synchronization edge $x_1$, the corresponding invocation of $F$ is guaranteed to complete before each invocation of $D$ is begun. Thus, $x_2$ is redundant in Figure 10.4 and can be removed from $E_s$ into the set $E_r$. It is easily verified that the path

$$p = ((F, G), (G, H), x_1, (B, C), (C, D))$$

is directed from $src(x_2)$ to $snk(x_2)$, and has a path delay (zero) that is equal to the delay on $x_2$.

In this section, we discuss an efficient algorithm to optimally remove redundant synchronization edges from a synchronization graph.

### 10.7.1    The Independence of Redundant Synchronizations

The following theorem establishes that the order in which we remove redundant synchronization edges is not important; therefore all the redundant synchronization edges can be removed together.

**Theorem 10.2:** Suppose that $G_s = (V, E_{int} \cup E_s)$ is a synchronization graph, $e_1$ and $e_2$ are distinct redundant synchronization edges in $G_s$ (i.e., these are edges that could be individually moved to $E_r$), and $G_s = (V, E_{int} \cup (E - \{e_1\}))$. Then $e_2$ is redundant in $G_s$. Thus, both $e_1$ and $e_2$ can be moved into $E_r$ *together*.

*Proof:* Since $e_2$ is redundant in $G_s$, there is a path $p \neq (e_2)$ in $G_s$ directed from $src(e_2)$ to $snk(e_2)$ such that

$$Delay(p) \leq delay(e_2). \tag{10-8}$$

Similarly, there is a path $p' \neq (e_1)$, contained in both $G_s$ and $\tilde{G}_s$, that is directed from $src(e_1)$ to $snk(e_1)$, and that satisfies

$$Delay(p') \leq delay(e_1). \tag{10-9}$$

Now, if $p$ does not contain $e_1$, then $p$ exists in $\tilde{G}_s$, and we are done. Otherwise, let $p' = (x_1, x_2, ..., x_n)$; observe that $p$ is of the form

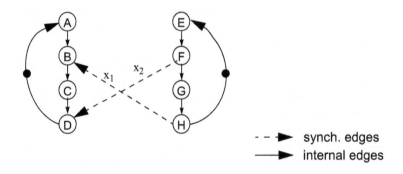

| | |
|---|---|
| - - ▶ | synch. edges |
| ──▶ | internal edges |

Figure 10.4. $x_2$ is an example of a redundant synchronization edge.

$$p = (y_1, y_2, \ldots, y_{k-1}, e_1, y_k, y_{k+1}, \ldots, y_m);$$

and define

$$p'' \equiv (y_1, y_2, \ldots, y_{k-1}, x_1, x_2, \ldots, x_n, y_k, y_{k+1}, \ldots, y_m).$$

Clearly, $p''$ is a path from $src(e_2)$ to $snk(e_2)$ in $\tilde{G}_s$. Also,

$$Delay(p'') = \sum delay(x_i) + \sum delay(y_i)$$

$$= Delay(p') + (Delay(p) - delay(e_1))$$

$$\leq Delay(p) \qquad\qquad \text{(from (10-9))}$$

$$\leq delay(e_2) \qquad\qquad \text{(from (10-8)).}$$

*QED.*

Theorem 10.2 tells us that we can avoid implementing synchronization for *all* redundant synchronization edges since the "redundancies" are not interdependent. Thus, an optimal removal of redundant synchronizations can be obtained by applying a straightforward algorithm that successively tests the synchronization edges for redundancy in some arbitrary sequence, and since computing the weight of the shortest path in a weighted directed graph is a tractable problem, we can expect such a solution to be practical.

### 10.7.2    Removing Redundant Synchronizations

Figure 10.5 presents an efficient algorithm, based on the ideas presented in the previous subsection, for optimal removal of redundant synchronization edges. In this algorithm, we first compute the path delay of a minimum-delay path from $x$ to $y$ for each ordered pair of vertices $(x, y)$; here, we assign a path delay of $\infty$ whenever there is no path from $x$ to $y$. This computation is equivalent to solving an instance of the well known *all points shortest paths problem* (see Section 3.13). Then, we examine each synchronization edge $e$ — in some arbitrary sequence — and determine whether or not there is a path from $src(e)$ to $snk(e)$ that does not contain $e$, and that has a path delay that does not exceed $delay(e)$. This check for redundancy is equivalent to the check that is performed by the *if* statement in *RemoveRedundantSynchs* because if $p$ is a path from $src(e)$ to $snk(e)$ that contains more than one edge and that contains $e$, then $p$ must contain a cycle $c$ such that $c$ does not contain $e$; and since all cycles must have positive path delay (from Lemma 8.1), the path delay of such a path $p$ must exceed $delay(e)$. Thus, if $e_0$ satisfies the inequality in the *if* statement of *RemoveRedundantSynchs*, and $p*$ is a path from $snk(e_0)$ to $snk(e)$ such that

$Delay(p^*) = \rho(snk(e_0), snk(e))$, then $p^*$ cannot contain $e$. This observation allows us to avoid having to recompute the shortest paths after removing a candidate redundant edge from $G_s$.

From the definition of a redundant synchronization edge, it is easily verified that the removal of a redundant synchronization edge does not alter any of the minimum-delay path values (path delays). That is, given a redundant synchronization edge $e_r$ in $G_s$, and two arbitrary vertices $x, y \in V$, if we let $G_s = (V, E_{int} \cup (E - \{e_r\}))$, then $\rho_{\hat{G}_s}(x, y) = \rho_{G_s}(x, y)$. Thus, none of the minimum-delay path values computed in Step 1 need to be recalculated after removing a redundant synchronization edge in Step 3.

Observe that the complexity of the function *RemoveRedundantSynchs* is dominated by Step 1 and Step 3. Since all edge delays are nonnegative, we can repeatedly apply Dijkstra's single-source shortest path algorithm (once for each vertex) to carry out Step 1 in $O(|V|^3)$ time; we discussed Dijkstra's algorithm in Section 3.13. A modification of Dijkstra's algorithm can be used to reduce the complexity of Step 1 to $O(|V|^2\log_2(|V|) + |V||E|)$ [CLR92]. In Step 3, $|E|$ is an upper bound for the number of synchronization edges, and in the worst case, each

**Function** RemoveRedundantSynchs
**Input**: A synchronization graph $G_s = E_{int} \cup E_s$
**Output**: The synchronization graph $G_s^* = (V, E_{int} \cup (E_s - E_r))$

1. Compute $\rho_{G_s}(x, y)$ for each ordered pair of vertices in $G_s$.

2. $E_r \leftarrow \varnothing$

3. **For** each $e \in E_s$

    **For** each output edge $e_o$ of $src(e)$ except for $e$

        **If** $delay(e_o) + \rho_{G_s}(snk(e_o), snk(e)) \leq delay(e)$

            $E_r \leftarrow E_r \cup \{e\}$
            Break          /* exit the innermost enclosing **For** loop */
        **Endif**
    **Endfor**
**Endfor**

4. **Return** $(V, E_{int} \cup (E_s - E_r))$.

Figure 10.5. An algorithm that optimally removes redundant synchronization edges.

vertex has an edge connecting it to every other member of $V$. Thus, the time-complexity of Step 3 is $O(|V||E|)$, and if we use the modification to Dijkstra's algorithm mentioned above for Step 1, then the time-complexity of *RemoveRedundantSynchs* is

$$O(|V|^2\log_2(|V|) + |V||E| + |V||E|) = O(|V|^2\log_2(|V|) + |V||E|).$$

### 10.7.3    Comparison with Shaffer's Approach

In [Sha89], Shaffer presents an algorithm that minimizes the number of directed synchronizations in the self-timed execution of an HSDFG under the (implicit) assumption that the execution of successive iterations of the HSDFG are not allowed to overlap. In Shaffer's technique, a construction identical to the synchronization graph is used except that there is no feedback edge connecting the last actor executed on a processor to the first actor executed on the same processor, and edges that have delay are ignored since only intraiteration dependencies are significant. Thus, Shaffer's synchronization graph is acyclic. *RemoveRedundantSynchs* can be viewed as an extension of Shaffer's algorithm to handle self-timed, iterative execution of an HSDFG; Shaffer's algorithm accounts for self-timed execution only within a graph iteration, and in general, it can be applied to iterative dataflow programs only if all processors are forced to synchronize between graph iterations.

### 10.7.4    An Example

In this subsection, we illustrate the benefits of removing redundant synchronizations through a practical example. Figure 10.6(a) shows an abstraction of a three channel, multiresolution quadrature mirror (QMF) filter bank, which has applications in signal compression [Vai93]. This representation is based on the general (not homogeneous) SDF model, and accordingly, each edge is annotated with the number of tokens produced and consumed by its source and sink actors. Actors $A$ and $F$ represent the subsystems that, respectively, supply and consume data to/from the filter bank system; $B$ and $C$ each represents a parallel combination of decimating high and low pass FIR analysis filters; $D$ and $E$ represent the corresponding pairs of interpolating synthesis filters. The amount of delay on the edge directed from $B$ to $E$ is equal to the sum of the filter orders of $C$ and $D$. For more details on the application represented by Figure 10.6(a), we refer the reader to [Vai93].

To construct a periodic parallel schedule, we must first determine the number of times $\mathbf{q}(N)$ that each actor $N$ must be invoked in the periodic schedule, as described in Section 3.6. Next, we must determine the precedence relationships between the actor invocations. In determining the exact precedence relationships, we must take into account the dependence of a given filter invocation on not only the invocation that produces the token that is "consumed" by the filter, but also on the invocations that produce the $n$ preceding tokens, where $n$ is the order of

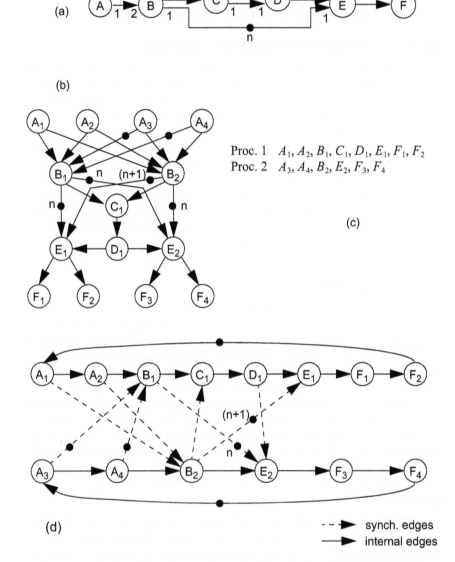

Proc. 1    $A_1, A_2, B_1, C_1, D_1, E_1, F_1, F_2$
Proc. 2    $A_3, A_4, B_2, E_2, F_3, F_4$

(c)

- - ▶   synch. edges
——▶   internal edges

Figure 10.6. (a) A multiresolution QMF filter bank used to illustrate the benefits of removing redundant synchronizations. (b) The precedence graph for (a). (c) A self-timed, two-processor, parallel schedule for (a). (d) The initial synchronization graph for (c).

the filter. Such dependence can easily be evaluated with an additional dataflow parameter on each actor input that specifies the number of *past tokens* that are accessed [Pri91][1]. Using this information, together with the invocation counts specified by $\mathbf{q}$, we obtain the precedence relationships specified by the graph of Figure 10.6(b), in which the $i$th invocation of actor $N$ is labeled $N_i$, and each edge $e$ specifies that invocation $snk(e)$ requires data produced by invocation $src(e)$ $delay(e)$ iteration periods after the iteration period in which the data is produced.

A self-timed schedule for Figure 10.6(b) that can be obtained from Hu's list scheduling method [Hu61] (described Section 6.3.2) is specified in Figure 10.6(c), and the synchronization graph that corresponds to the IPC graph of Figure 10.6(b) and Figure 10.6(c) is shown in Figure 10.6(d). All of the dashed edges in Figure 10.6(d) are synchronization edges. If we apply Shaffer's method, which considers only those synchronization edges that do not have delay, we can eliminate the need for explicit synchronization along only one of the 8 synchronization edges — edge $(A_1, B_2)$. In contrast, if we apply *RemoveRedundantSynchs*, we can detect the redundancy of $(A_1, B_2)$ as well as four additional redundant synchronization edges — $(A_3, B_1)$, $(A_4, B_1)$, $(B_2, E_1)$, and $(B_1, E_2)$. Thus, *RemoveRedundantSynchs* reduces the number of synchronizations from 8 down to 3 — a reduction of 62%. Figure 10.7 shows the synchronization graph of Figure 10.6(d) after all redundant synchronization edges are removed. It is easily verified that the synchronization edges that remain in this graph are not redundant; explicit synchronizations need only be implemented for these edges.

## 10.8    Making the Synchronization Graph Strongly Connected

In Section 10.5.1, we defined two different synchronization protocols — bounded buffer synchronization (BBS), which has a cost of 2 synchronization accesses per iteration period, and can be used whenever the associated edge is contained in a strongly connected component of the synchronization graph; and unbounded buffer synchronization (UBS), which has a cost of 4 synchronization accesses per iteration period. We pay the additional overhead of UBS whenever the associated edge is a feedforward edge of the synchronization graph.

---

1. It should be noted that some SDF-based design environments choose to forgo parallelization across multiple invocations of an actor in favor of simplified code generation and scheduling. For example, in the GRAPE system, this restriction has been justified on the grounds that it simplifies interprocessor data management, reduces code duplication, and allows the derivation of efficient scheduling algorithms that operate directly on general SDF graphs without requiring the use of the acyclic precedence graph (APG) [BELP94].

One alternative to implementing UBS for a feedforward edge $e$ is to add synchronization edges to the synchronization graph so that $e$ becomes encapsulated in a strongly connected component; such a transformation would allow $e$ to be implemented with BBS. However, extra synchronization accesses will be required to implement the new synchronization edges that are inserted. In this section, we show that by adding synchronization edges through a certain simple procedure, the synchronization graph can be transformed into a strongly connected graph in a way that the overhead of implementing the extra synchronization edges is always compensated by the savings attained by being able to avoid the use of UBS. That is, the conversion to a strongly connected synchronization graph ensures that the total number of synchronization accesses required (per iteration period) for the transformed graph is less than or equal to the number of synchronization accesses required for the original synchronization graph. Through a practical example, we show that this transformation can significantly reduce the number of required synchronization accesses. Also, we discuss a technique to compute the delay that should be added to each of the new edges added in the conversion to a strongly connected graph. This technique computes the delays in a way that the estimated throughput of the IPC graph is preserved with minimal increase in the shared memory storage cost required to implement the communication edges.

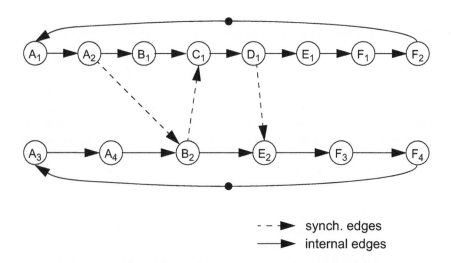

Figure 10.7. The synchronization graph of Figure 10.6(d) after all redundant synchronization edges are removed.

### 10.8.1    Adding Edges to the Synchronization Graph

Figure 10.8 presents an efficient algorithm for transforming a synchronization graph that is not strongly connected into a strongly connected graph. This algorithm simply "chains together" the source SCCs, and similarly, chains together the sink SCCs. The construction is completed by connecting the first SCC of the "source chain" to the last SCC of the sink chain with an edge that we call the **sink-source edge**. From each source or sink SCC, the algorithm selects a vertex that has minimum execution time to be the chain "link" corresponding to that SCC. Minimum execution time vertices are chosen in an attempt to minimize the amount of delay that must be inserted on the new edges to preserve the estimated throughput of the original graph. In Section 10.9, we discuss the selection of delays for the edges introduced by *Convert-to-SC-graph*.

It is easily verified that algorithm *Convert-to-SC-graph* always produces a strongly connected graph, and that a conversion to a strongly connected graph cannot be attained by adding fewer edges than the number of edges added by

**Function** *Convert-to-SC-graph*
**Input**: A synchronization graph $G$ that is not strongly connected.
**Output**: A strongly connected graph obtained by adding edges between the SCCs of $G$.

1. Generate an ordering $C_1, C_2, ..., C_m$ of the source SCCs of $G$, and similarly, generate an ordering $D_1, D_2, ..., D_n$ of the sink SCCs of $G$.
2. Select a vertex $v_1 \in C_1$ that minimizes $t(*)$ over $C_1$.
3. **For** $i = 2, 3..., m$
    • Select a vertex $v_i \in C_i$ that minimizes $t(*)$ over $C_i$.
    • Instantiate the edge $d_0(v_{i-1}, v_i)$.
**Endfor**
4. Select a vertex $w_1 \in D_1$ that minimizes $t(*)$ over $D_1$.
5. **For** $i = 2, 3..., n$
    • Select a vertex $w_i \in D_i$ that minimizes $t(*)$ over $D_i$.
    • Instantiate the edge $d_0(w_{i-1}, w_i)$.
**Endfor**
6. Instantiate the edge $d_0(w_m, v_1)$.

Figure 10.8. An algorithm for converting a synchronization graph that is not strongly connected into a strongly connected graph.

*Convert-to-SC-graph.* Figure 10.9 illustrates a possible solution obtained by algorithm *Convert-to-SC-graph*. Here, the black dashed edges are the synchronization edges contained in the original synchronization graph, and the grey dashed edges are the edges that are added by *Convert-to-SC-graph*. The dashed edge labeled $e_s$ is the sink-source edge.

Assuming the synchronization graph is connected, the number of feedforward edges $n_f$ must satisfy $(n_f \geq (n_c - 1))$, where $n_c$ is the number of SCCs. This follows from the fundamental graph theoretic fact that in a connected graph $(V^*, E^*)$, $|E^*|$ must be at least $(|V^*| - 1)$. Now, it is easily verified that the number of new edges introduced by *Convert-to-SC-graph* is equal to $(n_{src} + n_{snk} - 1)$, where $n_{src}$ is the number of source SCCs, and $n_{snk}$ is the number of sink SCCs. Thus, the number of synchronization accesses per iteration period, $S_+$, that is required to implement the edges introduced by *Convert-to-SC-graph* is $(2 \times (n_{src} + n_{snk} - 1))$, while the number of synchronization accesses, $S_-$, eliminated by *Convert-to-SC-graph* (by allowing the feedforward edges of the original synchronization graph to be implemented with BBS rather than UBS) equals $2n_f$. It follows that the net change $(S_+ - S_-)$ in the number of synchronization accesses satisfies

$$(S_+ - S_-) = 2(n_{src} + n_{snk} - 1) - 2n_f = 2(n_c - 1 - n_f) \leq 2(n_c - 1 - (n_c - 1))$$

and thus, $(S_+ - S_-) \leq 0$. We have established the following result.

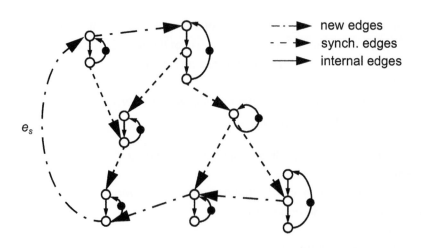

Figure 10.9. An illustration of a possible solution obtained by algorithm *Convert-to-SC-graph.*

**Theorem 10.3:** Suppose that $G$ is a synchronization graph, and $\hat{G}$ is the graph that results from applying algorithm *Convert-to-SC-graph* to $G$. Then the synchronization cost of $\hat{G}$ is less than or equal to the synchronization cost of $G$.

For example, without the edges added by *Convert-to-SC-graph* (the dashed grey edges) in Figure 10.9, there are 6 feedforward edges, which require 24 synchronization accesses per iteration period to implement. The addition of the 4 dashed edges requires 8 synchronization accesses to implement these new edges, but allows us to use BBS for the original feedforward edges, which leads to a savings of 12 synchronization accesses for the original feedforward edges. Thus, the net effect achieved by *Convert-to-SC-graph* in this example is a reduction of the total number of synchronization accesses by $(12 - 8) = 4$. As another example, consider Figure 10.10, which shows the synchronization graph topology (after redundant synchronization edges are removed) that results from a four-processor schedule of a synthesizer for plucked-string musical instruments in seven voices based on the Karplus-Strong technique. This algorithm was also discussed in Chapter 3, as an example application that was implemented on the ordered memory access architecture prototype. This graph contains $n_i = 6$ synchronization edges (the dashed edges), all of which are feedforward edges, so the synchronization cost is $4n_i = 24$ synchronization accesses per iteration period. Since the graph has one source SCC and one sink SCC, only one edge is added by *Convert-to-SC-graph*, and adding this edge reduces the synchronization cost to $2n_i + 2 = 14$ — a 42% savings. Figure 10.11 shows the topology of a possible solution computed by *Convert-to-SC-graph* on this example. Here, the dashed edges represent the synchronization edges in the synchronization graph returned by *Convert-to-SC-graph*.

## 10.9    Insertion of Delays

One important issue that remains to be addressed in the conversion of a synchronization graph $G_s$ into a strongly connected graph $\hat{G}_s$ is the proper insertion of delays so that $\hat{G}_s$ is not deadlocked, and does not have lower estimated throughput than $G_s$. The potential for deadlock and reduced estimated throughput arise because the conversion to a strongly connected graph must necessarily introduce one or more new fundamental cycles. In general, a new cycle may be delay-free, or its cycle mean may exceed that of the critical cycle in $G_s$. Thus, we may have to insert delays on the edges added by *Convert-to-SC-graph*. The location (edge) and magnitude of the delays that we add are significant since they affect the self-timed buffer bounds of the communication edges, as shown subsequently in Theorem 10.4. Since the self-timed buffer bounds determine the amount of memory that we allocate for the corresponding buffers, it is desirable to prevent deadlock and decrease in estimated throughput in a way that the sum of the self-timed buffer bounds over all communication edges is minimized. In this section, we outline a simple and efficient algorithm called *DetermineDelays* for addressing this problem. Algorithm *DetermineDelays* produces an optimal result if $G_s$ has only one source SCC or only one sink SCC; in other cases, the algorithm must be viewed as a heuristic. Our algorithm produces an optimal result if $G_s$ has only one source SCC or only one sink SCC; in other cases, the algorithm must be viewed as a heuristic. In practice, the assumptions under which we can expect an optimal result are frequently satisfied.

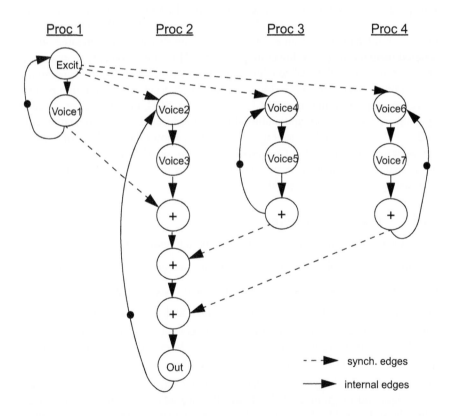

Figure 10.10. The synchronization graph, after redundant synchronization edges are removed, induced by a four-processor schedule of a music synthesizer based on the Karplus-Strong algorithm.

For simplicity in explaining the optimality result that has been established for Algorithm *DetermineDelays*, we first specify a restricted version of the algorithm that assumes only one sink SCC. After explaining the optimality of this restricted algorithm, we discuss how it can be modified to yield an optimal algorithm for the general single-source-SCC case, and finally, we discuss how it can be extended to provide a heuristic for arbitrary synchronization graphs.

Figure 10.12 outlines the restricted version of Algorithm *DetermineDelays* that applies when the synchronization graph $G_s$ has exactly one source SCC. Here, *BellmanFord* is assumed to be an algorithm that takes a synchronization graph $Z$ as input, and repeatedly applies the Bellman-Ford algorithm discussed in Section 3.13 to return the cycle mean of the critical cycle in $Z$; if one or more cycles exist that have zero path delay, then *BellmanFord* returns $\infty$ .

### 10.9.1    Analysis of *DetermineDelays*

In developing the optimality properties of Algorithm *DetermineDelays*, we will use the following definitions:

**Definition 10.3:** If $G = (V, E)$ is a DFG; $(e_0, e_1, ..., e_{n-1})$ is a sequence of

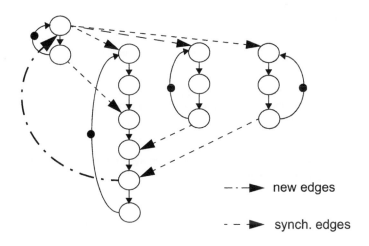

new edges

synch. edges

Figure 10.11. A possible solution obtained by applying *Convert-to-SC-graph* to the example of Figure 10.10.

distinct members of $E$ ; and

**Function** *DetermineDelays*

**Input**: Synchronization graphs $G_s = (V, E)$ and $\hat{G}_s$, where $\hat{G}_s$ is the graph computed by *Convert-to-SC-graph* when applied to $G_s$. The ordering of source SCCs generated in Step 2 of *Convert-to-SC-graph* is denoted $C_1, C_2, ..., C_m$. For $i = 1, 2, ...m-1$, $e_i$ denotes the edge instantiated by *Convert-to-SC-graph* from a vertex in $C_i$ to a vertex in $C_{i+1}$. The sink-source edge instantiated by *Convert-to-SC-graph* is denoted $e_0$.

**Output**: Nonnegative integers $d_o, d_1, ..., d_{m-1}$ such that the estimated throughput when $delay(e_i) = d_i$, $0 \le i \le m-1$, equals estimated throughput of $G_s$.

$X_0 = \hat{G}_s[e_0 \to \infty, ..., e_{m-1} \to \infty]$ /* set delays on each edge to be infinite */

$\lambda_{max} = BellmanFord(X_0)$       /* compute the max. cycle mean of $G_s$ */

$d_{ub} = \left\lceil \left( \sum_{x \in V} t(x) \right) / MCM \right\rceil$     /* an upper bound on the delay required for

any $e_i$ */

**For** $i = 0, 1, ..., m-1$

    $\delta_i = MinDelay(X_i, e_i, MCM, d_{ub})$

    $X_{i+1} = X_i[e_i \to \delta_i]$       /* fix the delay on $e_i$ to be $\delta_i$ */

**End For**

**Return** $\delta_o, \delta_1, ..., \delta_{m-1}$.

**Function** *MinDelay(X, e, λ, B )*

**Input**: A synchronization graph $X$, an edge $e$ in $X$, a positive real number $\lambda$, and a positive integer $B$.

**Output**: Assuming $X[e \to B]$ has estimated throughput no less than $\lambda^{-1}$, determine the minimum $d \in \{0, 1, ..., B\}$ such that the estimated throughput of $X[e \to d]$ is no less than $\lambda^{-1}$.

Perform a binary search in the range $[0, 1, ..., B]$ to find the minimum value of $r \in \{0, 1, ..., B\}$ such that $BellmanFord(X[e \to r])$ returns a value less than or equal to $\lambda$. Return this minimum value of $r$.

Figure 10.12. An algorithm for determining the delays on the edges introduced by algorithm *Convert-to-SC-graph*.

$$\Delta_0, \Delta_1, ..., \Delta_{n-1} \in \{0, 1, ..., \infty\},\tag{10-10}$$

then $G[e_0 \rightarrow \Delta_0, ..., e_{n-1} \rightarrow \Delta_{n-1}]$ denotes the DFG

$$(V, ((E - \{e_0, e_1, ..., e_{n-1}\}) \cup \{e_0', e_1', ..., e_{n-1}'\})),\tag{10-11}$$

where each $e_i'$ is defined by $src(e_i') = src(e_i)$, $snk(e_i') = snk(e_i)$, and $delay(e_i') = \Delta_i$. Thus, $G[e_0 \rightarrow \Delta_0, ..., e_{n-1} \rightarrow \Delta_{n-1}]$ is simply the DFG that results from "changing the delay" on each $e_i$ to the corresponding new delay value $\Delta_i$.

**Definition 10.4:** Suppose that $G$ is a synchronization graph that preserves $G_{ipc}$. An **IPC sink-source path** in $G$ is a minimum-delay path in $G$ directed from $snk(e)$ to $src(e)$, where $e$ is an IPC edge (in $G_{ipc}$).

Motivation for Algorithm *DetermineDelays* is based on the observations that the set of IPC sink-source paths introduced by *Convert-to-SC-graph* can be partitioned into $m$ nonempty subsets $P_0, P_1, ..., P_{m-1}$ such that each member of $P_i$ contains $e_0, e_1, ..., e_i$ [1] and contains no other members of $\{e_0, e_1, ..., e_{m-1}\}$, and similarly, the set of fundamental cycles introduced by *DetermineDelays* can be partitioned into $W_0, W_1, ..., W_{m-1}$ such that each member of $W_i$ contains $e_0, e_1, ..., e_i$ and contains no other members of $\{e_0, e_1, ..., e_{m-1}\}$.

By construction, a nonzero delay on any of the edges $e_0, e_1, ..., e_i$ "contributes to reducing the cycle means of all members of $W_i$". Algorithm *DetermineDelays* starts (iteration $i = 0$ of the *For* loop) by determining the minimum delay $\delta_0$ on $e_0$ that is required to ensure that none of the cycles in $W_0$ has a cycle mean that exceeds the maximum cycle mean $\lambda_{max}$ of $G_s$. Then (in iteration $i = 1$) the algorithm determines the minimum delay $\delta_1$ on $e_1$ that is required to guarantee that no member of $W_1$ has a cycle mean that exceeds $\lambda_{max}$, assuming that $delay(e_0) = \delta_0$.

Now, if $delay(e_0) = \delta_0$, $delay(e_1) = \delta_1$, and $\delta_1 > 0$, then for any positive integer $k \le \delta_1$, $k$ units of delay can be "transferred from $e_1$ to $e_0$" without violating the property that no member of $(W_0 \cup W_1)$ contains a cycle whose cycle mean exceeds $\lambda_{max}$. However, such a transformation increases the path delay of each member of $P_0$ while leaving the path delay of each member of $P_1$ unchanged, and therefore such a transformation cannot reduce the self-timed buffer bound of any IPC edge. Furthermore, apart from transferring delay from $e_1$ to $e_0$, the only other change that can be made to $delay(e_0)$ or $delay(e_1)$ — without introducing a member of $(W_0 \cup W_1)$ whose cycle mean exceeds $\lambda_{max}$ — is to increase one or both of these values by some positive integer amount(s). Clearly, such a change cannot reduce the self-timed buffer bound on any IPC

---

1. See Figure 10.12 for the specification of what the $e_i$ s represent.

edge.

Thus, we see that the values $\delta_0$ and $\delta_1$ computed by *DetermineDelays* for *delay*$(e_0)$ and *delay*$(e_1)$, respectively, optimally ensure that no member of $(W_0 \cup W_1)$ has a cycle mean that exceeds $\lambda_{max}$. After computing these values, *DetermineDelays* computes the minimum delay $\delta_2$ on $e_2$ that is required for all members of $W_2$ to have cycle means less than or equal to $\lambda_{max}$, assuming that *delay*$(e_0) = \delta_0$ and *delay*$(e_1) = \delta_1$. Given the "configuration" $(delay(e_0) = \delta_0, \; delay(e_1) = \delta_1, \; delay(e_2) = \delta_2)$, transferring delay from $e_2$ to $e_1$ increases the path delay of all members of $P_1$, while leaving the path delay of each member of $(P_0 \cup P_2)$ unchanged; and transferring delay from $e_2$ to $e_0$ increases the path delay across $(P_0 \cup P_1)$, while leaving the path delay across $P_2$ unchanged. Thus, by an argument similar to that given to establish the optimality of $(\delta_0, \delta_1)$ with respect to $(W_0 \cup W_1)$, we can deduce that (1) the values computed by *DetermineDelays* for the delays on $e_0, e_1, e_2$ guarantee that no member of $(W_0 \cup W_1 \cup W_2)$ has a cycle mean that exceeds $\lambda_{max}$; and (2) for any other assignment of delays $(\delta_0', \delta_1', \delta_2')$ to $(e_0, e_1, e_2)$ that preserves the estimated throughput across $(W_0 \cup W_1 \cup W_2)$, and for any IPC edge $e$ such that an IPC sink-source path of $e$ is contained in $(P_0 \cup P_1 \cup P_2)$, the self-timed buffer bound of $e$ under the assignment $(\delta_0', \delta_1', \delta_2')$ is greater than or equal to self-timed buffer bound of $e$ under the assignment $(\delta_0, \delta_1, \delta_2)$ computed by iterations $i = 0, 1, 2$ of *DetermineDelays*.

After extending this analysis successively to each of the remaining iterations $i = 3, 4, ..., m - 1$ of the *for* loop in *DetermineDelays*, we arrive at the following result.

**Theorem 10.4:** Suppose that $G_s$ is a synchronization graph that has exactly one sink SCC; let $\hat{G}_s$ and $(e_0, e_1, ..., e_{m-1})$ be as in Figure 10.12; let $(d_0, d_1, ..., d_{m-1})$ be the result of applying *DetermineDelays* to $G_s$ and $\hat{G}_s$; and let $(d_0', d_1', ..., d_{m-1}')$ be any sequence of $m$ nonnegative integers such that $\hat{G}_s[e_0 \rightarrow d_0', ..., e_{m-1} \rightarrow d_{m-1}']$ has the same estimated throughput as $G_s$. Then

$$\Phi(\hat{G}_s[e_0 \rightarrow d_0', ..., e_{m-1} \rightarrow d_{m-1}']) \geq \Phi(\hat{G}_s[e_0 \rightarrow d_0, ..., e_{m-1} \rightarrow d_{m-1}]),$$

where $\Phi(X)$ denotes the sum of the self-timed buffer bounds over all IPC edges in $G_{ipc}$ induced by the synchronization graph $X$.

### 10.9.2    Delay Insertion Example

Figure 10.13 illustrates a solution obtained from *DetermineDelays*. Here we assume that $t(v) = 1$, for each vertex $v$, and we assume that the set of IPC edges is $\{e_a, e_b\}$ (for clarity, we are assuming in this example that the IPC edges are present in the given synchronization graph). The grey dashed edges are the edges added by *Convert-to-SC-graph*. We see that $\lambda_{max}$ is determined by the cycle in the sink SCC of the original graph, and inspection of this cycle yields $\lambda_{max} = 4$. Also, we see that the set $W_0$ — the set of fundamental cycles that contain $e_0$, and do not contain $e_1$ — consists of a single cycle $c_0$ that contains three edges. By inspection of this cycle, we see that the minimum delay on $e_0$ required to guarantee that its cycle mean does not exceed $\lambda_{max}$ is 1. Thus, the $i = 0$ iteration of the *For* loop in *DetermineDelays* computes $\delta_0 = 1$. Next, we see that $W_1$ consists of a single cycle that contains five edges, and we see that two delays must be present on this cycle for its cycle mean to be less than or equal to $\lambda_{max}$. Since one delay has been placed on $e_0$, *DetermineDelays* computes $\delta_1 = 1$ in the $i = 1$ iteration of the *For* loop. Thus, the solution determined by *DetermineDelays* for Figure 10.13 is $(\delta_0, \delta_1) = (1, 1)$; the resulting self-timed buffer bounds of $e_a$ and $e_b$ are, respectively, 1 and 2; and $\Phi = 2 + 1 = 3$.

Now $(2, 0)$ is an alternative assignment of delays on $(e_0, e_1)$ that preserves the estimated throughput of the original graph. However, in this assignment, we see that the self-timed buffer bounds of $e_a$ and $e_b$ are identically equal to 2, and thus, $\Phi = 4$, one greater than the corresponding sum from the delay assignment $(1, 1)$ computed by *DetermineDelays*. Thus, if $G_s$ denotes the graph returned by *Convert-to-SC-graph* for the example of Figure 10.13, we have that

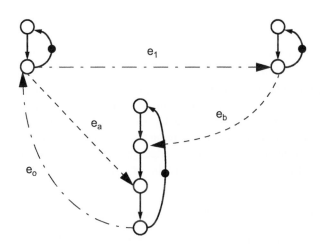

Figure 10.13. An example used to illustrate a solution obtained by algorithm *DetermineDelays*.

$$\Phi(\hat{G}_s[e_0 \to \delta_0, e_1 \to \delta_1]) < \Phi(\hat{G}_s[e_0 \to 2, e_1 \to 0]), \qquad (10\text{-}12)$$

where $\Phi(X)$ denotes the sum of the self-timed buffer bounds over all IPC edges in $X$.

### 10.9.3    Extending the Algorithm

Algorithm *DetermineDelays* can easily be modified to optimally handle general graphs that have only one *source* SCC. Here, the algorithm specification remains essentially the same, with the exception that for $i = 1, 2, ..., (m-1)$, $e_i$ denotes the edge directed from a vertex in $D_{m-i}$ to a vertex in $D_{m-i+1}$, where $D_1, D_2, ..., D_m$ is the ordering of sink SCCs generated in Step 2 of the corresponding invocation of *Convert-to-SC-graph* ($e_0$ still denotes the sink-source edge instantiated by *Convert-to-SC-graph*). By adapting the reasoning behind Theorem 10.4, it is easily verified that when it is applicable, this modified algorithm always yields an optimal solution.

As far as we are aware, there is no straightforward extension of *DetermineDelays* to general graphs (multiple source SCCs and multiple sink SCCs) that is guaranteed to yield optimal solutions. The fundamental problem for the general case is the inability to derive the partitions $W_0, W_1, ..., W_{m-1}$ ($P_0, P_1, ..., P_{m-1}$) of the fundamental cycles (IPC sink-source paths) introduced by *Convert-to-SC-graph* such that each $W_i$ ($P_i$) contains $e_0, e_1, ..., e_i$, and contains no other members of $E_s \equiv \{e_0, e_1, ..., e_{m-1}\}$, where $E_s$ is the set of edges added by *Convert-to-SC-graph*. The existence of such partitions was crucial to our development of Theorem 10.4 because it implied that once the minimum values for $e_0, e_1, ..., e_i$ are successively computed, "transferring" delay from some $e_i$ to some $e_j$, $j < i$, is never beneficial. Figure 10.14 shows an example of a synchronization graph that has multiple source SCCs and multiple sink SCCs, and that does not induce a partition of the desired form for the fundamental cycles.

However, *DetermineDelays* can be extended to yield heuristics for the general case in which the original synchronization graph $G_s$ contains more than one source SCC *and* more than one sink SCC. For example, if $(a_1, a_2, ..., a_k)$ denote edges that were instantiated by *Convert-to-SC-graph* "between" the source SCCs — with each $a_i$ representing the $i$ th edge created — and similarly, $(b_1, b_2, ..., b_l)$ denote the sequence of edges instantiated between the sink SCCs, then algorithm *DetermineDelays* can be applied with the modification that $m = k + l + 1$, and

$$(e_0, e_1, ..., e_{m-1}) \equiv (e_s, a_1, a_2, ..., a_k, b_l, b_{l-1}, ..., b_1), \qquad (10\text{-}13)$$

where $e_s$ is the sink-source edge from *Convert-to-SC-graph*.

The derivation of alternative heuristics for general synchronization graphs appears to be an interesting direction for further research. It should be noted, though, that practical synchronization graphs frequently contain either a single source SCC or a single SCC, or both — such as the example of Figure 10.10 — so that algorithm *DetermineDelays,* together with its counterpart for graphs that have a single source SCC, form a widely-applicable solution for optimally determining the delays on the edges created by *Convert-to-SC-graph.*

## 10.9.4     Complexity

If we assume that there exist constants $T$ and $D$ such that $t(v) \leq T$, for all $v$, and $delay(e) \leq D$ for all edges $e$, then the complexity of *BellmanFord* is $O(|V||E|\log_2(|V|))$ (see Section 3.13.2); and we have

$$\lambda_{max} \geq \frac{1}{D}$$

and

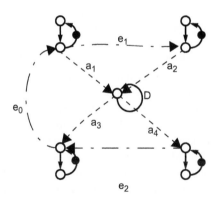

Figure 10.14. A synchronization graph, after processing by *Convert-to-SC-graph*, such that there is no $m$-way partition $W_0, W_1, ..., W_{m-1}$ of the fundamental cycles introduced by *Convert-to-SC-graph* that satisfies both (1). Each $W_i$ contains $e_0, e_1, ..., e_i$ and (2). Each $W_i$ does not contain any member of $e_{i+1}, e_{i+2}, ..., e_{m-1}$. Here, the fundamental cycles introduced by *Convert-to-SC-graph* (the grey dashed edges are the edges instantiated by *Convert-to-SC-graph*) are $(e_0, a_1, a_3)$, $(e_0, e_1, a_2, a_3)$, $(e_0, e_1, a_2, a_4, e_2)$, and $(e_0, a_1, a_4, a_2)$. It is easily verified that these cycles cannot be decomposed into a partition of the above form even if we are allowed to reorder the $e_i$ s.

$$\sum t(v) \le T|V|,$$

so that $d_{ub} \le DT|V|$. Thus, each invocation of *MinDelay* runs in

$$O(\log_2(DT|V|)|V||E|\log_2(|V|)) = O(|V||E|(\log_2(|V|))^2) \text{ time.} \quad (10\text{-}14)$$

It follows that *DetermineDelays* — and any of the variations of *DetermineDelays* defined above — is $O(m|V||E|(\log_2(|V|))^2)$, where $m$ is the number of edges instantiated by *Convert-to-SC-graph*. Since $m = (n_{src} + n_{snk} - 1)$, where $n_{src}$ is the number of source SCCs, and $n_{snk}$ is the number of sink SCCs, it is obvious that $m < |V|$. With this observation, and the observation that $|E| \le |V|^2$, we have that *DetermineDelays* and its variations are $O(|V|^4(\log_2(|V|))^2)$. Furthermore, it is easily verified that the time complexity of *DetermineDelays* dominates that of *Convert-to-SC-graph,* so the time complexity of applying *Convert-to-SC-graph* and *DetermineDelays* in succession is also $O(|V|^4(\log_2(|V|))^2)$.

Although the issue of deadlock does not explicitly arise in algorithm *DetermineDelays,* the algorithm does guarantee that the output graph is not deadlocked, assuming that the input graph is not deadlocked. This is because (from Lemma 8.1) deadlock is equivalent to the existence of a cycle that has zero path delay, and is thus equivalent to an infinite maximum cycle mean. Since *DetermineDelays* does not increase the maximum cycle mean, it follows that the algorithm cannot convert a graph that is not deadlocked into a deadlocked graph.

## 10.9.5    Related Work

Converting a mixed grain HSDFG that contains feedforward edges into a strongly connected graph has been studied by Zivojnovic, Koerner, and Meyr [ZKM94] in the context of retiming when the assignment of actors to processors is fixed beforehand. In this case, the objective is to retime the input graph so that the number of communication edges that have nonzero delay is maximized, and the conversion is performed to constrain the set of possible retimings in such a way that an integer linear programming formulation can be developed. The technique generates two dummy vertices that are connected by an edge; the sink vertices of the original graph are connected to one of the dummy vertices, while the other dummy vertex is connected to each source. It is easily verified that in a self-timed execution, this scheme requires at least four more synchronization accesses per graph iteration than the method that we have proposed. We can obtain further relative savings if we succeed in detecting one or more beneficial resynchronization opportunities. The effect of Zivojnovic's retiming algorithm on synchronization overhead is unpredictable since, on one hand, a communication edge becomes "easier to make redundant" when its delay increases, while on the other hand, the edge becomes less useful in making other communication edges redundant since the path delay of all paths that contain the edge increase.

## 10.10   Summary

This chapter has developed two software strategies for minimizing synchronization overhead when implementing self-timed, iterative dataflow programs. These techniques rely on a graph-theoretic analysis framework based on two data structures called the interprocessor communication graph and the synchronization graph. This analysis framework allows us to determine the effects on throughput and buffer sizes of modifying the points in the target program at which synchronization functions are carried out, and we have shown how this framework can be used to extend an existing technique — removal of redundant synchronization edges — for noniterative programs to the iterative case, and to develop a new method for reducing synchronization overhead — the conversion of a synchronization graph into a strongly connected graph so that a more efficient synchronization protocol can be used.

As in Chapter 8, the main premise of the techniques discussed in the chapter is that estimates are available for the execution times of actors such that the actual execution time of an actor exhibits large variation from its corresponding estimate only with very low frequency. Accordingly, our techniques have been devised to guarantee that if the actual execution time of each actor invocation is always equal to the corresponding execution time estimate, then the throughput of an implementation that incorporates our synchronization minimization techniques is never less than the throughput of a corresponding unoptimized implementation — that is, we never accept an opportunity to reduce synchronization overhead if it constrains execution in such a way that throughput is decreased. Thus, the techniques discussed in this section are particularly relevant to embedded DSP applications, where the price of synchronization is high, and accurate execution time estimates are often available, but guarantees on these execution times do not exist due to infrequent events such as cache misses, interrupts, and error handling.

In the next two chapters, we discuss a third software-based technique called *resynchronization*, for reducing synchronization overhead in application-specific multiprocessors.

## 10.10 Summary

# 11

## RESYNCHRONIZATION

This chapter discusses a technique, called **resynchronization**, for reducing synchronization overhead in application-specific multiprocessor implementations. The technique applies to arbitrary collections of dedicated, programmable or configurable processors, such as combinations of programmable DSPs, ASICS, and FPGA subsystems. Resynchronization is based on the concept of redundant synchronization operations, which was defined in the previous chapter. The objective of resynchronization is to introduce new synchronizations in such a way that the number of original synchronizations that consequently become redundant is significantly more than number of new synchronizations.

## 11.1     Definition of Resynchronization

Intuitively, resynchronization is the process of adding one or more new synchronization edges and removing the redundant edges that result. Figure 11.1(a) illustrates how this concept can be used to reduce the total number of synchronizations in a multiprocessor implementation. Here, the dashed edges represent synchronization edges. Observe that if the new synchronization edge $d_0(C, H)$ is inserted, then two of the original synchronization edges — $(B, G)$ and $(E, J)$ — become redundant. Since redundant synchronization edges can be removed from the synchronization graph to yield an equivalent synchronization graph, we see that the net effect of adding the synchronization edge $d_0(C, H)$ is to reduce the number of synchronization edges that need to be implemented by 1. Figure 11.1(b) shows the synchronization graph that results from inserting the *resynchronization edge* $d_0(C, H)$ into Figure 11.1(a), and then removing the redundant synchronization edges that result.

Definition 11.1 gives a formal definition of resynchronization. This considers resynchronization only "across" feedforward edges. Resynchronization that includes inserting edges into SCCs is also possible; however, in general, such resynchronization may increase the estimated throughput (see Theorem 11.1

at the end of Section 11.2). Thus, for our objectives, it must be verified that each new synchronization edge introduced in an SCC does not decrease the estimated throughput. To avoid this complication, which requires a check of significant complexity $(O(|V||E|\log_2(|V|))$, where $(V, E)$ is the modified synchronization graph, using the Bellman Ford algorithm described in Section 3.13.2) *for each* candidate resynchronization edge, we focus only on "feedforward" resynchronization in this chapter. Future research will address combining the insights developed here for feedforward resynchronization with efficient techniques to estimate the impact that a given *feedback* resynchronization edge has on the estimated throughput.

Opportunities for feedforward resynchronization are particularly abundant in the dedicated hardware implementation of dataflow graphs. If each actor is mapped to a separate piece of hardware, as in the VLSI dataflow arrays of Kung, Lewis, and Lo [KLL87], then for any application graph that is acyclic, every communication channel between two units will have an associated feedforward synchronization edge. Due to increasing circuit integration levels, such isomorphic mapping of dataflow subsystems into hardware is becoming attractive for a growing family of applications. Feedforward synchronization edges often arise naturally in multiprocessor software implementations as well. A software example is reviewed in detail in Section 11.5.

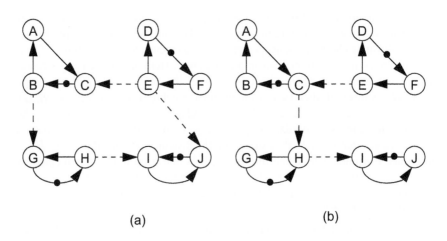

(a)                                    (b)

Figure 11.1. An example of resynchronization.

**Definition 11.1:** Suppose that $G = (V, E)$ is a synchronization graph, and $F \equiv \{e_1, e_2, ..., e_n\}$ is the set of all feedforward edges in $G$. A **resynchronization** of $G$ is a finite set $R \equiv \{e_1', e_2', ..., e_m'\}$ of edges that are not necessarily contained in $E$, but whose source and sink vertices are in $V$, such that a) $e_1', e_2', ..., e_m'$ are feedforward edges in the HSDFG $G^* \equiv (V, ((E - F) + R))$; and b) $G^*$ preserves $G$ — that is, $\rho_{G^*}(src(e_i), snk(e_i)) \leq delay(e_i)$ for all $i \in \{1, 2, ..., n\}$. Each member of $R$ that is not in $E$ is called a **resynchronization edge** of the resynchronization $R$, $G^*$ is called the **resynchronized graph** associated with $R$, and this graph is denoted by $\Psi(R, G)$.

If we let $G$ denote the graph in Figure 11.1, then the set of feedforward edges is $F = \{(B, G), (E, J), (E, C), (H, I)\}$; $R = \{d_0(C, H), (E, C), (H, I)\}$ is a resynchronization of $G$; Figure 11.1(b) shows the HSDFG

$$G^* = (V, ((E - F) + R));$$

and from Figure 11.1(b), it is easily verified that $F$, $R$, and $G^*$ satisfy conditions (a) and (b) of Definition 11.1.

Typically, resynchronization is meaningful only in the context of synchronization graphs that are not **deadlocked** — that is, synchronization graphs that do not contain any delay-free cycles, or equivalently, that have infinite estimated throughput. In the remainder of this chapter and throughout Chapter 12, we are concerned only with deadlock-free synchronization graphs. Thus, unless otherwise stated, we assume the absence of delay-free synchronization graph cycles. In practice, this assumption is not a problem, since delay-free cycles can be detected efficiently [Kar78].

## 11.2    Properties of Resynchronization

This section reviews a number of useful properties of synchronization redundancy and resynchronization that we will apply throughout the developments of this chapter and Chapter 12.

**Fact 11.1:** Suppose that $G = (V, E)$ is a synchronization graph and $s$ is a redundant synchronization edge in $G$. Then there exists a simple path $p$ in $G$ directed from $src(s)$ to $snk(s)$ such that $p$ does not contain $s$, and $Delay(p) \leq delay(s)$.

*Proof:* Let $G' \equiv (V, (E - \{s\}))$ denote the synchronization graph that results when we remove $s$ from $G$. Then from Definition 10.2, there exists a path $p'$ in $G'$ directed from $src(s)$ to $snk(s)$ such that

$$Delay(p') \leq delay(s). \tag{11-1}$$

Now observe that every edge in $G'$ is also contained in $G$, and thus, $G$ contains the path $p'$. If $p'$ is a simple path, then we are done. Otherwise, $p'$ can be expressed as a concatenation

$$\langle (q_0, C_1, q_1, C_2, ..., q_{n-1}, C_n, q_n) \rangle, \; n \geq 1, \tag{11-2}$$

where each $q_i$ is a simple path, at least one $q_i$ is non-empty, and each $C_j$ is a (not necessarily simple) cycle. Since valid synchronization graphs cannot contain delay-free-cycles (Section 11.1), we must have $Delay(C_k) \geq 1$ for $1 \leq k \leq n$. Thus, since each $C_i$ originates and terminates at the same actor, the path $p'' = \langle q_0, q_1, ..., q_n \rangle$ is a simple path directed from $src(s)$ to $snk(s)$ such that $Delay(p'') < Delay(p')$. Combining this last inequality with (11-1) yields

$$Delay(p'') < delay(s). \tag{11-3}$$

Furthermore, since $p'$ is contained in $G$, it follows from the construction of $p''$, that $p''$ must also be contained in $G$.

Finally, since $p'$ is contained in $G'$, $G'$ does not contain $s$, and the set of edges contained in $p''$ is a subset of the set of edges contained in $p'$, we have that $p''$ does not contain $s$. *QED.*

**Lemma 11.1:** Suppose that $G$ and $G'$ are synchronization graphs such that $G'$ preserves $G$, and $p$ is a path in $G$ from actor $x$ to actor $y$. Then there is a path $p'$ in $G'$ from $x$ to $y$ such that $Delay(p') \leq Delay(p)$, and $tr(p) \subseteq tr(p')$, where $tr(\varphi)$ denotes the set of actors traversed by the path $\varphi$.

Thus, if a synchronization graph $G'$ preserves another synchronization graph $G$ and $p$ is a path in $G$ from actor $x$ to actor $y$, then there is at least one path $p'$ in $G'$ such that 1) the path $p'$ is directed from $x$ to $y$; 2) the cumulative delay on $p'$ does not exceed the cumulative delay on $p$; and 3) every actor that is traversed by $p$ is also traversed by $p'$ (although $p'$ may traverse one or more actors that are not traversed by $p$).

For example in Figure 11.1(a), if we let $x = B$, $y = I$, and

$$p = ((B, G), (G, H), (H, I)),$$

then the path

$$p' = ((B, A), (A, C), (C, H), (H, G), (G, H), (H, I)) \tag{11-4}$$

in Figure 11.1(b) confirms Lemma 11.1 for this example. Here

$$tr(p) = \{B, G, H, I\} \text{ and } tr(p') = \{A, B, C, G, H, I\}.$$

*Proof of Lemma 11.1:* Let $p = (e_1, e_2, ..., e_n)$. By definition of the *preserves* relation, each $e_i$ that is not a synchronization edge in $G$ is contained in $G'$. For each $e_i$ that is a synchronization edge in $G$, there must be a path $p_i$ in $G'$ from $src(e_i)$ to $snk(e_i)$ such that $Delay(p_i) \leq delay(e_i)$. Let $e_{i_1}, e_{i_2}, ..., e_{i_m}$, $i_1 < i_2 < ... < i_m$, denote the set of $e_i$s that are synchronization edges in $G$, and define the path $\tilde{p}$ to be the concatenation

$$\langle (e_1, e_2, ..., e_{i_1-1}), p_1, (e_{i_1+1}, ..., e_{i_2-1}), p_2, ..., \qquad . \qquad (11\text{-}5)$$
$$(e_{i_{m-1}+1}, ..., e_{i_m-1}), p_m, (e_{i_m+1}, ..., e_n)\rangle$$

Clearly, $\tilde{p}$ is a path in $G'$ from $x$ to $y$, and since $Delay(p_i) \le delay(e_i)$ holds whenever $e_i$ is a synchronization edge, it follows that $Delay(\tilde{p}) \le Delay(p)$. Furthermore, from the construction of $\tilde{p}$, it is apparent that every actor that is traversed by $p$ is also traversed by $\tilde{p}$. *QED*.

The following lemma states that if a resynchronization contains a resynchronization edge $e$ such that there is a delay-free path in the original synchronization graph from the source of $e$ to the sink of $e$, then $e$ must be redundant in the resychronized graph.

**Lemma 11.2:** Suppose that $G$ is a synchronization graph; $R$ is a resynchronization of $G$; and $(x, y)$ is a resynchronization edge such that $\rho_G(x, y) = 0$. Then $(x, y)$ is redundant in $\Psi(R, G)$. Thus, a minimal resynchronization (fewest number of elements) has the property that $\rho_G(x', y') > 0$ for each resynchronization edge $(x', y')$.

*Proof:* Let $p$ denote a minimum-delay path from $x$ to $y$ in $G$. Since $(x, y)$ is a resynchronization edge, $(x, y)$ is not contained in $G$, and thus, $p$ traverses at least three actors. From Lemma 11.1, it follows that there is a path $p'$ in $\Psi(R, G)$ from $x$ to $y$ such that

$$Delay(p') = 0, \qquad\qquad (11\text{-}6)$$

and $p'$ traverses at least three actors. Thus,

$$Delay(p') \le delay((x, y)) \qquad\qquad (11\text{-}7)$$

and $p' \ne ((x, y))$. Furthermore, $p'$ cannot properly contain $(x, y)$. To see this, observe that if $p'$ contains $(x, y)$ but $p' \ne ((x, y))$, then from (11-6), it follows that there exists a delay-free cycle in $G$ (that traverses $x$), and hence that our assumption of a deadlock-free schedule (Section 11.1) is violated. Thus, we conclude that $(x, y)$ is redundant in $\Psi(R, G)$. *QED*.

As a consequence of Lemma 11.1, the estimated throughput of a given synchronization graph is always less than or equal to that of every synchronization graph that it preserves.

**Theorem 11.1:** If $G$ is a synchronization graph, and $G'$ is a synchronization graph that preserves $G$, then $\lambda_{max}(G') \ge \lambda_{max}(G)$.

*Proof:* Suppose that $C$ is a critical cycle in $G$. Lemma 11.1 guarantees that there is a cycle $C'$ in $G'$ such that a) $Delay(C') \le Delay(C)$, and b) the set of actors that are traversed by $C$ is a subset of the set of actors traversed by $C'$. Now clearly, b) implies that

$$\sum_{v \text{ is traversed by } C'} t(v) \geq \sum_{v \text{ is traversed by } C} t(v), \qquad (11\text{-}8)$$

and this observation together with a) implies that the cycle mean of $C'$ is greater than or equal to the cycle mean of $C$. Since $C$ is a critical cycle in $G$, it follows that $\lambda_{max}(G') \geq \lambda_{max}(G)$. *QED.*

Thus, any saving in synchronization cost obtained by rearranging synchronization edges may come at the expense of a decrease in estimated throughput. As implied by Definition 11.1, we avoid this complication by restricting our attention to feedforward synchronization edges. Clearly, resynchronization that rearranges only feedforward synchronization edges cannot decrease the estimated throughput since no new cycles are introduced and no existing cycles are altered. Thus, with the form of resynchronization that is addressed in this chapter, any decrease in synchronization cost that we obtain is not diminished by a degradation of the estimated throughput.

## 11.3    Relationship to Set Covering

We refer to the problem of finding a resynchronization with the fewest number of elements as the **resynchronization problem**. In Section 11.4, it is formally shown that the resynchronization problem is NP-hard, which means that it is unlikely that efficient algorithms can be devised to solve the problem exactly, and thus, for practical use, we should search for good heuristic solutions [GJ79]. In this section, we explain the intuition behind this result. To establish the NP-hardness of the resynchronization problem, we examine a special case that occurs when there are exactly two SCCs, which we call the **pairwise resynchronization problem**, and we derive a polynomial-time reduction from the classic *set covering problem* [CLR92], a well-known NP-hard problem, to the pairwise resynchronization problem. In the set-covering problem, one is given a finite set $X$ and a family $T$ of subsets of $X$, and asked to find a minimal (fewest number of members) subfamily $T_s \subseteq T$ such that

$$\bigcup_{t \in T_s} t = X.$$

A subfamily of $T$ is said to *cover* $X$ if each member of $X$ is contained in some member of the subfamily. Thus, the set-covering problem is the problem of finding a minimal cover.

**Definition 11.2:** Given a synchronization graph $G$, let $(x_1, x_2)$ be a synchronization edge in $G$, and let $(y_1, y_2)$ be an ordered pair of actors in $G$. We say that

$(y_1, y_2)$ **subsumes** $(x_1, x_2)$ in $G$ if

$$\rho(x_1, y_1) + \rho(y_2, x_2) \leq delay((x_1, x_2)).$$

Thus, every synchronization edge subsumes itself, and intuitively, if $(x_1, x_2)$ is a synchronization edge, then $(y_1, y_2)$ subsumes $(x_1, x_2)$ if and only if a zero-delay synchronization edge directed from $y_1$ to $y_2$ makes $(x_1, x_2)$ redundant.

The following fact is easily verified from Definitions 11.1 and 11.2.

**Fact 11.2:** Suppose that $G$ is a synchronization graph that contains exactly two SCCs, $F$ is the set of feedforward edges in $G$, and $F'$ is a resynchronization of $G$. Then for each $e \in F$, there exists $e' \in F'$ such that $(src(e'), snk(e'))$ subsumes $e$ in $G$.

An intuitive correspondence between the pairwise resynchronization problem and the set covering problem can be derived from Fact 11.2. Suppose that $G$ is a synchronization graph with exactly two SCCs, $C_1$ and $C_2$, such that each feedforward edge is directed from a member of $C_1$ to a member of $C_2$. We start by viewing the set $F$ of feedforward edges in $G$ as the finite set that we wish to cover, and with each member $p$ of $\{(x, y) | (x \in C_1, y \in C_2)\}$, we associate the subset of $F$ defined by $\chi(p) \equiv \{e \in F | (p \text{ subsumes } e)\}$. Thus, $\chi(p)$ is the set of feedforward edges of $G$ whose corresponding synchronizations can be eliminated if we implement a zero-delay synchronization edge directed from the first vertex of the ordered pair $p$ to the second vertex of $p$. Clearly then, $\{e_1', e_2', ..., e_n'\}$ is a resynchronization if and only if each $e \in F$ is contained in at least one $\chi((src(e_i'), snk(e_i')))$ — that is, if and only if $\{\chi((src(e_i'), snk(e_i'))) | 1 \leq i \leq n\}$ covers $F$. Thus, solving the pairwise resynchronization problem for $G$ is equivalent to finding a minimal cover for $F$ given the family of subsets $\{\chi(x, y) | (x \in C_1, y \in C_2)\}$.

Figure 11.2 helps to illustrate this intuition. Suppose that we are given the set $X = \{x_1, x_2, x_3, x_4\}$, and the family of subsets $T = \{t_1, t_2, t_3\}$, where $t_1 = \{x_1, x_3\}$, $t_2 = \{x_1, x_2\}$, and $t_3 = \{x_2, x_4\}$. To construct an instance of the pairwise resynchronization problem, we first create two vertices and an edge directed between these vertices *for each* member of $X$; we label each of the edges created in this step with the corresponding member of $X$. Then for each $t \in T$, we create two vertices $vsrc(t)$ and $vsnk(t)$. Next, for each relation $x_i \in t_j$ (there are six such relations in this example), we create two delayless edges — one directed from the source of the edge corresponding to $x_i$ and directed to $vsrc(t_j)$, and another directed from $vsnk(t_j)$ to the sink of the edge corresponding to $x_i$. This last step has the effect of making each pair $(vsrc(t_i), vsnk(t_i))$ subsume exactly those edges that correspond to members of $t_i$; in other words, after this construction, $\chi((vsrc(t_i), vsnk(t_i))) = t_i$, for each $i$. Finally, for each edge created in the previous step, we create a corresponding feedback edge oriented in the opposite direction, and having a unit delay.

Figure 11.2(a) shows the synchronization graph that results from this construction process. Here, it is assumed that each vertex corresponds to a separate processor; the associated unit delay, self loop edges are not shown to avoid clutter. Observe that the graph contains two SCCs — the SCC

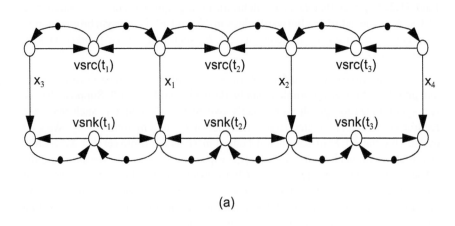

(a)

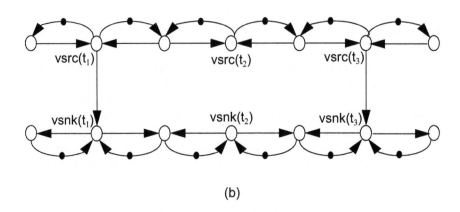

(b)

Figure 11.2. (a) An instance of the pairwise resynchronization problem that is derived from an instance of the set-covering problem; (b) the HSDFG that results from a solution to this instance of pairwise resynchronization.

($\{src(x_i)\} \cup \{vsrc(t_i)\}$) and the SCC ($\{snk(x_i)\} \cup \{vsnk(t_i)\}$) — and that the set of feedforward edges is the set of edges that correspond to members of $X$. Now, recall that a major correspondence between the given instance of set covering and the instance of pairwise resynchronization defined by Figure 11.2(a) is that $\chi((vsrc(t_i), vsnk(t_i))) = t_i$, for each $i$. Thus, if we can find a minimal resynchronization of Figure 11.2(a) such that each edge in this resynchronization is directed from some $vsrc(t_k)$ to the corresponding $vsnk(t_k)$, then the associated $t_k$'s form a minimum cover of $X$. For example, it is easy, albeit tedious, to verify that the resynchronization illustrated in Figure 11.2(b),

$$\{d_0(vsrc(t_1), vsnk(t_1)), d_0(vsrc(t_3), vsnk(t_3))\},$$

is a minimal resynchronization of Figure 11.2(a), and from this, we can conclude that $\{t_1, t_3\}$ is a minimal cover for $X$. From inspection of the given sets $X$ and $T$, it is easily verified that this conclusion is correct.

This example illustrates how an instance of pairwise resynchronization can be constructed (in polynomial time) from an instance of set covering, and how a solution to this instance of pairwise resynchronization can easily be converted into a solution of the set covering instance. The formal proof of the NP-hardness of pairwise resynchronization that is given in the following section is a generalization of the example in Figure 11.2.

## 11.4    Intractability of Resynchronization

In this section, the NP completeness of the resynchronization problem is established. This result is derived by reducing an arbitrary instance of the set-covering problem, a well-known NP-hard problem, to an instance of the pairwise resynchronization problem, which is a special case of the resynchronization problem that occurs when there are exactly two SCCs. The intuition behind this reduction is explained in Section 11.3 above.

Suppose that we are given an instance $(X, T)$ of set covering, where $X$ is a finite set, and $T$ is a family of subsets of $X$ that covers $X$. Without loss of generality, we assume that

$T$ does *not* contain a proper nonempty subset $T'$ that satisfies

$$\left( \bigcup_{t \in (T-T')} t \right) \cap \left( \bigcup_{t \in T'} t \right) = \varnothing. \tag{11-9}$$

We can assume this without loss of generality because if this assumption does not hold, then we can apply the construction below to each "independent subfamily" separately, and then combine the results to get a minimal cover for $X$.

The following steps specify how we construct an HSDFG from $(X, T)$. Except where stated otherwise, no delay is placed on the edges that are instantiated.

1. For each $x \in X$, instantiate two vertices $vsrc(x)$ and $vsnk(x)$, and instantiate an edge $e(x)$ directed from $vsrc(x)$ to $vsnk(x)$.

2. For each $t \in T$

    (a) Instantiate two vertices $vsrc(t)$ and $vsnk(t)$.

    (b) For each $x \in t$

        • Instantiate an edge directed from $vsrc(x)$ to $vsrc(t)$.

        • Instantiate an edge directed from $vsrc(t)$ to $vsrc(x)$, and place one delay on this edge.

        • Instantiate an edge directed from $vsnk(t)$ to $vsnk(x)$.

        • Instantiate an edge directed from $vsnk(x)$ to $vsnk(t)$, and place one delay on this edge.

3. For each vertex $v$ that has been instantiated, instantiate an edge directed from $v$ to itself, and place one delay on this edge.

Observe from our construction, that whenever $x \in X$ is contained in $t \in T$, there is an edge directed from $vsrc(x)$ ($vsnk(t)$) to $vsrc(t)$ ($vsnk(x)$), and there is also an edge (having unit delay) directed from $vsrc(t)$ ($vsnk(x)$) to $vsrc(x)$ ($vsnk(t)$). Thus, from the assumption stated in (11-9), it follows that $\{vsrc(z)|z \in (X \cup T)\}$ forms one SCC, $\{vsnk(z)|z \in (X \cup T)\}$ forms another SCC, and $F \equiv \{e(x)|x \in X\}$ is the set of feedforward edges.

Let $G$ denote the HSDFG that we have constructed, and as in Section 11.3, define $\chi(p) \equiv \{e \in F|(p \text{ subsumes } (src(e), snk(e)))\}$ for each ordered pair of vertices $p = (y_1, y_2)$ such that $y_1$ is contained in the source SCC of $G$, and $y_2$ is contained in the sink SCC of $G$. Clearly, $G$ gives an instance of the pairwise resynchronization problem.

**Observation 1:** By construction of $G$, observe that

$\{x \in X|((vsrc(t), vsnk(t)) \text{ subsumes } (vsrc(x), vsnk(x)))\} = t$, for all $t \in T$. Thus, for all $t \in T$, $\chi(vsrc(t), vsnk(t)) = \{e(x)|x \in t\}$.

**Observation 2:** For each $x \in X$, all input edges of $vsrc(x)$ have unit delay on them. It follows that for any vertex $y$ in the sink SCC of $G$, $\chi(vsrc(x), y) \subseteq \{e \in F|src(e) = vsrc(x)\} = \{e(x)\}$.

**Observation 3:** For each $t \in T$, the only vertices in $G$ that have a delay-free path to $vsrc(t)$ are those vertices contained in $\{vsrc(x)|x \in t\}$. It follows that for any vertex $y$ in the sink SCC of $G$,

$$\chi(vsrc(t), y) \subseteq \chi(vsrc(t), vsnk(t)) = \{e(x) | x \in t\}.$$

Now suppose that $F' = \{f_1, f_2, ..., f_m\}$ is a minimal resynchronization of $G$. For each $i \in \{1, 2, ..., m\}$, exactly one of the following two cases must apply:

Case 1: $vsrc(f_i) = vsrc(x)$ for some $x \in X$. In this case, we pick an arbitrary $t \in T$ that contains $x$, and we set $v_i = vsrc(t)$ and $w_i = vsnk(t)$. From Observation 2, it follows that

$$\chi((src(f_i), snk(f_i))) \subseteq \{e(x)\} \subseteq \chi(v_i, w_i).$$

Case 2: $vsrc(f_i) = vsrc(t)$ for some $t \in T$. We set $v_i = vsrc(t)$ and $w_i = vsnk(t)$. From Observation 3, we have

$$\chi((src(f_i), snk(f_i))) \subseteq \chi(v_i, w_i).$$

**Observation 4:** From our definition of the $v_i$s and $w_i$s, $\{d_o(v_i, w_i) | (i \in \{1, 2, ..., m\})\}$ is a minimal resynchronization of $G$. Also, each $(v_i, w_i)$ is of the form $(vsrc(t), vsnk(t))$, where $t \in T$.

Now, for each $i \in \{1, 2, ..., m\}$, we define

$$Z_i \equiv \{x \in X | (v_i, w_i) \text{ subsumes } (vsrc(x), vsnk(x))\}.$$

**Proposition 1:** $\{Z_1, Z_2, ..., Z_m\}$ covers $X$.

*Proof:* From Observation 4, we have that for each $Z_i$, there exists a $t \in T$ such that $Z_i = \{x \in X | (vsrc(t), vsnk(t)) \text{ subsumes } (vsrc(x), vsnk(x))\}$. Thus, each $Z_i$ is a member of $T$. Also, since $\{d_o(v_i, w_i) | (i \in \{1, 2, ..., m\})\}$ is a resynchronization of $G$, each member of $\{(vsrc(x), vsnk(x)) | x \in X\}$ must be preserved by some $(v_i, w_i)$, and thus each $x \in X$ must be contained in some $Z_i$. QED.

**Proposition 2:** $\{Z_1, Z_2, ..., Z_m\}$ is a minimal cover for $X$.

*Proof:* (By contraposition). Suppose there exists a cover $\{Y_1, Y_2, ..., Y_{m'}\}$ (among the members of $T$) for $X$, with $m' < m$. Then, each $x \in X$ is contained in some $Y_j$, and from Observation 1, $(vsrc(Y_j), vsnk(Y_j))$ subsumes $e(x)$. Thus, $\{(vsrc(Y_i), vsnk(Y_i)) | (i \in \{1, 2, ..., m'\})\}$ is a resynchronization of $G$. Since $m' < m$, it follows that $F' = \{f_1, f_2, ..., f_m\}$ is not a minimal resynchronization of $G$. QED.

In summary, we have shown how to convert an arbitrary instance $(X, T)$ of the set-covering problem into an instance $G$ of the pairwise resynchronization problem, and we have shown how to convert a solution $F' = \{f_1, f_2, ..., f_m\}$ of this instance of pairwise resynchronization into a solution $\{Z_1, Z_2, ..., Z_m\}$ of

$(X, T)$. It is easily verified that all of the steps involved in deriving $G$ from $(X, T)$, and in deriving $\{Z_1, Z_2, ..., Z_m\}$ from $F'$ can be performed in polynomial time. Thus, from the NP hardness of set covering [CLR92], we can conclude that the pairwise resynchronization problem is NP hard.

## 11.5    Heuristic Solutions

### 11.5.1    Applying Set Covering Techniques to Pairs of SCCs

A heuristic framework for the pairwise resynchronization problem emerges naturally from the relationship that was established in Section 11.3 between set-covering and pairwise resynchronization. Given an arbitrary algorithm *COVER* that solves the set-covering problem, and given an instance of pairwise resynchronization that consists of two SCCs $C_1$ and $C_2$, and a set $S$ of feedforward synchronization edges directed from members of $C_1$ to members of $C_2$, this heuristic framework first computes the subset

$$\chi((u, v))=\{e \in S | (\rho_G(src(e), u) = 0) + (\rho_G(v, snk(e)) \le delay(e))\}$$

for each ordered pair of actors $(u, v)$ that is contained in the set

$$T \equiv \{(u', v') | (u' \text{ is in } C_1 \text{ and } v' \text{ is in } C_2)\},$$

and then applies the algorithm *COVER* to the instance of set covering defined by the set $S$ together with the family of subsets $\{\chi((u', v')) | ((u', v') \in T)\}$. If $\Xi$ denotes the solution returned by *COVER*, then a resynchronization for the given instance of pairwise resynchronization can be derived by

$$\{d_0(u, v) | \chi((u, v)) \in \Xi\}.$$

This resynchronization is the solution returned by the heuristic framework.

From the correspondence between set-covering and pairwise resynchronization that is outlined in Section 11.3, it follows that the quality of a resynchronization obtained by the heuristic framework is determined entirely by the quality of the solution computed by the set-covering algorithm that is employed; that is, if the solution computed by *COVER* is $X\%$ worse ($X\%$ more subfamilies) than an optimal set-covering solution, then the resulting resynchronization will be $X\%$ worse ($X\%$ more synchronization edges) than an optimal resynchronization of the given instance of pairwise resynchronization.

The application of the heuristic framework for pairwise resynchronization to each pair of SCCs, in some arbitrary order, in a general synchronization graph yields a heuristic framework for the general resynchronization problem. However, a major limitation of this extension to general synchronization graphs arises from its inability to consider resynchronization opportunities that involve paths that traverse more than two SCCs, and paths that contain more than one feedforward synchronization edge.

Thus, in general, the quality of the solutions obtained by this approach will be worse than the quality of the solutions that are derived by the particular set covering heuristic that is employed, and roughly, this discrepancy can be expected to increase as the number of SCCs increases relative to the number of synchronization edges in the original synchronization graph.

For example, Figure 11.3 shows the synchronization graph that results from a six-processor schedule of a synthesizer for plucked-string musical instruments in 11 voices based on the Karplus-Strong technique. Here, *exc* represents the excitation input, each $v_i$ represents the computation for the $i$ th voice, and the actors marked with "+" signs specify adders. Execution time estimates for the actors are shown in the table at the bottom of the figure. In this example, the only pair of distinct SCCs that have more than one synchronization edge between them is the pair consisting of the SCC containing {*exc*, $v_1$} and the SCC containing $v_2$, $v_3$, five addition actors, and the actor labeled *out*. Thus, the best result that can be derived from the heuristic extension for general synchronization graphs described above is a resynchronization that optimally rearranges the synchronization edges between these two SCCs in isolation, and leaves all other synchronization edges unchanged. Such a resynchronization is illustrated in Figure 11.4. This synchronization graph has a total of nine synchronization edges, which

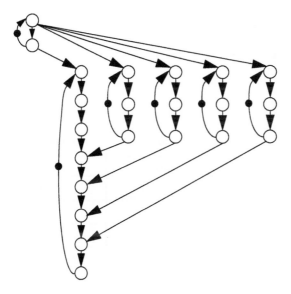

Figure 11.4. The synchronization graph that results from applying the heuristic framework based on pairwise resynchronization to the example of Figure 11.3.

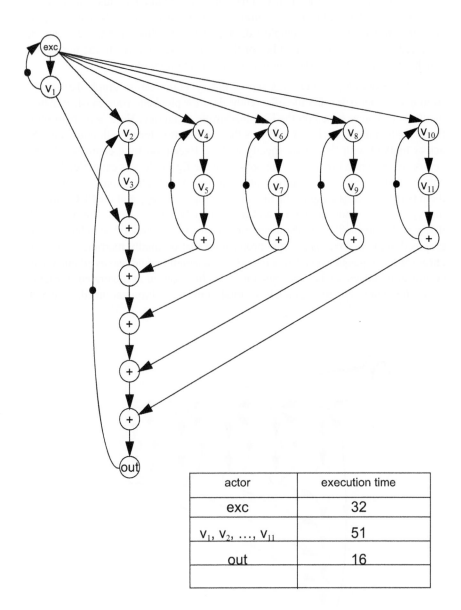

| actor | execution time |
|-------|----------------|
| exc | 32 |
| $v_1, v_2, \ldots, v_{11}$ | 51 |
| out | 16 |
| | |

Figure 11.3. The synchronization graph that results from a six-processor schedule of a music synthesizer based on the Karplus-Strong technique.

is only one less than the number of synchronization edges in the original graph. In contrast, it is shown in the following subsection that with a more flexible approach to resynchronization, the total synchronization cost of this example can be reduced to only five synchronization edges.

## 11.5.2    A More Flexible Approach

This subsection presents a more global approach to resynchronization, called Algorithm Global-resynchronize, which overcomes the major limitation of the pairwise approach discussed in Section 11.5.1. Algorithm Global-resynchronize is based on the simple greedy approximation algorithm for set-covering that repeatedly selects a subset that covers the largest number of *remaining elements*, where a remaining element is an element that is not contained in any of the subsets that have already been selected. In [Joh74, Lov75] it is shown that this set-covering technique is guaranteed to compute a solution whose cardinality is no greater than $(\ln(|X|) + 1)$ times that of the optimal solution, where $X$ is the set that is to be covered.

To adapt this set-covering technique to resynchronization, we construct an instance of set covering by choosing the set $X$, the set of elements to be covered, to be the set of feedforward synchronization edges, and choosing the family of subsets to be

$$T \equiv \{\chi(v_1, v_2) \,|\, (((v_1, v_2) \notin E) \text{ and } (\rho_G(v_2, v_1) = \infty))\}, \qquad (11\text{-}10)$$

where $G = (V, E)$ is the input synchronization graph. The constraint $\rho_G(v_2, v_1) = \infty$ in (11-10) ensures that inserting the resynchronization edge $(v_1, v_2)$ does not introduce a cycle, and thus that it does not introduce deadlock or reduce the estimated throughput.

Algorithm Global-resynchronize assumes that the input synchronization graph is reduced (a reduced synchronization graph can be derived efficiently, for example, by using the redundant synchronization removal technique discussed in the previous chapter). The algorithm determines the family of subsets specified by (11-10), chooses a member of this family that has maximum cardinality, inserts the corresponding delayless resynchronization edge, removes all synchronization edges that it subsumes, and updates the values $\rho_G(x, y)$ for the new synchronization graph that results. This entire process is then repeated on the new synchronization graph, and it continues until it arrives at a synchronization graph for which the computation defined by (11-10) produces the empty set — that is, the algorithm terminates when no more resynchronization edges can be added. Figure 11.5 gives a pseudocode specification of this algorithm (with some straightforward modifications to improve the running time).

To analyze the complexity of Algorithm Global-resynchronize, the following definition is useful.

**Function** *Global-resynchronize*
**Input**: a reduced synchronization graph $G = (V, E)$
**Output**: an alternative reduced synchronization graph that preserves $G$.

Compute $\rho_G(x, y)$ for all actor pairs $x, y \in V$
*complete* = FALSE
**While** not (*complete*)
    *best* = *NULL*, $M = 0$
    **For** $x \in V$
        **For** $y \in V$
            **If** $(\rho_G(y, x) = \infty)$ and $(x, y) \notin E$
                $\chi^* = \chi((x, y))$
                **If** $(|\chi^*| > M)$
                    $M = |\chi^*|$
                    *best* = $(x, y)$
                **Endif**
            **Endif**
        **Endfor**
    **Endfor**
    **If** (*best* = *NULL*)
        *complete* = *TRUE*
    **Else**
        $E = E - \chi(best) + \{d_0(best)\}$
        $G = (V, E)$
        **For** $x, y \in V$             /* update $\rho_G$ */
            $\rho_{new}(x, y) =$
            $min(\{\rho_G(x, y), \rho_G(x, src(best)) + \rho_G(snk(best), y)\})$
        **Endfor**
        $\rho_G = \rho_{new}$
    **Endif**
**Endwhile**
**Return** $G$
**End function**

Figure 11.5. A heuristic for resynchronization.

**Definition 11.3:** Suppose that $G$ is a synchronization graph. The **delayless connectivity** of $G$, denoted $DC(G)$, is the number of distinct ordered vertex-pairs $(x, y)$ in $G$ that satisfy $\rho_G(x, y) = 0$. That is,

$$DC(G) = \left|\hat{S}(G)\right|, \text{ where } \hat{S}(G) = \{(x, y) | (\rho_G(x, y) = 0)\}. \tag{11-11}$$

The following lemma shows that as long as the input synchronization graph is reduced, the resynchronization operations performed in Algorithm Global-resynchronize always yield a reduced synchronization graph.

**Lemma 11.3:** Suppose that $G = (V, E)$ is a reduced synchronization graph; and $(x, y)$ is an ordered pair of vertices in $G$ such that $(x, y) \notin E$, $(\rho_G(y, x) = \infty)$, and $|\chi(x, y)| \geq 1$. Let $G'$ denote the synchronization graph obtained by inserting $d_0(x, y)$ into $G$ and removing all members of $\chi(x, y)$; that is, $G' = (V, E')$, where

$$E' = (E - \chi(x, y)) + \{d_0(x, y)\}.$$

Then $G'$ is a reduced synchronization graph. In other words, $G'$ does not contain any redundant synchronizations. Furthermore, $DC(G') > DC(G)$.

*Proof:* We prove the first part of this lemma by contraposition. Suppose that there exists a redundant synchronization edge $s$ in $G'$, and first suppose that $s = (x, y)$. Then from Fact 11.1, there exists a path in $G'$ directed from $x$ to $y$ such that $Delay(p) = 0$, and

$$p \text{ does not contain } (x, y). \tag{11-12}$$

Also, observe that from the definition of $E'$,

$$(E' - (x, y)) \subseteq E, \tag{11-13}$$

It follows from (11-12) and (11-13) that $G$ also contains the path $p$.

Now let $(x', y')$ be an arbitrary member of $\chi(x, y)$. Then

$$\rho_G(x', x) + \rho_G(y, y') \leq delay(x', y'). \tag{11-14}$$

Since $G$ contains the path $p$, we have $\rho_G(x, y) = 0$, and thus, from the triangle inequality (3-4) together with (11-14),

$$\rho_G(x', y') \leq \rho_G(x', x) + \rho_G(x, y) + \rho_G(y, y') \leq delay(x', y'). \tag{11-15}$$

We conclude that $(x', y')$ is redundant in $G$, which violates the assumption that $G$ is reduced.

If, on the other hand, $s \neq (x, y)$, then from Fact 11.1, there exists a simple path $p_s \neq (s)$ in $G'$ directed from $src(s)$ to $snk(s)$ such that

$$Delay(p_s) \le delay(s). \tag{11-16}$$

Also, it follows from (11-13) that $G$ contains $s$. Since $G$ is reduced, the path $p_s$ must contain the edge $(x, y)$ (otherwise $s$ would be redundant in $G$). Thus, $p_s$ can be expressed as a concatenation $p_s = \langle (p_1, ((x, y)), p_2) \rangle$, where either $p_1$ or $p_2$ may be empty, but not both. Furthermore, since $p_s$ is a simple path, neither $p_1$ nor $p_2$ contains $(x, y)$. Hence, from (11-13), we are guaranteed that both $p_1$ and $p_2$ are also contained in $G$.

Now from (11-16), we have

$$Delay(p_1) + Delay(p_2) \le delay(s). \tag{11-17}$$

Furthermore, from the definition of $p_1$ and $p_2$,

$$\rho_G(src(s), x) \le Delay(p_1) \text{ and } \rho_G(y, snk(s)) \le Delay(p_2). \tag{11-18}$$

Combining (11-17) and (11-18) yields

$$\rho_G(src(s), x) + \rho_G(y, snk(s)) \le delay(s), \tag{11-19}$$

which implies that $s \in \chi(x, y)$. But this violates the assumption that $G'$ does not contain any edges that are subsumed by $(x, y)$ in $G$. This concludes the proof of the first part of Lemma *11.3*.

It remains to be shown that $DC(G') > DC(G)$. Now, from Lemma 11.1 and Definition 11.3, it follows that

$$\hat{S}(G) \subseteq \hat{S}(G'). \tag{11-20}$$

Also, from the first part of Lemma 11.3, which has already been proven, we know that $G'$ is reduced. Thus, from Lemma 11.2, we have

$$(x, y) \notin \hat{S}(G). \tag{11-21}$$

But, clearly from the construction of $G'$, $\rho_{G'}(x, y) = 0$, and thus,

$$(x, y) \in \hat{S}(G'). \tag{11-22}$$

From (11-20), (11-21), and (11-22), it follows that $\hat{S}(G)$ is a proper subset of $\hat{S}(G')$. Hence, $DC(G') > DC(G)$. *QED.*

Clearly from Lemma 11.3, each time a Algorithm Global-resynchronize performs a resynchronization operation (an iteration of the **while** loop of Figure 11.5), the number of ordered vertex pairs $(x, y)$ that satisfy $\rho_G(x, y) = 0$ is increased by at least one. Thus, the number of iterations of the **while** loop in Figure 11.5 is bounded above by $|V|^2$. The complexity of one iteration of the **while** loop is dominated by the computation in the pair of nested **for** loops. The computation of one iteration of the inner **for** loop is dominated by the time required to compute $\chi(x, y)$ for a specific actor pair $(x, y)$. Assuming $\rho_G(x', y')$ is available for all $x', y' \in V$, the time to compute $\chi(x, y)$ is $O(s_c)$, where $s_c$ is the number of feedforward synchronization edges in the current synchronization graph. Since the number of feedforward synchronization edges never increases from one iteration of the **while** loop to the next, it follows that the time-complexity of the overall algorithm is $O(s|V|^4)$, where $s$ is the number of feedforward synchronization edges in the input synchronization graph. In practice, however, the number of resynchronization steps (**while** loop iterations) is usually much lower than $|V|^2$ since the constraints on the introduction of cycles severely limit the number of resynchronization steps. Thus, the $O(s|V|^4)$ bound can be viewed as a very conservative estimate.

### 11.5.3    Unit-Subsumption Resynchronization Edges

At first, it may seem that it only makes sense to continue the **while** loop of Algorithm Global-resynchronize as long as a resynchronization edge can be found that subsumes at least two existing synchronization edges. However, in general it may be advantageous to continue the resynchronization process even if each resynchronization candidate subsumes at most one synchronization edge. This is because although such a resynchronization candidate does not lead to an immediate reduction in synchronization cost, its insertion may lead to future resynchronization opportunities in which the number of synchronization edges can be reduced.

Figures 11.6 and 11.7 illustrate a simple example. In the synchronization graph shown in Figure 11.6(a), there are 5 synchronization edges, $(A, B)$, $(B, C)$, $(D, F)$, $(G, F)$, and $(A, E)$. Self-loop edges incident to actors $A$, $B$, $C$, and $F$ (each of these four actors executes on a separate processor) are omitted from the illustration for clarity. It is easily verified that no resynchronization candidate in Figure 11.6(a) subsumes more than one synchronization edge. If we terminate the resynchronization process at this point, we must accept a synchronization cost of 5 synchronization edges.

However, suppose that we insert the resynchronization edge $(B, D)$, which subsumes $(B, C)$, and then we remove the subsumed edge $(B, C)$. Then we arrive at the synchronization graph of Figure 11.6(b). In this graph, resynchronization candidates exist that subsume up to two synchronization edges each. For example, insertion of the resynchronization edge $(F, E)$, allows us to remove synchronization edges $(G, F)$ and $(A, E)$. The resulting synchronization graph, shown in Figure 11.7(a), contains only four synchronization edges.

Alternatively, from Figure 11.6(b), we could insert the resynchronization edge $(C, E)$ and remove both $(D, F)$ and $(A, E)$. This gives us the synchronization graph of Figure 11.7(b), which also contains four synchronization edges. This is the solution derived by an actual implementation of Algorithm Global-resynchronize [BSL96b] when it is applied to the graph of Figure 11.6(a).

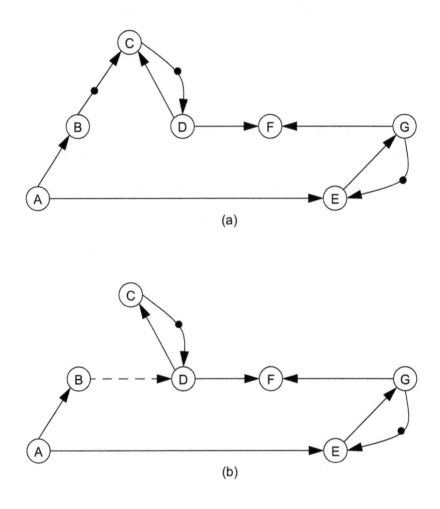

(a)

(b)

Figure 11.6. An example in which inserting a resynchronization edge that sub-sumes only one existing synchronization edge eventually leads to a reduction in the total number of synchronizations.

### 11.5.4     Example

Figure 11.8 shows the optimized synchronization graph that is obtained when Algorithm Global-resynchronize is applied to the example of Figure 11.3 (using the implementation discussed in [BSL96b]). Observe that the total number

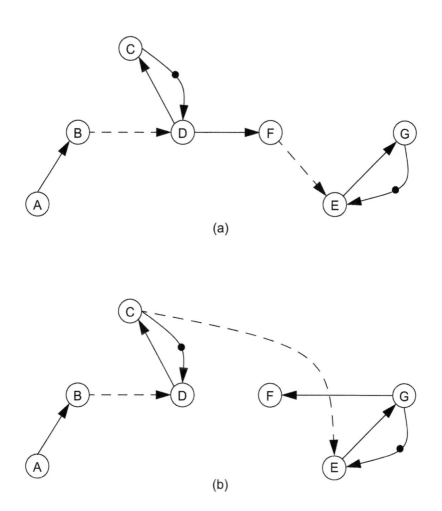

(a)

(b)

Figure 11.7. A continuation of the example in Figure 11.6.

of synchronization edges has been reduced from 10 to 5. The total number of "resynchronization steps" (number of while-loop iterations) required by the heuristic to complete this resynchronization is 7.

Table 11.1 shows the relative throughput improvement delivered by the optimized synchronization graph of Figure 11.8 over the original synchronization graph as the shared memory access time varies from 1 to 10 processor clock cycles. The assumed synchronization protocol is FFS, and the back-off time for each simulation is obtained by the experimental procedure discussed in Section 10.5. The second and fourth columns show the *average iteration period* for the original synchronization graph and the resynchronized graph, respectively. The average iteration period, which is the reciprocal of the average throughput, is the average number of time units required to execute an iteration of the synchronization graph. From the sixth column, we see that the resynchronized graph consistently attains a throughput improvement of 22% to 26%. This improvement includes the effect of reduced overhead for maintaining synchronization variables and reduced contention for shared memory. The third and fifth columns of

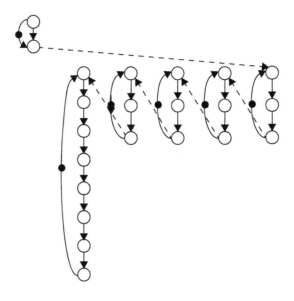

Figure 11.8. The optimized synchronization graph that is obtained when Algorithm Global-resynchronize is applied to the example of Figure 11.3.

Table 11.1 show the average number of shared memory accesses per iteration of the synchronization graph. Here we see that the resynchronized solution consistently obtains at least a 30% improvement over the original synchronization graph. Since accesses to shared memory typically require significant amounts of energy, particularly for a multiprocessor system that is not integrated on a single chip, this reduction in the average rate of shared memory accesses is especially useful when low power consumption is an important implementation issue.

## 11.5.5    Simulation Approach

The simulation is written in C making use of a package called CSIM [Sch88] that allows concurrently running processes to be modeled. Each CSIM process is "created," after which it runs concurrently with the other processes in the simulation. Processes communicate and synchronize through *events* and

| Mem. acc-ess time | Original graph | | Resynchronized graph | | Decrease in iter. period |
|---|---|---|---|---|---|
| | Iter. period | Memory accesses/ period | Iter. period | Memory accesses/ period | |
| 1 | 250 | 67 | 195 | 43 | 22% |
| 2 | 292 | 66 | 216 | 43 | 26% |
| 3 | 335 | 64 | 249 | 43 | 26% |
| 4 | 368 | 63 | 273 | 40 | 26% |
| 5 | 408 | 63 | 318 | 43 | 22% |
| 6 | 459 | 63 | 350 | 43 | 24% |
| 7 | 496 | 63 | 385 | 43 | 22% |
| 8 | 540 | 63 | 420 | 43 | 22% |
| 9 | 584 | 63 | 455 | 43 | 22% |
| 10 | 655 | 65 | 496 | 43 | 24% |

Table 11.1. Performance comparison between the resynchronized solution and the original synchronization graph for the example of Figure 11.3.

*mailboxes* (which are FIFO queues of events between two processes). Time delays are specified by the function *hold*. Holding for an appropriate time causes the process to be put into an event queue, and the process "wakes up" when the simulation time has advanced by the amount specified by the hold statement. Passage of time is modeled in this fashion. In addition, CSIM allows specification of *facilities*, which can be accessed by only one process at a time. Mutual exclusion of access to shared resources is modeled in this fashion.

For the multiprocessor simulation, each processor is made into a process, and synchronization is attained by sending and receiving messages from mailboxes. The shared bus is made into a facility. Polling of the mailbox for checking the presence of data is done by first reserving the bus, then checking for the message count on that particular mailbox; if the count is greater than zero, data can be read from shared memory, or else the processor backs off for a certain duration, and then resumes polling.

When a processor sends data, it increments a counter in shared memory, and then writes the data value. When a processor receives, it first polls the corresponding counter, and if the counter is nonzero, it proceeds with the read; otherwise, it backs off for some time and then polls the counter again. Experimentally determined back-off times are used for each value of the memory access time. For a send, the processor checks if the corresponding buffer is full or not. For the simulation, all buffers are sized equal to 5; these sizes can of course be jointly minimized to reduce buffer memory. Polling time is defined as the time required to access the bus and check the counter value.

## 11.6    Chainable Synchronization Graphs

In this section, it is shown that although optimal resynchronization is intractable for general synchronization graphs, a broad class of synchronization graphs exists for which optimal resynchronizations can be computed using an efficient polynomial-time algorithm.

### 11.6.1    Chainable Synchronization Graph SCCs

**Definition 11.4:** Suppose that $C$ is an SCC in a synchronization graph $G$, and $x$ is an actor in $C$. Then $x$ is an **input hub** of $C$ if for each feedforward synchronization edge $e$ in $G$ whose sink actor is in $C$, we have $\rho_C(x, snk(e)) = 0$. Similarly, $x$ is an **output hub** of $C$ if for each feedforward synchronization edge $e$ in $G$ whose source actor is in $C$, we have $\rho_C(src(e), x) = 0$. We say that $C$ is **linkable** if there exist actors $x, y$ in $C$ such that $x$ is an input hub, $y$ is an output hub, and $\rho_C(x, y) = 0$. A synchronization graph is **chainable** if each SCC is linkable.

For example, consider the SCC in Figure 11.9(a), and assume that the dashed edges represent the synchronization edges that connect this SCC with other SCCs. This SCC has exactly one input hub, actor $A$ , and exactly one output hub, actor $F$, and since $\rho(A, F) = 0$ , it follows that the SCC is linkable. However, if we remove the edge $(C, F)$, then the resulting graph (shown in Figure 11.9(b)) is not linkable since it does not have an output hub. A class of linkable SCCs that occur commonly in practical synchronization graphs are those SCCs that correspond to only one processor, such as the SCC shown in Figure 11.9(c). In such cases, the first actor executed on the processor is always an input hub and the last actor executed is always an output hub.

In the remainder of this section, we assume that for each linkable SCC, an input hub $x$ and output hub $y$ are selected such that $\rho(x, y) = 0$, and these

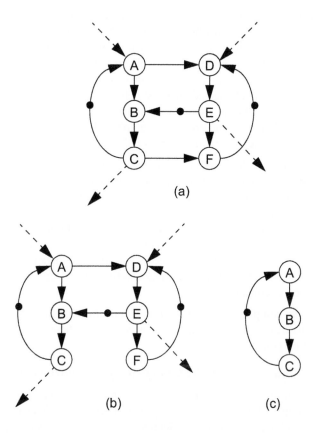

Figure 11.9. An illustration of input and output hubs for synchronization graph.

actors are referred to as the **selected input hub** and the **selected output hub** of the associated SCC. Which input hub and output hub are chosen as the "selected" ones makes no difference to our discussion of the techniques in this section as long they are selected so that $\rho(x, y) = 0$.

An important property of linkable synchronization graphs is that if $C_1$ and $C_2$ are distinct linkable SCCs, then all synchronization edges directed from $C_1$ to $C_2$ are subsumed by the single ordered pair $(l_1, l_2)$, where $l_1$ denotes the selected output hub of $C_1$ and $l_2$ denotes the selected input hub of $C_2$. Furthermore, if there exists a path between two SCCs $C_1', C_2'$ of the form $((o_1, i_2), (o_2, i_3), ..., (o_{n-1}, i_n))$, where $o_1$ is the selected output hub of $C_1'$, $i_n$ is the selected input hub of $C_2'$, and there exist distinct SCCs $Z_1, Z_2, ..., Z_{n-2} \notin \{C_1', C_2'\}$ such that for $k = 2, 3, ..., (n-1)$, $i_k, o_k$ are respectively the selected input hub and the selected output hub of $Z_{k-1}$, then all synchronization edges between $C_1'$ and $C_2'$ are redundant.

From these properties, an optimal resynchronization for a chainable synchronization graph can be constructed efficiently by computing a topological sort of the SCCs, instantiating a zero delay synchronization edge from the selected output hub of the $i$ th SCC in the topological sort to the selected input hub of the $(i+1)$ th SCC, for $i = 1, 2, ..., (n-1)$, where $n$ is the total number of SCCs, and then removing all of the redundant synchronization edges that result. For example, if this algorithm is applied to the chainable synchronization graph of Figure 11.10(a), then the synchronization graph of Figure 11.10(b) is obtained, and the number of synchronization edges is reduced from 4 to 2.

This chaining technique can be viewed as a form of pipelining, where each SCC in the output synchronization graph corresponds to a pipeline stage. As discussed in Chapter 6, pipelining can be used to increase the throughput in multiprocessor DSP implementations through improved parallelism. However, in the form of pipelining that is associated with chainable synchronization graphs, the load of each processor is unchanged, and the estimated throughput is not affected (since no new cyclic paths are introduced), and thus, the benefit to the *overall* throughput of the chaining technique arises chiefly from the optimal reduction of synchronization overhead.

The time-complexity of the optimal algorithm discussed above for resynchronizing chainable synchronization graphs is $O(v^2)$, where $v$ is the number of synchronization graph actors.

## 11.6.2      Comparison to the Global-Resynchronize Heuristic

It is easily verified that the original synchronization graph for the music synthesis example of Section 11.5.2, shown in Figure 11.3, is chainable. Thus, the chaining technique presented in Section 11.6.1 is guaranteed to produce an optimal resynchronization for this example, and since no feedback synchronization edges are present, the number of synchronization edges in the resynchro-nized solution is guaranteed to be equal to one less than the number of SCCs in the original synchronization graph; that is, the optimized synchronization graph contains $6 - 1 = 5$ synchronization edges. From Figure 11.8, we see that this is precisely the number of synchronization edges in the synchronization graph that results from the implementation of Algorithm Global-resynchronize that was dis-cussed in Section 11.5.4.

However, Algorithm Global-resynchronize does not always produce opti-mal results for chainable synchronization graphs. For example, consider the syn-chronization graph shown in Figure 11.11(a), which corresponds to an eight-processor schedule in which each of the following subsets of actors are assigned to a separate processor — $\{I\}$, $\{J\}$, $\{G, K\}$, $\{C, H\}$, $\{D\}$, $\{E, L\}$, $\{A, F\}$, and $\{B\}$. The dashed edges are synchronization edges, and the remaining edges connect actors that are assigned to the same processor. The total number of syn-chronization edges is 14. Now it is easily verified that actor $K$ is both an input hub and an output hub for the SCC $\{C, G, H, J, K\}$, and similarly, actor $L$ is both an input and output hub for the SCC $\{A, D, E, F, L\}$. Thus, we see that the overall synchronization graph is chainable. It is easily verified that the chaining technique developed in Section 11.6.1 uniquely yields the optimal resynchroniza-

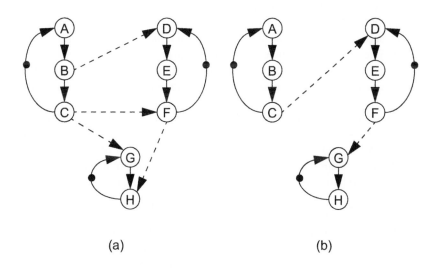

(a)                                    (b)

Figure 11.10. An illustration of an algorithm for optimal resynchronization of chain-able synchronization graphs. The dashed edges are synchronization edges.

tion illustrated in Figure 11.11(b), which contains only 11 synchronization edges.

In contrast, the quality of the resynchronization obtained for Figure 11.11(a) by Algorithm Global-resynchronize depends on the order in which the actors are traversed by each of the two nested **for** loops in Figure 11.5. For example, if both loops traverse the actors in alphabetical order, then Global-resynchronize obtains the suboptimal solution shown in Figure 11.11(c), which contains 12 synchronization edges.

However, actor traversal orders exist for which Global-resynchronize achieves optimal resynchronizations of Figure 11.11(a). One such ordering is

$$K, D, C, B, E, F, G, H, I, J, L, A; \tag{11-23}$$

if both **for** loops traverse the actors in this order, then Global-resynchronize yields the same resynchronized graph that is computed uniquely by the chaining technique of Section 11.6.1 (Figure 11.11(b)). It is an open question whether or not given an arbitrary chainable synchronization graph, actor traversal orders always exist with which Global-resynchronize arrives at optimal resynchronizations. Furthermore, even if such traversal orders are always guaranteed to exist, it is doubtful that they can, in general, be computed efficiently.

### 11.6.3    A Generalization of the Chaining Technique

The chaining technique developed in Section 11.6.1 can be generalized to optimally resynchronize a somewhat broader class of synchronization graphs. This class consists of all synchronization graphs for which each source SCC has an output hub (but not necessarily an input hub), each sink SCC has an input hub (but not necessarily an output hub), and each internal SCC is linkable. In this case, the internal SCCs are pipelined as in the previous algorithm, and then for each source SCC, a synchronization edge is inserted from one of its output hubs to the selected input hub of the first SCC in the pipeline of internal SCCs, and for each sink SCC, a synchronization edge is inserted to one of its input hubs from the selected output hub of the last SCC in the pipeline of internal SCCs. If there are no internal SCCs, then the sink SCCs are pipelined by selecting one input hub from each SCC, and joining these input hubs with a chain of synchronization edges. Then a synchronization edge is inserted from an output hub of each source SCC to an input hub of the first SCC in the chain of sink SCCs.

### 11.6.4    Incorporating the Chaining Technique

In addition to guaranteed optimality, another important advantage of the chaining technique for chainable synchronization graphs is its relatively low time-complexity ($O(v^2)$ versus $O(sv^4)$ for Global-resynchronize), where $v$ is the number of synchronization graph actors, and $s$ is the number of feedforward synchronization edges. The primary disadvantage is, of course, its restricted

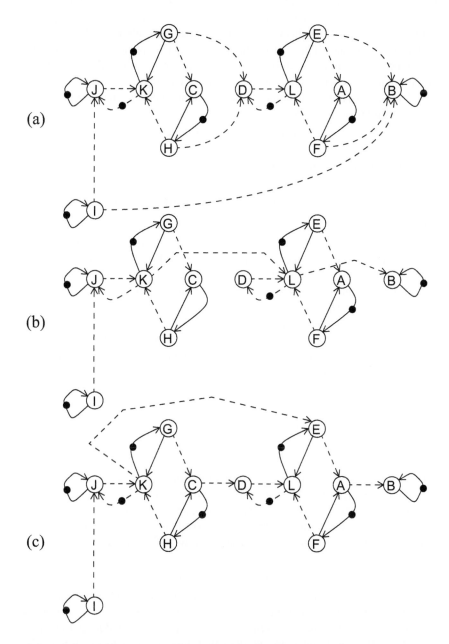

Figure 11.11. A chainable synchronization graph for which Algorithm Global-resynchronize fails to produce an optimal solution.

applicability. An obvious solution is to first check if the general form of the chaining technique (described above in Section 11.6.3) can be applied, apply the chaining technique if the check returns an affirmative result, or apply Algorithm Global-resynchronize if the check returns a negative result. The check must determine whether or not each source SCC has an output hub, each sink SCC has an input hub, and each internal SCC is linkable. This check can be performed in $O(n^3)$ time, where $n$ is the number of actors in the input synchronization graph, using a straightforward algorithm.

A useful direction for further investigation is a deeper integration of the chaining technique with algorithm Global-resynchronize for general (not necessarily chainable) synchronization graphs.

## 11.7     Resynchronization of Constraint Graphs for Relative Scheduling

Filo, Ku and De Micheli have studied synchronization rearrangement in the context of minimizing the controller area for hardware synthesis of synchronization digital circuitry [FKM92, FKCM93] and significant differences in the underlying analytical models prevent these techniques from applying to the context of self-timed HSDF implementation. In the graphical hardware model of [FKM92], called the **constraint graph** model, each vertex corresponds to a separate hardware device and edges have arbitrary weights that specify sequencing constraints. When the source vertex has bounded execution time, a positive weight $w(e)$ (*forward constraint*) imposes the constraint

$$start(snk(e)) \geq w(e) + start(src(e)),\qquad\qquad(11\text{-}24)$$

while a negative weight (*backward constraint*) implies

$$start(snk(e)) \leq w(e) + start(src(e)).\qquad\qquad(11\text{-}25)$$

If the source vertex has unbounded execution time, the forward and backward constraints are relative to the *completion* time of the source vertex. In contrast, in the synchronization graph model, multiple actors can reside on the same processing element (implying zero synchronization cost between them), and the timing constraints always correspond to the case where $w(e)$ is positive and equal to the execution time of $src(e)$.

The implementation models, and associated implementation cost functions are also significantly different. A constraint graph is implemented using a scheduling technique called **relative scheduling** [KM92], which can roughly be viewed as intermediate between self-timed and fully-static scheduling. In relative scheduling, the constraint graph vertices that have unbounded execution time, called **anchors**, are used as reference points against which all other vertices are scheduled: for each vertex $v$, an offset $f_i$ is specified for each anchor $a_i$ that affects the activation of $v$, and $v$ is scheduled to occur once $f_i$ clock cycles have elapsed from the completion of $a_i$, for each $i$.

In the implementation of a relative schedule, each anchor has attached control circuitry that generates offset signals, and each vertex has a synchronization circuit that asserts an *activate* signal when all relevant offset signals are present. The resynchronization optimization is driven by a cost function that estimates the total area of the synchronization circuitry, where the offset circuitry area estimate for an anchor is a function of the maximum offset, and the synchronization circuitry estimate for a vertex is a function of the number of offset signals that must be monitored.

As a result of the significant differences in both the scheduling models and the implementation models, the techniques developed for resynchronizing constraint graphs do not extend in any straightforward manner to the resynchronization of synchronization graphs for self-timed multiprocessor implementation, and the solutions that we have discussed for synchronization graphs are significantly different in structure from those reported in [FKM92]. For example, the fundamental relationships that have been established between set covering and the resynchronization of self-timed HSDF schedules have not emerged in the context of constraint graphs.

## 11.8    Summary

This chapter has discussed a post-optimization called resynchronization for self-timed, multiprocessor implementations of DSP algorithms. The goal of resynchronization is to introduce new synchronizations in such a way that the number of additional synchronizations that become redundant exceeds the number of new synchronizations that are added, and thus the net synchronization cost is reduced.

It was shown that optimal resynchronization is intractable by deriving a reduction from the classic set-covering problem. However, a broad class of systems was defined for which optimal resynchronization can be performed in polynomial time. This chapter also discussed a heuristic algorithm for resynchronization of general systems that emerges naturally from the correspondence to set covering. The performance of an implementation of this heuristic was demonstrated on a multiprocessor schedule for a music synthesis system. The results demonstrate that the heuristic can efficiently reduce synchronization overhead and improve throughput significantly.

# 12

# LATENCY-CONSTRAINED
# RESYNCHRONIZATION

Chapter 11 introduced the concept of resynchronization, a post-optimization for static multiprocessor schedules in which extraneous synchronization operations are introduced in such a way that the number of original synchronizations that consequently become *redundant* significantly exceeds the number of additional synchronizations. Redundant synchronizations are synchronization operations whose corresponding sequencing requirements are enforced completely by other synchronizations in the system. The amount of run-time overhead required for synchronization can be reduced significantly by eliminating redundant synchronizations [Sha89, BSL97]. Thus, effective resynchronization reduces the net synchronization overhead in the implementation of a multiprocessor schedule, and improves the overall throughput.

However, since additional serialization is imposed by the new synchronizations, resynchronization can produce significant increase in latency. In Chapter 11, we discussed fundamental properties of resynchronization and we studied the problem of optimal resynchronization under the assumption that arbitrary increases in latency can be tolerated ("maximum-throughput resynchronization"). Such an assumption is valid, for example, in a wide variety of simulation applications. This chapter discusses the problem of computing an optimal resynchronization among all resynchronizations that do not increase the latency beyond a prespecified upper bound $L_{max}$. This study of resynchronization is based in the context of self-timed execution of iterative dataflow specifications, which is an implementation model that has been applied extensively for digital signal processing systems.

Latency constraints become important in interactive applications such as video conferencing, games, and telephony, where latency beyond a certain point becomes annoying to the user. This chapter demonstrates how to obtain the benefits of resynchronization while maintaining a specified latency constraint

## 12.1    Elimination of Synchronization Edges

This section introduces a number of useful properties that pertain to the process by which resynchronization can make certain synchronization edges in the original synchronization graph become redundant. The following definition is fundamental to these properties.

**Definition 12.1:** If $G$ is a synchronization graph, $s$ is a synchronization edge in $G$ that is not redundant, $R$ is a resynchronization of $G$, and $s$ is not contained in $R$, then we say that $R$ **eliminates** $s$. If $R$ eliminates $s$, $s' \in R$, and there is a path $p$ from $src(s)$ to $snk(s)$ in $\Psi(R, G)$ such that $p$ contains $s'$ and $Delay(p) \leq delay(s)$, then we say that $s'$ **contributes to the elimination of** $s$.

A synchronization edge $s$ can be eliminated if a resynchronization creates a path $p$ from $src(s)$ to $snk(s)$ such that $Delay(p) \leq delay(s)$. In general, the path $p$ may contain more than one resynchronization edge, and thus, it is possible that none of the resynchronization edges allows us to eliminate $s$ "by itself." In such cases, it is the contribution of all of the resynchronization edges within the path $p$ that enables the elimination of $s$. This motivates the choice of terminology in Definition 12.1. An example is shown in Figure 12.1.

The following two facts follow immediately from Definition 12.1.

**Fact 12.1:** Suppose that $G$ is a synchronization graph, $R$ is a resynchronization of $G$, and $r$ is a resynchronization edge in $R$. If $r$ does not contribute to the elimination of any synchronization edges, then $(R - \{r\})$ is also a resynchronization of $G$. If $r$ contributes to the elimination of one and only one synchronization edge $s$, then $(R - \{r\} + \{s\})$ is a resynchronization of $G$.

**Fact 12.2:** Suppose that $G$ is a synchronization graph, $R$ is a resynchronization of $G$, $s$ is a synchronization edge in $G$, and $s'$ is a resynchronization edge in $R$ such that $delay(s') > delay(s)$. Then $s'$ does not contribute to the elimination of $s$.

For example, let $G$ denote the synchronization graph in Figure 12.2(a). Figure 12.2(b) shows a resynchronization $R$ of $G$. In the resynchronized graph of Figure 12.2(b), the resynchronization edge $(x_4, y_3)$ does not contribute to the elimination of any of the synchronization edges of $G$, and thus Fact 12.1 guarantees that $R' \equiv R - \{(x_4, y_3)\}$, illustrated in Figure 12.2(c), is also a resynchronization of $G$. In Figure 12.2(c), it is easily verified that $(x_5, y_4)$ contributes to the elimination of exactly one synchronization edge — the edge $(x_5, y_5)$, and from Fact 12.1, we have that $R'' \equiv R' - \{(x_5, y_4)\} + \{(x_5, y_5)\}$, illustrated in Figure

12.2(d), is a also resynchronization of $G$.

## 12.2     Latency-Constrained Resynchronization

As discussed in Section 11.2, resynchronization cannot decrease the estimated throughput since it manipulates only the feedforward edges of a synchro-

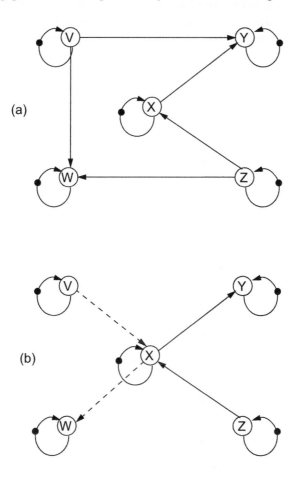

Figure 12.1. An illustration of Definition 12.1. Here each processor executes a single actor. A resynchronization of the synchronization graph in (a) is illustrated in (b). In this resynchronization, the resynchronization edges $(V, X)$ and $(X, W)$ both contribute to the elimination of $(V, W)$.

nization graph. Frequently in real-time DSP systems, latency is also an important issue, and although resynchronization does not degrade the estimated throughput, it generally does increase the latency. This section defines the *latency-constrained resynchronization problem* for self-timed multiprocessor systems.

**Definition 12.2:** Suppose $G_0$ is an application graph, $G$ is a synchronization graph that results from a multiprocessor schedule for $G_0$, $x$ is an execution source (an actor that has no input edges or has nonzero delay on all input edges) in $G$, and $y$ is an actor in $G$ other than $x$. We define the **latency** from $x$ to $y$ by

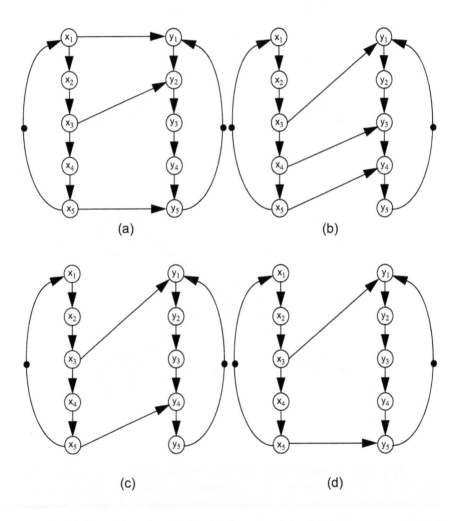

Figure 12.2. Properties of resynchronization.

$L_G(x, y) \equiv end(y, 1 + \rho_{G_0}(x, y))$ [1]. We refer to $x$ as the **latency input** associated with this measure of latency, and we refer to $y$ as the **latency output**.

Intuitively, the latency is the time required for the first invocation of the latency input to influence the associated latency output, and thus the latency corresponds to the critical path in the dataflow implementation to the first output invocation that is influenced by the input. This interpretation of the latency as the critical path is widely used in VLSI signal processing [Kun88, Mad95].

In general, the latency can be computed by performing a simple simulation of the ASAP execution for $G$ through the $(1 + \rho_{G_0}(x, y))$ th execution of $y$. Such a simulation can be performed as a functional simulation of an HSDFG ?Gsim? that has the same topology (vertices and edges) as $G$, and that maintains the simulation time of each processor in the values of data tokens. Each initial token (delay) in ?Gsim? is initialized to have the value 0, since these tokens are all present at time 0. Then, a data-driven simulation of ?Gsim? is carried out. In this simulation, an actor may execute whenever it has sufficient data, and the value of the output token produced by the invocation of any actor $z$ in the simulation is given by

$$max(\{v_1, v_2, ..., v_n\}) + t(z), \tag{12-1}$$

where $\{\{v_1, v_2, ..., v_n\}\}$ is the set of token values consumed during the actor execution. In such a simulation, the $i$ th token value produced by an actor $z$ gives the completion time of the $i$ th invocation of $z$ in the ASAP execution of $G$. Thus, the latency can be determined as the value of the $(1 + \rho_{G_0}(x, y))$ th output token produced by $y$. With careful implementation of the functional simulator described above, the latency can be determined in $O(d \times max(\{|V|, s\}))$ time, where $d = 1 + \rho_{G_0}(x, y)$, and $s$ denotes the number of synchronization edges in $G$. The simulation approach described above is similar to approaches described in [TTL95].

For a broad class of synchronization graphs, latency can be analyzed even more efficiently during resynchronization. This is the class of synchronization graphs in which the first invocation of the latency output is influenced by the first invocation of the latency input. Equivalently, it is the class of graphs that contain at least one delayless path in the corresponding application graph directed from the latency input to the latency output. For transparent synchronization graphs, we can directly apply well-known longest-path based techniques for computing latency.

---

1. Recall from Chapter 5 that $start(v, k)$ and $end(v, k)$ denote the time at which invocation $k$ of actor $v$ commences and completes execution. Also, note that $start(x, 1) = 0$ since $x$ is an execution source.

**Definition 12.3:** Suppose that $G_0$ is an application graph, $x$ is a source actor in $G_0$, and $y$ is an actor in $G_0$ that is not identical to $x$. If $\rho_{G_0}(x, y) = 0$, then we say that $G_0$ is **transparent** with respect to latency input $x$ and latency output $y$. If $G$ is a synchronization graph that corresponds to a multiprocessor schedule for $G_0$, we also say that $G$ is **transparent**.

If a synchronization graph is transparent with respect to a latency input/output pair, then the latency can be computed efficiently using longest path calculations on an *acyclic* graph that is derived from the input synchronization graph $G$. This acyclic graph, which we call the **first-iteration graph** of $G$, denoted $fi(G)$, is constructed by removing all edges from $G$ that have nonzero-delay; adding a vertex $\upsilon$, which represents the beginning of execution; setting $t(\upsilon) = 0$; and adding delayless edges from $\upsilon$ to each source actor (other than $\upsilon$) of the partial construction until the only source actor that remains is $\upsilon$. Figure 12.3 illustrates the derivation of $fi(G)$.

Given two vertices $x$ and $y$ in $fi(G)$ such that there is a path in $fi(G)$ from $x$ to $y$, we denote the sum of the execution times along a path from $x$ to $y$ that has maximum cumulative execution time by $T_{fi(G)}(x, y)$. That is,

$$T_{fi(G)}(x, y) = max\left( \sum_{p \text{ traverses } z} t(z) \middle| (p \text{ is a path from } x \text{ to } y \text{ in } fi(G)) \right). \quad (12\text{-}2)$$

If there is no path from $x$ to $y$, then we define $T_{fi(G)}(x, y)$ to be $-\infty$. Note that for all $x, y$ $T_{fi(G)}(x, y) < +\infty$, since $fi(G)$ is acyclic. The values $T_{fi(G)}(x, y)$ for

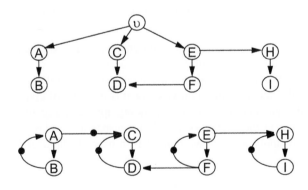

Figure 12.3. An example used to illustrate the construction of $fi(G)$. The graph on the bottom is $fi(G)$ if $G$ is the top graph.

all pairs $x, y$ can be computed in $O(n^3)$ time, where $n$ is the number of actors in $G$, by using a simple adaptation of the Floyd-Warshall algorithm described in Section 3.13.3.

**Fact 12.3:** Suppose that $G_0$ is a an HSDFG that is transparent with respect to latency input $x$ and latency output $y$, $G_s$ is the synchronization graph that results from a multiprocessor schedule for $G_0$, and $G$ is a resynchronization $G_s$. Then $\rho_G(x, y) = 0$, and thus $T_{fi(G)}(x, y) \geq 0$ (i.e., $T_{fi(G)}(x, y) \neq -\infty$).

*Proof:* Since $G_0$ is transparent, there is a delayless path $p$ in $G_0$ from $x$ to $y$. Let $(u_1, u_2, ..., u_n)$, where $x = u_1$ and $y = u_n$, denote the sequence of actors traversed by $p$. From the semantics of the HSDFG $G_0$, it follows that for $1 \leq i < n$, either $u_i$ and $u_{i+1}$ execute on the same processor, with $u_i$ scheduled earlier than $u_{i+1}$, or there is a zero-delay synchronization edge in $G_s$ directed from $u_i$ to $u_{i+1}$. Thus, for $1 \leq i < n$, we have $\rho_{G_s}(u_i, u_{i+1}) = 0$, and thus, that $\rho_{G_s}(x, y) = 0$. Since $G$ is a resynchronization of $G_s$, it follows from Lemma 11.1 that $\rho_G(x, y) = 0$. *QED.*

The following theorem gives an efficient means for computing the latency $L_G$ for transparent synchronization graphs.

**Theorem 12.1:** Suppose that $G$ is a synchronization graph that is transparent with respect to latency input $x$ and latency output $y$. Then $L_G(x, y) = T_{fi(G)}(\upsilon, y)$.

*Proof:* By induction, we show that for every actor $w$ in $fi(G)$,

$$end(w, 1) = T_{fi(G)}(\upsilon, w), \tag{12-3}$$

which clearly implies the desired result.

First, let $mt(w)$ denote the maximum number of actors that are traversed by a path in $fi(G)$ (over all paths in $fi(G)$) that starts at $\upsilon$ and terminates at $w$. If $mt(w) = 1$, then clearly $w = \upsilon$. Since both the LHS and RHS of (12-3) are identically equal to $t(\upsilon) = 0$ when $w = \upsilon$, we have that (12-3) holds whenever $mt(w) = 1$.

Now suppose that (12-3) holds whenever $mt(w) \leq k$, for some $k \geq 1$, and consider the scenario $mt(w) = k + 1$. Clearly, in the self-timed (ASAP) execution of $G$, invocation $w_1$, the first invocation of $w$, commences as soon as all invocations in the set

$$Z = \{z_1 | (z \in P_w)\}$$

have completed execution, where $z_1$ denotes the first invocation of actor $z$, and $P_w$ is the set of predecessors of $w$ in $fi(G)$. All members $z \in P_w$ satisfy $mt(z) \leq k$, since otherwise $mt(w)$ would exceed $(k + 1)$. Thus, from the induction hypothesis, we have

$$start(w, 1) = max(end(z, 1)|(z \in P_w)) = max(T_{fi(G)}(\upsilon, z)|(z \in P_w)),$$

which implies that

$$end(w, 1) = max(T_{fi(G)}(\upsilon, z)|(z \in P_w)) + t(w). \qquad (12\text{-}4)$$

But, by definition of $T_{fi(G)}$, the RHS of (12-4) is clearly equal to $T_{fi(G)}(\upsilon, w)$, and thus we have that $end(w, 1) = T_{fi(G)}(\upsilon, w)$.

We have shown that (12-3) holds for $mt(w) = 1$, and that whenever it holds for $mt(w) = k \geq 1$, it must hold for $mt(w) = (k+1)$. Thus, (12-3) holds for all values of $mt(w)$. *QED.*

In the context of resynchronization, the main benefit of transparent synchronization graphs is that the change in latency induced by adding a new synchronization edge (a "resynchronization operation") can be computed in $O(1)$ time, given $T_{fi(G)}(a, b)$ for all actor pairs $(a, b)$. We will discuss this further in Section 12.5.

Since many practical application graphs contain delayless paths from input to output and these graphs admit a particularly efficient means for computing latency, the first implementation of latency-constrained resynchronization was targeted to the class of transparent synchronization graphs [BSL96a]. However, the overall resynchronization framework described in this chapter does not depend on any particular method for computing latency, and thus, it can be fully applied to general graphs (with a moderate increase in complexity) using the ASAP simulation approach mentioned above. This framework can also be applied to subclasses of synchronization graphs other than transparent graphs for which efficient techniques for computing latency are discovered.

**Definition 12.4:** An instance of the **latency-constrained resynchronization problem** consists of a synchronization graph $G$ with latency input $x$ and latency output $y$, and a *latency constraint* $L_{max} \geq L_G(x, y)$. A solution to such an instance is a resynchronization $R$ such that 1) $L_{\Psi(R, G)}(x, y) \leq L_{max}$, and 2) no resynchronization of $G$ that results in a latency less than or equal to $L_{max}$ has smaller cardinality than $R$.

Given a synchronization graph $G$ with latency input $x$ and latency output $y$, and a latency constraint $L_{max}$, we say that a resynchronization $R$ of $G$ is a **latency-constrained resynchronization (LCR)** if $L_{\Psi(R, G)}(x, y) \leq L_{max}$. Thus, the latency-constrained resynchronization problem is the problem of determining a minimal LCR.

Generally, resynchronization can be viewed as complementary to the *Convert-to-SC-graph* optimization defined in Chapter 11: resynchronization is performed first, followed by *Convert-to-SC-graph*. Under severe latency constraints, it may not be possible to accept the solution computed by *Convert-to-SC-graph*, in which case the feedforward edges that emerge from the resynchronized solution must be implemented with FFS. In such a situation, *Convert-to-SC-graph* can be attempted on the original (before resynchronization) graph to see if it achieves a better result than resynchronization without *Convert-to-SC-graph*. However, for transparent synchronization graphs that have only one source SCC *and* only one sink SCC, the latency is not affected by *Convert-to-SC-graph*, and thus, for such systems, resynchronization and *Convert-to-SC-graph* are fully complementary. This is fortunate since such systems arise frequently in practice.

Trade-offs between latency and throughput have been studied by Potkonjac and Srivastava in the context of transformations for dedicated implementation of linear computations [PS94]. Because this work is based on synchronous implementations, it does not address the synchronization issues and opportunities that we encounter in the self-timed dataflow context.

## 12.3    Intractability of LCR

This section shows that the latency-constrained resynchronization problem is NP-hard even for the very restricted subclass of synchronization graphs in which each SCC corresponds to a single actor, and all synchronization edges have zero delay.

As with the maximum-throughput resynchronization problem, discussed in Chapter 11, the intractability of this special case of latency-constrained resynchronization can be established by a reduction from set-covering. To illustrate this reduction, we suppose that we are given the set $X = \{x_1, x_2, x_3, x_4\}$, and the family of subsets $T = \{t_1, t_2, t_3\}$, where $t_1 = \{x_1, x_3\}$, $t_2 = \{x_1, x_2\}$, and $t_3 = \{x_2, x_4\}$. Figure 12.4 illustrates the instance of latency-constrained resynchronization that we derive from the instance of set-covering specified by $(X, T)$. Here, each actor corresponds to a single processor and the self-loop edge for each actor is not shown. The numbers beside the actors specify the actor execution times, and the latency constraint is $L_{max} = 103$. In the graph of Figure 12.4, which we denote by $G$, the edges labeled $ex_1, ex_2, ex_3, ex_4$ correspond respectively to the members $x_1, x_2, x_3, x_4$ of the set $X$ in the set-covering instance, and the vertex pairs (resynchronization candidates) $(v, st_1), (v, st_2), (v, st_3)$ correspond to the members of $T$. For each relation $x_i \in t_j$, an edge exists that is directed from $st_j$ to $sx_i$. The latency input and latency output are defined to be *in* and *out* respectively, and it is assumed that $G$ is transparent.

The synchronization graph that results from an optimal resynchronization of $G$ is shown in Figure 12.5, with redundant resynchronization edges removed. Since the resynchronization candidates $(v, st_1), (v, st_3)$ were chosen to obtain the solution shown in 12.5, this solution corresponds to the solution of $(X, T)$ that consists of the subfamily $\{t_1, t_3\}$.

A correspondence between the set-covering instance $(X, T)$ and the instance of latency-constrained resynchronization defined by Figure 12.4 arises from two properties of the construction described above:

**Observation 5:**

$(x_i \in t_j$ in the set-covering instance $) \Leftrightarrow ((v, st_j)$ subsumes $ex_i$ in $G)$

**Observation 6:** If $R$ is an optimal LCR of $G$, then each resynchronization edge in $R$ is of the form

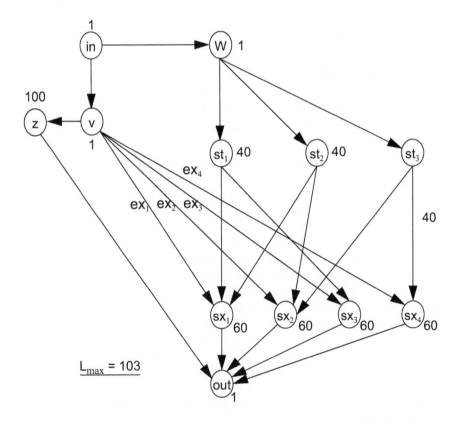

Figure 12.4. An instance of latency-constrained resynchronization that is derived from an instance of the set-covering problem.

$$(v, st_i), i \in \{1, 2, 3\}, \text{ or of the form } (st_j, sx_i), x_i \notin t_j. \tag{12-5}$$

The first observation is immediately apparent from inspection of Figure 12.4. A proof of the second observation follows.

*Proof of Observation 6*: We must show that no other resynchronization edges can be contained in an optimal LCR of $G$. Figure 12.6 specifies arguments with which we can discard all possibilities other than those given in (12-5). In the matrix shown in Figure 12.6(a), each entry specifies an index into the list of arguments given in Figure 12.6(b). For each of the six categories of arguments,

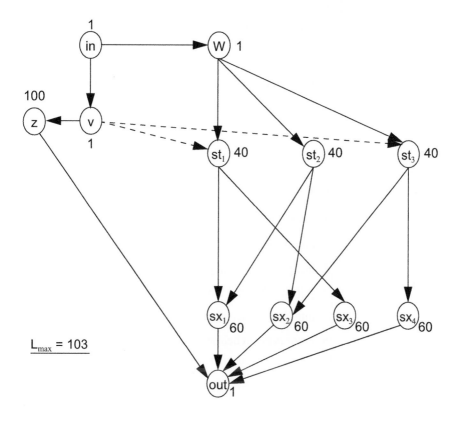

Figure 12.5.    The synchronization graph that results from a solution to the instance of latency-constrained resynchronization shown in Figure 12.4.

(a)

|  | $v$ | $w$ | $z$ | $in$ | $out$ | $st_i$ | $sx_i$ |
|---|---|---|---|---|---|---|---|
| $v$ | 5 | 3 | 1 | 2 | 4 | OK | 1 |
| $w$ | 3 | 5 | 6 | 2 | 4 | 1 | 4 |
| $z$ | 2 | 3 | 5 | 2 | 1 | 3 | 3 |
| $in$ | 1 | 1 | 4 | 5 | 4 | 4 | 4 |
| $out$ | 2 | 2 | 2 | 2 | 5 | 2 | 2 |
| $st_j$ | 3 | 2 | 3 | 2 | 4 | 3/5 | OK[a] |
| $sx_j$ | 2 | 2 | 3 | 2 | 1 | 2/3 | 3/5 |

a. Assuming that $x_j \notin t_i$; otherwise 1 applies.

(b)

Legend of entries in the table above

1. Exists in $G$.

2. Introduces a cycle.

3. Increases the latency beyond $L_{max}$.

4. $\rho_G(a_1, a_2) = 0$ (Lemma 11.2).

5. Introduces a delayless self-loop.

6. Proof is given below.

Figure 12.6. Arguments that support Observation 6.

except for #6, the reasoning is either obvious or easily understood from inspection of Figure 12.4. A proof of argument #6 follows shortly within this same section.

For example, edge $(v, z)$ cannot be a resynchronization edge in $R$ because the edge already exists in the original synchronization graph; an edge of the form $(sx_j, w)$ cannot be in $R$ because there is a path in $G$ from $w$ to each $sx_i$; $(z, w) \notin R$ since otherwise there would be a path from $in$ to $out$ that traverses $v, z, w, st_1, sx_1$, and thus, the latency would be increased to at least $204$; $(in, z) \notin R$ from Lemma 11.2 since $\rho_G(in, z) = 0$; and $(v, v) \notin R$ since otherwise there would be a delayless self-loop. Three of the entries in Figure 12.6 point to multiple argument categories. For example, if $x_j \in t_i$, then $(sx_j, st_i)$ introduces a cycle, and if $x_j \notin t_i$ then $(sx_j, st_i)$ cannot be contained in $R$ because it would increase the latency beyond $L_{max}$.

The entries in Figure 12.6 marked $OK$ are simply those that correspond to (12-5), and thus we have justified Observation 6. $QED.$

In the proof of Observation 6, we deferred the proof of Argument #6 for Figure 12.6. A proof of this argument follows.

*Proof of Argument #6 in Figure 12.6:* By contraposition, we show that $(w, z)$ cannot contribute to the elimination of any synchronization edge of $G$, and thus, from Fact 12.1, it follows from the optimality of $R$ that $(w, z) \notin R$. Suppose that $(w, z)$ contributes to the elimination of some synchronization edge $s$. Then

$$\rho_{\tilde{G}}(src(s), w) = \rho_{\tilde{G}}(z, snk(s)) = 0, \tag{12-6}$$

where

$$\tilde{G} \equiv \Psi(R, G). \tag{12-7}$$

From the matrix in Figure 12.6, we see that no resynchronization edge can have $z$ as the source vertex. Thus, $snk(s) \in \{z, out\}$. Now, if $snk(s) = z$, then $s = (v, z)$, and thus, from (12-6), there is a zero delay path from $v$ to $w$ in $\tilde{G}$. However, the existence of such a path in $\tilde{G}$ implies the existence of a path from $in$ to $out$ that traverses actors $v, w, st_1, sx_1$, which in turn implies that $L_{\tilde{G}}(in, out) \geq 104$, and thus that $R$ is not a valid LCR.

On the other hand, if $snk(s) = out$, then $src(s) \in \{z, sx_1, sx_2, sx_3, sx_4\}$. Now, from (12-6), $src(s) = z$ implies the existence of a zero delay path from $z$ to $w$ in $\tilde{G}$, which implies the existence of a path from $in$ to $out$ that traverses $v, w, z, st_1, sx_1$, which in turn implies that $L_{max} \geq 204$. On the other hand, if $src(s) = sx_i$ for some $i$, then since from Figure 12.6, there are no resynchronization edges that have an $sx_i$ as the source, it follows from (12-6) that there must be a zero delay path in $\tilde{G}$ from $out$ to $w$. The existence of such a path, however, implies the existence of a cycle in $\tilde{G}$ since $\rho_G(w, out) = 0$. Thus, $snk(s) = out$ implies that $R$ is not an LCR. QED.

The following observation states that a resynchronization edge of the form $(st_j, sx_i)$ contributes to the elimination of exactly one synchronization edge, which is the edge $ex_i$.

**Observation 7:** Suppose that $R$ is an optimal LCR of $G$ and suppose that $e = (st_j, sx_i)$ is a resynchronization edge in $R$, for some $i \in \{1, 2, 3, 4\}, j \in \{1, 2, 3\}$ such that $x_i \notin t_j$. Then $e$ contributes to the elimination of one and only one synchronization edge — $ex_i$.

*Proof:* Since $R$ is an optimal LCR, we know that $e$ must contribute to the elimination of at least one synchronization edge (from Fact 12.1). Let $s$ be some synchronization edge such that $e$ contributes to the elimination of $s$. Then

$$\rho_{R(G)}(src(s), src(e)) = \rho_{R(G)}(snk(e), snk(s)) = 0. \tag{12-8}$$

Now from Figure 12.6, it is apparent that there are no resynchronization edges in $R$ that have $sx_i$ or $out$ as their source actor. Thus, from (12-8), $snk(s) = sx_i$ or $snk(s) = out$. Now, if $snk(s) = out$, then $src(s) = sx_k$ for some $k \neq i$, or $src(s) = z$. However, since no resynchronization edge has a member of $\{sx_1, sx_2, sx_3, sx_4\}$ as its source, we must (from 12-8) rule out $src(s) = sx_k$. Similarly, if $src(s) = z$, then from (12-8) there exists a zero delay path in $R(G)$ from $z$ to $st_j$, which in turn implies that $L_{R(G)}(in, out) > 140$. But this is not possible since the assumption that $R$ is an LCR guarantees that $L_{R(G)}(in, out) \leq 103$. Thus, we conclude that $snk(s) \neq out$, and thus, that $snk(s) = sx_i$.

Now $(snk(s) = sx_i)$ implies that (a) $s = ex_i$ or (b) $s = (st_k, sx_i)$ for some $k$ such that $x_i \in t_k$ (recall that $x_i \notin t_j$, and thus, that $k \neq j$). If $s = (st_k, sx_i)$, then from (12-8), $\rho_{R(G)}(st_k, st_j) = 0$. It follows that for any member $x_l \in t_j$, there is a zero delay path in $R(G)$ that traverses $st_k$, $st_j$ and $sx_l$. Thus, $s = (st_k, sx_i)$ does not hold since otherwise $L_{R(G)}(in, out) \geq 140$.

Thus, we are left with only possibility (a) — $s = ex_i$. QED.

Now, suppose that we are given an optimal LCR $R$ of $G$. From Observation 7 and Fact 12.1, we have that for each resynchronization edge $(st_j, sx_i)$ in $R$, we can replace this resynchronization edge with $ex_i$ and obtain another optimal LCR. Thus, from Observation 6, we can efficiently obtain an optimal LCR $R'$ such that all resynchronization edges in $R'$ are of the form $(v, st_i)$.

For each $x_i \in X$ such that

$$\exists t_j | ((x_i \in t_j) \textbf{ and } ((v, st_j) \in R')), \tag{12-9}$$

we have that $ex_i \notin R'$. This is because $R'$ is assumed to be optimal, and thus, $\Psi(R, G)$ contains no redundant synchronization edges. For each $x_i \in X$ for which (12-9) does not hold, we can replace $ex_i$ with any $(v, st_j)$ that satisfies $x_i \in t_j$, and since such a replacement does not affect the latency, we know that the result will be another optimal LCR for $G$. In this manner, if we repeatedly replace each $ex_i$ that does not satisfy (12-9) then we obtain an optimal LCR $R''$ such that

$$\text{each resynchronization edge in } R'' \text{ is of the form } (v, st_i), \qquad (12\text{-}10)$$

and for each $x_i \in X$, there exists a resynchronization edge $(v, t_j)$ in $R''$ such that

$$x_i \in t_j. \qquad (12\text{-}11)$$

It is easily verified that the set of synchronization edges eliminated by $R''$ is $\{ex_i | x_i \in X\}$. Thus, the set $T' \equiv \{t_j | (v, t_j) \text{ is a resynchronization edge in } R''\}$ is a cover for $X$, and the cost (number of synchronization edges) of the resynchronization $R''$ is $(N - |X| + |T'|)$, where $N$ is the number of synchronization edges in the original synchronization graph. Now, it is also easily verified (from Figure 12.4) that given an arbitrary cover $T_a$ for $X$, the resynchronization defined by

$$R_a \equiv (R'' - \{(v, t_j) | (t_j \in T')\}) + \{(v, t_j) | (t_j \in T_a)\} \qquad (12\text{-}12)$$

is also a valid LCR of $G$, and that the associated cost is $(N - |X| + |T_a|)$. Thus, it follows from the optimality of $R''$ that $T'$ must be a minimal cover for $X$, given the family of subsets $T$.

To summarize, we have shown how from the particular instance $(X, T)$ of set-covering, we can construct a synchronization graph $G$ such that from a solution to the latency-constrained resynchronization problem instance defined by $G$, we can efficiently derive a solution to $(X, T)$. This example of the reduction from set-covering to latency-constrained resynchronization is easily generalized to an arbitrary set-covering instance $(X', T')$. The generalized construction of the initial synchronization graph $G$ is specified by the steps listed in Figure 12.7.

The main task in establishing a general correspondence between latency-constrained resynchronization and set-covering is generalizing Observation 6 to apply to all constructions that follow the steps in Figure 12.7. This generalization is not conceptually difficult (although it is rather tedious) since it is easily verified that all of the arguments in Figure 12.6 hold for the general construction. Similarly, the reasoning that justifies converting an optimal LCR for the construction into an optimal LCR of the form implied by (12-10) and (12-11) extends in a straightforward fashion to the general construction.

## 12.4     Two-Processor Systems

This section shows that although latency-constrained resynchronization for transparent synchronization graphs is NP-hard, the problem becomes tractable for systems that consist of only two processors — that is, synchronization graphs in which there are two SCCs and each SCC is a simple cycle. This reveals a pattern of complexity that is analogous to the classic nonpreemptive processor scheduling problem with deterministic execution times, in which the problem is also intractable for general systems, but an efficient greedy algorithm suffices to yield optimal solutions for two-processor systems in which the execution times of all tasks are identical [Cof76, Hu61]. However, for latency-constrained resynchronization, the tractability for two-processor systems does not depend on any constraints on the task (actor) execution times. Two-processor optimality results in multiprocessor scheduling have also been reported in the context of a stochastic model for parallel computation in which tasks have random execution times and communication patterns [Nic89].

In an instance of the **two-processor latency-constrained resynchronization (2LCR) problem,** we are given a set of *source processor actors* $x_1, x_2, ..., x_p$, with associated execution times $\{t(x_i)\}$, such that each $x_i$ is the $i$ th actor scheduled on the processor that corresponds to the source SCC of the

- Instantiate actors $v, w, z, in, out$, with execution times $1$, $1$, $100$, $1$, and $1$, respectively, and instantiate all of the edges in Figure 12.4 that are contained in the subgraph associated with these five actors.
- For each $t \in T'$, instantiate an actor labeled $st$ that has execution time $40$.
- For each $x \in X'$
     Instantiate an actor labeled $sx$ that has execution time $60$.
     Instantiate the edge $ex \equiv d_0(v, sx)$.
     Instantiate the edge $d_0(sx, out)$.
- For each $t \in T'$
     Instantiate the edge $d_0(w, st)$.
     For each $x \in t$, instantiate the edge $d_0(st, sx)$.
- Set $L_{max} = 103$.

Figure 12.7. A procedure for constructing an instance $I_{lr}$ of latency-constrained resynchronization from an instance $I_{sc}$ of set-covering such that a solution to $I_{lr}$ yields a solution to $I_{sc}$.

synchronization graph; a set of *sink processor actors* $y_1, y_2, ..., y_q$, with associated execution times $\{t(y_i)\}$, such that each $y_i$ is the $i$th actor scheduled on the processor that corresponds to the sink SCC of the synchronization graph; a set of non-redundant synchronization edges $S = \{s_1, s_2, ..., s_n\}$ such that for each $s_i$, $src(s_i) \in \{x_1, x_2, ..., x_p\}$ and $snk(s_i) \in \{y_1, y_2, ..., y_q\}$; and a latency constraint $L_{max}$, which is a positive integer. A solution to such an instance is a minimal resynchronization $R$ that satisfies $L_{\Psi(R, G)}(x_1, y_q) \le L_{max}$. In the remainder of this section, we denote the synchronization graph corresponding to the generic instance of 2LCR by $\tilde{G}$.

We assume that $delay(s_i) = 0$ for all $s_i$, and we refer to the subproblem that results from this restriction as **delayless 2LCR**. This section demonstrates an algorithm that solves the delayless 2LCR problem in $O(N^2)$ time, where $N$ is the number of vertices in $\tilde{G}$. An extension of this algorithm to the general 2LCR problem (arbitrary delays can be present) is also given.

### 12.4.1    Interval Covering

An efficient polynomial-time solution to delayless 2LCR can be derived by reducing the problem to a special case of set-covering called **interval covering**, in which we are given an ordering $w_1, w_2, ..., w_N$ of the members of $X$ (the set that must be covered), such that the collection of subsets $T$ consists entirely of subsets of the form $\{w_a, w_{a+1}, ..., w_b\}$, $1 \le a \le b \le N$. Thus, while general set-covering involves covering a set from a collection of subsets, interval covering amounts to covering an interval from a collection of subintervals.

Interval covering can be solved in $O(|X||T|)$ time by a simple procedure that first selects the subset $\{w_1, w_2, ..., w_{b_1}\}$, where

$$b_1 = max(\{b|(w_1, w_b \in t) \text{ for some } t \in T\});$$

then selects any subset of the form $\{w_{a_2}, w_{a_2+1}, ..., w_{b_2}\}$, $a_2 \le b_1 + 1$, where

$$b_2 = max(\{b|(w_{b_1+1}, w_b \in t) \text{ for some } t \in T\});$$

then selects any subset of the form $\{w_{a_3}, w_{a_3+1}, ..., w_{b_3}\}$, $a_3 \le b_2 + 1$, where

$$b_3 = max(\{b|(w_{b_2+1}, w_b \in t) \text{ for some } t \in T\});$$

and so on until $b_n = N$.

### 12.4.2    Two-Processor Latency-Constrained Resynchronization

To reduce delayless 2LCR to interval covering, we start with the following observations.

**Observation 8:** Suppose that $R$ is a resynchronization of $\tilde{G}$, $r \in R$, and $r$ contributes to the elimination of synchronization edge $s$. Then $r$ subsumes $s$. Thus, the set of synchronization edges that $r$ contributes to the elimination of is simply the set of synchronization edges that are subsumed by $r$.

*Proof:* This follows immediately from the restriction that there can be no resynchronization edges directed from a $y_j$ to an $x_i$ (feedforward resynchronization), and thus in $\Psi(R, \tilde{G})$, there can be at most one synchronization edge in any path directed from $src(s)$ to $snk(s)$. *QED.*

**Observation 9:** If $R$ is a resynchronization of $\tilde{G}$, then

$$L_{\Psi(R, \tilde{G})}(x_1, y_q) = max(\{t_{pred}(src(s')) + t_{succ}(snk(s')) | s' \in R\}), \text{ where}$$

$$t_{pred}(x_i) \equiv \sum_{j \le i} t(x_j) \text{ for } i = 1, 2, ..., p, \text{ and } t_{succ}(y_i) \equiv \sum_{j \ge i} t(y_j) \text{ for } i = 1, 2, ..., q.$$

*Proof:* Given a synchronization edge $(x_a, y_b) \in R$, there is exactly one delayless path in $R(\tilde{G})$ from $x_1$ to $y_q$ that contains $(x_a, y_b)$ and the set of vertices traversed by this path is $\{x_1, x_2, ..., x_a, y_b, y_{b+1}, ..., y_q\}$. The desired result follows immediately. *QED.*

Now, corresponding to each of the source processor actors $x_i$ that satisfies $t_{pred}(x_i) + t(y_q) \le L_{max}$ we define an ordered pair of actors (a "resynchronization candidate") by

$$v_i \equiv (x_i, y_j), \text{ where } j = min(\{k | (t_{pred}(x_i) + t_{succ}(y_k) \le L_{max})\}). \qquad (12\text{-}13)$$

Consider the example shown in Figure 12.8. Here, we assume that $t(z) = 1$ for each actor $z$, and $L_{max} = 10$. From (12-13), we have

$$v_1 = (x_1, y_1), v_2 = (x_2, y_1), v_3 = (x_3, y_2), v_4 = (x_4, y_3),$$

$$v_5 = (x_5, y_4), v_6 = (x_6, y_5), v_7 = (x_7, y_6), v_8 = (x_8, y_7). \qquad (12\text{-}14)$$

If $v_i$ exists for a given $x_i$, then $d_0(v_i)$ can be viewed as the best resynchronization edge that has $x_i$ as the source actor, and thus, to construct an optimal LCR, we can select the set of resynchronization edges entirely from among the $v_i$ s. This is established by the following two observations.

**Observation 10:** Suppose that $R$ is an LCR of $\tilde{G}$, and suppose that $(x_a, y_b)$ is a delayless synchronization edge in $R$ such that $(x_a, y_b) \ne v_a$. Then $(R - \{(x_a, y_b)\} + \{d_0(v_a)\})$ is an LCR of $R$.

*Proof:* Let $v_a = (x_a, y_c)$ and $R' = (R - \{(x_a, y_b)\} + \{d_0(v_a)\})$, and observe that $v_a$ exists, since

$$((x_a, y_b) \in R) \Rightarrow (t_{pred}(x_a) + t_{succ}(y_b) \le L_{max}) \Rightarrow (t_{pred}(x_a) + t(y_q) \le L_{max}) (12\text{-}15)$$

From Observation 8 and the assumption that $(x_a, y_b)$ is delayless, the set of synchronization edges that $(x_a, y_b)$ contributes to the elimination of is simply the set of synchronization edges that are subsumed by $(x_a, y_b)$. Now, if $s$ is a synchronization edge that is subsumed by $(x_a, y_b)$, then

$$\rho_{\tilde{G}}(src(s), x_a) + \rho_{\tilde{G}}(y_b, snk(s)) \leq delay(s). \tag{12-16}$$

From the definition of $v_a$, we have that $c \leq b$, and thus, that $\rho_{\tilde{G}}(y_c, y_b) = 0$. It follows from (12-16) that

$$\rho_{\tilde{G}}(src(s), x_a) + \rho_{\tilde{G}}(y_c, snk(s)) \leq delay(s), \tag{12-17}$$

and thus, that $v_a$ subsumes $s$. Hence, $v_a$ subsumes all synchronization edges that $(x_a, y_b)$ contributes to the elimination of, and we can conclude that $R'$ is a valid resynchronization of $\tilde{G}$.

From the definition of $v_a$, we know that $t_{pred}(x_a) + t_{succ}(y_c) \leq L_{max}$, and

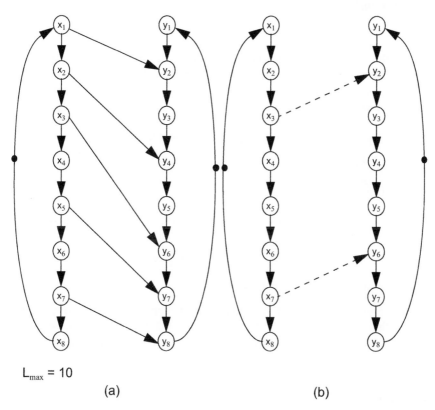

$L_{max} = 10$

(a)                                    (b)

Figure 12.8. An instance of delayless, two-processor latency-constrained resynchronization. In this example, the execution times of all actors are identically equal to unity.

thus since $R$ is an LCR, we have from Observation 9 that $R'$ is an LCR. *QED.*

From Fact 12.2 and the assumption that the members of $S$ are all delay-less, an optimal LCR of $\tilde{G}$ consists only of delayless synchronization edges. Thus, from Observation 10, we know that there exists an optimal LCR that consists only of members of the form $d_0(v_i)$. Furthermore, from Observation 9, we know that a collection $V$ of $v_i$s is an LCR if and only if

$$\bigcup_{v \in V} \chi(v) = \{s_1, s_2, ..., s_n\},$$

where $\chi(v)$ is the set of synchronization edges that are subsumed by $v$. The following observation completes the correspondence between 2LCR and interval covering.

**Observation 11:** Let $s_1', s_2', ..., s_n'$ be the ordering of $s_1, s_2, ..., s_n$ specified by

$$(x_a = src(s_i'), x_b = src(s_j'), a < b) \Rightarrow (i < j). \tag{12-18}$$

That is the $s_i'$'s are ordered according to the order in which their respective source actors execute on the source processor. Suppose that for some $j \in \{1, 2, ..., p\}$, some $m > 1$, and some $i \in \{1, 2, ..., n - m\}$, we have $s_i' \in \chi(v_j)$ and $s_{i+m}' \in \chi(v_j)$. Then $s_{i+1}', s_{i+2}', ..., s_{i+m-1}' \in \chi(v_j)$.

In Figure 12.8(a), the ordering specified by (12-18) is

$$s_1' = (x_1, y_2), s_2' = (x_2, y_4), s_3' = (x_3, y_6), s_4' = (x_5, y_7), s_5' = (x_7, y_8), \tag{12-19}$$

and thus from (12-14), we have

$$\chi(v_1) = \{s_1'\}, \chi(v_2) = \{s_1', s_2'\}, \chi(v_3) = \{s_1', s_2', s_3'\}, \chi(v_4) = \{s_2', s_3'\},$$
$$\chi(v_5) = \{s_2', s_3', s_4'\}, \chi(v_6) = \{s_3', s_4'\},$$
$$\chi(v_7) = \{s_3', s_4', s_5'\}, \chi(v_8) = \{s_4', s_5'\}, \tag{12-20}$$

which is clearly consistent with Observation 11.

*Proof of Observation 11:* Let $v_j = (x_j, y_l)$, and suppose $k$ is a positive integer such that $i < k < i + m$. Then from (12-18), we know that $\rho_{\tilde{G}}(src(s_k'), src(s_{i+m}')) = 0$. Thus, since $s_{i+m}' \in \chi(v_j)$, we have that

$$\rho_{\tilde{G}}(src(s_k'), x_j) = 0. \tag{12-21}$$

Now clearly

$$\rho_{\tilde{G}}(snk(s_i'), snk(s_k')) = 0, \tag{12-22}$$

since otherwise $\rho_{\tilde{G}}(snk(s_k'), snk(s_i')) = 0$ and thus (from 12-18) $s_k'$ subsumes $s_i'$, which contradicts the assumption that the members of $S$ are not redundant. Finally, since $s_i' \in \chi(v_j)$, we know that $\rho_{\tilde{G}}(y_l, snk(s_i')) = 0$. Combining this with (12-22) yields

$$\rho_{\tilde{G}}(y_l, snk(s_k')) = 0, \tag{12-23}$$

and (12-21) and (12-23) together yield that $s_k' \in \chi(v_j)$. *QED.*

From Observation 11 and the preceding discussion, we conclude that an optimal LCR of $\tilde{G}$ can be obtained by the following steps.
(a) Construct the ordering $s_1', s_2', ..., s_n'$ specified by (12-18).
(b) For $i = 1, 2, ..., p$, determine whether or not $v_i$ exists, and if it exists, compute $v_i$.
(c) Compute $\chi(v_j)$ for each value of $j$ such that $v_j$ exists.
(d) Find a minimal cover $C$ for $S$ given the family of subsets $\{\chi(v_j)|v_j \text{ exists}\}$.
(e) Define the resynchronization $R = \{v_j|\chi(v_j) \in C\}$.

Steps (a), (b), and (e) can clearly be performed in $O(N)$ time, where $N$ is the number of vertices in $\tilde{G}$. If the algorithm outlined in Section 12.4.1 is employed for step (d), then from the discussion in Section 12.4.1 and Observation 12(e) in Section 12.4.3, it can be easily verified that the time complexity of step (d) is $O(N^2)$. Step (c) can also be performed in $O(N^2)$ time using the observation that if $v_i = (x_i, y_j)$, then

$$\chi(v_i) \equiv \{(x_a, y_b) \in S | a \leq i \text{ and } b \geq j\},$$

where $S = \{s_1, s_2, ..., s_n\}$ is the set of synchronization edges in $\tilde{G}$. Thus, we have the following result.

**Theorem 12.2:** Polynomial-time solutions (quadratic in the number of synchronization graph vertices) exist for the delayless, two-processor latency-constrained resynchronization problem.

Note that solutions more efficient than the $O(N^2)$ approach described above may exist.

From (12-20), we see that there are two possible solutions that can result if we apply Steps (a)-(e) to Figure 12.8(a) and use the technique described earlier for interval covering. These solutions correspond to the interval covers $C_1 = \{\chi(v_3), \chi(v_7)\}$ and $C_2 = \{\chi(v_3), \chi(v_8)\}$. The synchronization graph that results from the interval cover $C_1$ is shown in Figure 12.8(b).

## 12.4.3    Taking Delays into Account

If delays exist on one or more edges of the original synchronization graph, then the correspondence defined in the previous subsection between 2LCR and interval covering does not necessarily hold. For example, consider the synchronization graph in Figure 12.9. Here, the numbers beside the actors specify execution times; a "D" on top of an edge specifies a unit delay; the latency input and latency output are respectively $x_1$ and $y_8$; and the latency constraint is $L_{max} = 12$. It is easily verified that $v_i$ exists for $i = 1, 2, ..., 6$, and from (12-13), we obtain

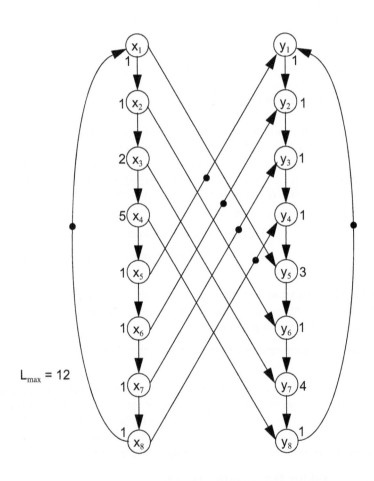

Figure 12.9. A synchronization graph with unit delays on some of the synchronization edges.

$$v_1 = (x_1, y_3), \; v_2 = (x_2, y_4), \; v_3 = (x_3, y_6), \; v_4 = (x_4, y_8),$$
$$v_5 = (x_5, y_8), \; v_6 = (x_6, y_8). \qquad (12\text{-}24)$$

Now if we order the synchronization edges as specified by (12-18), then

$$s_i' = (x_i, y_{i+4}) \text{ for } i = 1, 2, 3, 4,$$
$$\text{and } s_i' = (x_i, y_{i-4}) \text{ for } i = 5, 6, 7, 8, \qquad (12\text{-}25)$$

and if the correspondence between delayless 2LCR and interval covering defined in the previous section were to hold for general 2LCR, then we would have that

each subset $\chi(v_i)$ is of the form $\{s_a', s_{a+1}', \dots, s_b'\}, \; 1 \le a \le b \le 8$.     (12-26)

However, computing the subsets $\chi(v_i)$, we obtain

$$\chi(v_1) = \{s_1', s_7', s_8'\}, \chi(v_2) = \{s_1', s_2', s_8'\}, \chi(v_3) = \{s_2', s_3'\}$$
$$\chi(v_4) = \{s_4'\}, \chi(v_5) = \{s_4', s_5'\}, \chi(v_6) = \{s_4', s_5', s_6'\}, \qquad (12\text{-}27)$$

and these subsets are clearly not all consistent with the form specified in (12-26). Thus, the algorithm developed in Section 12.4.2 does not apply directly to handle delays.

However, the technique developed in Section 12.4.2 can be extended to solve the general 2LCR problem in polynomial time. This extension is based on separating the subsumption relationships between the $v_i$'s and the synchronization edges into two categories: if $v_i = (x_i, y_j)$ subsumes the synchronization edge $s = (x_k, y_l)$ then we say that $v_i$ **1-subsumes** $s$ if $i < k$, and we say that $v_i$ **2-subsumes** $s$ if $i \ge k$. For example in Figure 12.8(a), $v_1 = (x_1, y_3)$ 1-subsumes both $(x_7, y_3)$ and $(x_8, y_4)$, and $v_5 = (x_5, y_8)$ 2-subsumes $(x_4, y_8)$ and $(x_5, y_1)$.

**Observation 12:** Assuming the same notation for a generic instance of 2LRC that was defined in the previous subsection, the initial synchronization graph $\tilde{G}$ satisfies the following conditions:

(a) Each synchronization edge has at most one unit of delay ($delay(s_i) \in \{0, 1\}$).

(b) If $(x_i, y_j)$ is a zero-delay synchronization edge and $(x_k, y_l)$ is a unit-delay synchronization edge, then $i < k$ and $j > l$.

(c) If $v_i$ 1-subsumes a unit-delay synchronization edge $(x_i, y_j)$, then $v_i$ also 1-subsumes all unit-delay synchronization edges $s$ that satisfy $src(s) = x_{i+n}, n > 0$.

(d) If $v_i$ 2-subsumes a unit-delay synchronization edge $(x_i, y_j)$, then $v_i$ also 2-subsumes all unit-delay synchronization edges $s$ that satisfy $src(s) = x_{i-n}, n > 0$.

(e) If $(x_i, y_j)$ and $(x_k, y_l)$ are both distinct zero-delay synchronization edges or they are both distinct unit-delay synchronization edges, then $i \ne k$ and

$(i < k) \Leftrightarrow (j < l)$.

(f) If $(x_i, y_j)$ 1-subsumes a unit delay synchronization edge $(x_k, y_l)$, then $l \geq j$.

*Proof outline:* From Fact 12.3, we know that $\rho(x_1, y_q) = 0$. Thus, there exists at least one delayless synchronization edge in $\tilde{G}$. Let $e$ be one such delayless synchronization edge. Then it is easily verified from the structure of $\tilde{G}$ that for all $x_i, y_j$, there exists a path $p_{i,j}$ in $\tilde{G}$ directed from $x_i$ to $y_j$ such that $p_{i,j}$ contains $e$, $p_{i,j}$ contains no other synchronization edges, and $Delay(p_{i,j}) \leq 2$. It follows that any synchronization edge $e'$ whose delay exceeds unity would be redundant in $\tilde{G}$. Thus, part (a) follows from the assumption that none of the synchronization edges in $\tilde{G}$ are redundant.

The other parts can be verified easily from the structure of $\tilde{G}$, including the assumption that no synchronization edge in $\tilde{G}$ is redundant. We omit the details.

Resynchronizations for instances of general 2LCR can be partitioned into two categories — **category A** consists of all resynchronizations that contain at least one synchronization edge having nonzero delay, and **category B** consists of all resynchronizations that consist entirely of delayless synchronization edges. An optimal category A solution (a category A solution whose cost is less than or equal to the cost of all category A solutions) can be derived by simply applying the optimal solution described in Subsection 12.4.2 to "rearrange" the delayless resynchronization edges, and then replacing all synchronization edges that have nonzero delay with a single unit delay synchronization edge directed from $x_p$, the last actor scheduled on the source processor to $y_1$, the first actor scheduled on the sink processor. We refer to this approach as **Algorithm A**.

An example is shown in Figure 12.10. When general 2LCR is applied to the instance of Figure 12.10(a), the constraint that all synchronization edges have zero delay is too restrictive to permit a globally optimal solution. Here, the latency constraint is assumed to be $L_{max} = 2$. Under this constraint, it is easily seen that no zero-delay resynchronization edges can be added without violating the latency constraint. However, if we allow resynchronization edges that have delay, then we can apply Algorithm A to achieve a cost of two synchronization edges. The resulting synchronization graph, with redundant synchronization edges removed, is shown in Figure 12.10(b). Observe that this resynchronization is an LCR since only delayless synchronization edges affect the latency of a transparent synchronization graph.

Now suppose that $\tilde{G}$ (our generic instance of 2LCR) contains at least one unit-delay synchronization edge, suppose that $G_b$ is an optimal category B solution for $\tilde{G}$, and let $R_b$ denote the set of resynchronization edges in $G_b$. Let $ud(\tilde{G})$ denote the set of synchronization edges in $\tilde{G}$ that have unit delay, and let

$(x_{k_1}, y_{l_1}), (x_{k_2}, y_{l_2}), \ldots, (x_{k_M}, y_{l_M})$ denote the ordering of the members of $ud(\tilde{G})$ that corresponds to the order in which the source actors execute on the source processor — that is, $(i < j) \Rightarrow (k_i < k_j)$. Note from Observation 12(a) that $ud(\tilde{G})$ is the set of all synchronization edges in $\tilde{G}$ that are not delayless. Also,

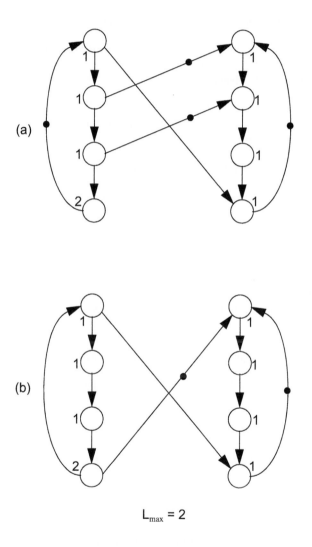

$$L_{max} = 2$$

Figure 12.10. An example in which constraining all resynchronization edges to be delayless makes it impossible to derive an optimal resynchronization.

let $1subs(\tilde{G}, G_b)$ denote the set of unit-delay synchronization edges in $\tilde{G}$ that are 1-subsumed by resynchronization edges in $G_b$. That is,

$$1subs(\tilde{G}, G_b) \equiv \{s \in ud(\tilde{G}) | (\exists((z_1, z_2) \in R_b)) \text{ s.t } ((z_1, z_2) \text{ 1-subsumes } s \text{ in} \tilde{G})\}$$

If $1subs(\tilde{G}, G_b)$ is not empty, define

$$r = min(\{j | (x_{k_j}, y_{l_j}) \in 1subs(\tilde{G}, G_b)\}).\tag{12-28}$$

Suppose $(x_m, y_{m'}) \in 1subs(\tilde{G}, G_b)$. Then by definition of $r$, $m' \geq l_r$, and thus $\rho_{\tilde{G}}(y_{l_r}, y_{m'}) = 0$. Furthermore, since $x_m$ and $x_1$ execute on the same processor, $\rho_{\tilde{G}}(x_m, x_1) \leq 1$. Hence

$$\rho_{\tilde{G}}(x_m, x_1) + \rho_{\tilde{G}}(y_{l_r}, y_{m'}) \leq 1 = delay(x_m, y_{m'}),$$

so we have that $(x_1, y_{l_r})$ subsumes $(x_m, y_{m'})$ in $\tilde{G}$. Since $(x_m, y_{m'})$ is an arbitrary member of $ud(\tilde{G})$, we conclude that

Every member of $1subs(\tilde{G}, G_b)$ is subsumed by $(x_1, y_{l_r})$. (12-29)

Now, if $\Gamma \equiv (ud(\tilde{G}) - 1subs(\tilde{G}, G_b))$ is not empty, then define

$$u = max(\{j | (x_{k_j}, y_{l_j}) \in \Gamma\}),\tag{12-30}$$

and suppose $(x_m, y_{m'}) \in \Gamma$. By definition of $u$, $m \leq k_u$ and thus $\rho_{\tilde{G}}(x_m, x_{k_u}) = 0$. Furthermore, since $y_{m'}$ and $y_q$ execute on the same processor, $\rho_{\tilde{G}}(y_q, y_{m'}) \leq 1$. Hence,

$$\rho_{\tilde{G}}(x_m, x_{k_u}) + \rho_{\tilde{G}}(y_q, y_{m'}) \leq 1 = delay(x_m, y_{m'}),$$

and we have that

Every member of $\Gamma$ is subsumed by $(x_{k_u}, y_q)$. (12-31)

Observe also that from the definitions of $r$ and $u$, and from Observation 12(c),

$$((1subs(\tilde{G}, G_b) \neq \varnothing) \textbf{ and } (\Gamma \neq \varnothing)) \Rightarrow (u = r - 1);\tag{12-32}$$

$$(1subs(\tilde{G}, G_b) = \varnothing) \Rightarrow (u = M);\tag{12-33}$$

and

$$(\Gamma = \varnothing) \Rightarrow (r = 1).\tag{12-34}$$

Now we define the synchronization graph $Z(\tilde{G})$ by

$$Z(\tilde{G}) = (V, (E - ud(\tilde{G})) + P),\tag{12-35}$$

where $V$ and $E$ are the sets of vertices and edges in $\tilde{G}$; $P = \{d_0(x_1, y_{l_r}), d_0(x_{k_u}, y_q)\}$, if both $1subs(\tilde{G}, G_b)$ and $\Gamma$ are nonempty; $P = \{d_0(x_1, y_{l_r})\}$ if $\Gamma$ is empty; and $P = \{d_0(x_{k_u}, y_q)\}$ if $1subs(\tilde{G}, G_b)$ is empty.

**Theorem 12.3:** $G_b$ is a resynchronization of $Z(\tilde{G})$.

*Proof:* The set of synchronization edges in $Z(\tilde{G})$ is $E_0 + P$, where $E_0$ is the set of delayless synchronization edges in $\tilde{G}$. Since $G_b$ is a resynchronization of $\tilde{G}$, it suffices to show that for each $e \in P$,

$$\rho_{G_b}(src(e), snk(e)) = 0. \tag{12-36}$$

If $1subs(\tilde{G}, G_b)$ is non-empty then from (12-28) (the definition of $r$) and Observation 12(f), there must be a delayless synchronization edge $e'$ in $G_b$ such that $snk(e') = y_w$ for some $w \le l_r$. Thus,

$$\rho_{G_b}(x_1, y_{l_r}) \le \rho_{G_b}(x_1, src(e')) + \rho_{G_b}(snk(e'), y_{l_r}) = 0 + \rho_{G_b}(y_w, y_{l_r}) = 0,$$

and we have that (12-36) is satisfied for $e = (x_1, y_{l_r})$.

Similarly if $\Gamma$ is non-empty, then from (12-30) (the definition of $u$) and from the definition of *2-subsumes*, there exists a delayless synchronization edge $e'$ in $G_b$ such that $src(e') = x_w$ for some $w \ge k_u$. Thus,

$$\rho_{G_b}(x_{k_u}, y_q) \le \rho_{G_b}(x_{k_u}, src(e')) + \rho_{G_b}(snk(e'), y_q) = \rho_{G_b}(x_{k_u}, x_w) + 0 = 0;$$

hence, we have that (12-36) is satisfied for $e = (x_{k_u}, y_q)$.

From the definition of $P$, it follows that (12-36) is satisfied for every $e \in P$.

**Corollary 12.1:** The latency of $Z(\tilde{G})$ is not greater than $L_{max}$. That is, $L_{Z(\tilde{G})}(x_1, y_q) \le L_{max}$.

*Proof:* From Theorem 12.3, we know that $G_b$ preserves $Z(\tilde{G})$. Thus, from Lemma 11.1, it follows that $L_{Z(\tilde{G})}(x_1, y_q) \le L_{G_b}(x_1, y_q)$. Furthermore, from the assumption that $G_b$ is an optimal category B LCR, we have $L_{G_b}(x_1, y_q) \le L_{max}$. We conclude that $L_{Z(\tilde{G})}(x_1, y_q) \le L_{max}$.

Theorem 12.3, along with (12-32)-(12-34), tells us that an optimal category B LCR of $\tilde{G}$ is always a resynchronization of

(1) a synchronization graph of the form

$$(V, ((E - ud(\tilde{G})) + \{d_0(x_1, y_{l_\alpha}), d_0(x_{k_{\alpha-1}}, y_q)\})), \quad 1 < \alpha \le M, \tag{1}$$

or

(2) of the graph $(V, ((E - ud(\tilde{G})) + \{d_0(x_1, y_{l_i})\}))$, $\qquad$ (2)

or

(3) of the graph $(V, ((E - ud(\tilde{G})) + \{d_0(x_{k_M}, y_q)\}))$. $\qquad$ (3)

Thus, from Corollary 12.1, an optimal resynchronization can be computed by examining each of the $(M+1) = (|ud(G)| + 1)$ synchronization graphs defined by (1)-(3), computing an optimal LCR for each of these graphs whose latency is no greater than $L_{max}$, and returning one of the optimal LCRs that has the fewest number of synchronization edges. This is straightforward since these graphs contain only delayless synchronization edges, and thus the algorithm of Section 12.4.2 can be used.

Recall the example of Figure 12.8(a). Here,

$$ud(\tilde{G}) = \{(x_5, y_1), (x_6, y_2), (x_7, y_3), (x_8, y_4)\},$$

and the set of synchronization graphs that correspond to (1)-(3) are shown in Figures 12.11 and 12.12. The latencies of the graphs in Figure 12.11(a)-(c) and Figure 12.12(a-b) are respectively 14, 13, 12, 13, and 14. Since $L_{max} = 12$, we only need to compute an optimal LCR for the graph of Figure 12.11(c) (from Corollary 12.1). This is done by first removing redundant edges from the graph (yielding the graph in Figure 12.13(b)) and then applying the algorithm developed in Section 12.4.2. For the synchronization graph of Figure 12.13(b), and $L_{max} = 12$, it is easily verified that the set of $v_i$ s is

$$v_1 = (x_1, y_3), v_2 = (x_2, y_4), v_3 = (x_3, y_6), v_4 = (x_4, y_8),$$
$$v_5 = (x_5, y_8), v_6 = (x_6, y_8).$$

If we let

$$s_1 = (x_1, y_3), s_2 = (x_2, y_6), s_3 = (x_3, y_7), s_4 = (x_6, y_8), \qquad (12\text{-}37)$$

then we have

$$, \chi(v_1) = \{s_1\}, \chi(v_2) = \{s_2\}, \chi(v_3) = \{s_2, s_3\},$$
$$\chi(v_4) = \chi(v_5) = \varnothing, \chi(v_6) = \{s_4\}. \qquad (12\text{-}38)$$

From (12-38), the algorithm outlined in Section 12.4.1 for interval covering can be applied to obtain an optimal resynchronization. This results in the resynchronization $R = \{v_1, v_3, v_6\}$. The resulting synchronization graph is shown in Figure 12.13(c). Observe that the number of synchronization edges has been reduced from $8$ to $3$, while the latency has increased from $10$ to $L_{max} = 12$. Also, none of the original synchronization edges in $\tilde{G}$ are retained

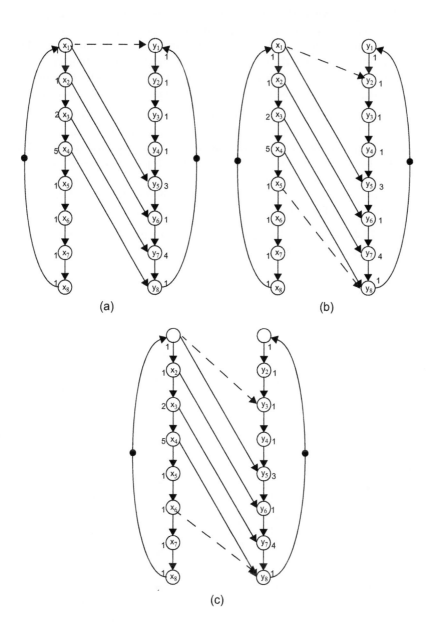

Figure 12.11. An illustration of Algorithm B.

in the resynchronization.

We say that **Algorithm B** for general 2LCR is the approach of constructing the $(|ud(\tilde{G})| + 1)$ synchronization graphs corresponding to (1)-(3), computing an optimal LCR for each of these graphs whose latency is no greater than $L_{max}$, and returning one of the optimal LCRs that has the fewest number of synchronization edges. We have shown previously that Algorithm B leads to an optimal LCR *under the constraint that all resynchronization edges have zero delay*.

Thus, given an instance of a general 2LCR, a globally optimal solution can be derived by applying Algorithm A and Algorithm B and retaining the best of the resulting two solutions. The time complexity of this two-phased approach is dominated by the complexity of Algorithm B, which is $O(|ud(\tilde{G})|N^2)$ (a factor of $|ud(\tilde{G})|$ greater than the complexity of the technique for delayless 2LCR that was developed in Section 12.4.2), where $N$ is the number of vertices in $\tilde{G}$. Since $|ud(\tilde{G})| \leq N$ from Observation 12(e), the complexity is $O(N^3)$.

**Theorem 12.4:** Polynomial-time solutions exist for the general two-processor latency-constrained resynchronization problem.

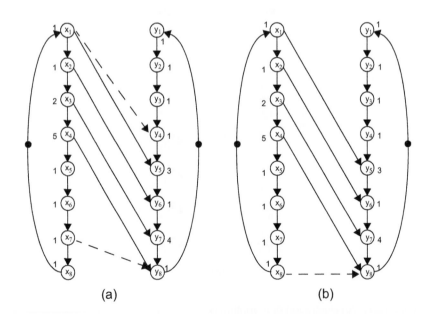

(a)                                      (b)

Figure 12.12. Continuation of the illustration of Algorithm B in Figure 12.11.

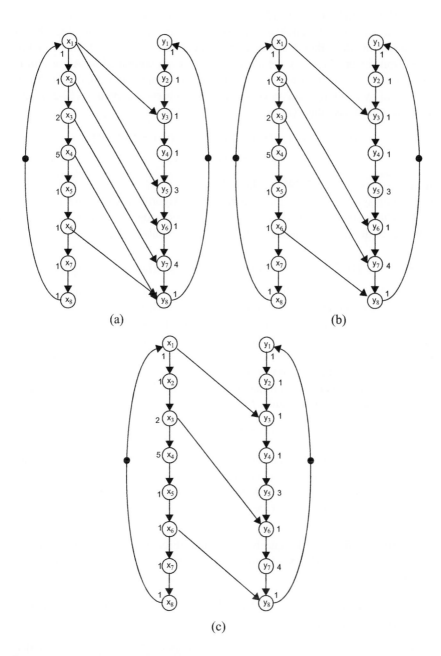

Figure 12.13.  Optimal LCR example.

The example in Figure 12.10 shows how it is possible for Algorithm A to produce a better result than Algorithm B. Conversely, the ability of Algorithm B to outperform Algorithm A can be demonstrated through the example of Figure 12.9. From Figure 12.13(c), we know that the result computed by Algorithm B has a cost of 3 synchronization edges. The result computed by Algorithm A can be derived by applying interval covering to the subsets specified in (12-27) with all of the unit-delay edges $(s_5', s_6', s_7', s_8')$ removed:

$$\chi(v_1) = \{s_1'\}, \chi(v_2) = \{s_1', s_2'\}, \chi(v_3) = \{s_2', s_3'\}$$
$$\chi(v_4) = \chi(v_5) = \chi(v_6) = \{s_4'\}. \tag{12-39}$$

A minimal cover for (12-39) is achieved by $\{\chi(v_2), \chi(v_3), \chi(v_4)\}$, and the corresponding synchronization graph computed by Algorithm A is shown in Figure 12.14. This solution has a cost of 4 synchronization edges, which is one greater than that of the result computed by Algorithm B for this example.

## 12.5    A Heuristic for General Synchronization Graphs

In Section 11.5, we discussed a heuristic called Global-resynchronize for the maximum-throughput resynchronization problem, which is the problem of determining an optimal resynchronization under the assumption that arbitrary increases in latency can be tolerated. In this section, we extend Algorithm Global-resynchronize to derive an efficient heuristic that addresses the latency-constrained resynchronization problem for general synchronization graphs. Given an input synchronization graph $G$, Algorithm Global-resynchronize operates by first computing the family of subsets

$$T \equiv \{\chi(v_1, v_2) | (((v_1, v_2) \notin E) \text{ and } (\rho_G(v_2, v_1) = \infty))\}. \tag{12-40}$$

After computing the family of subsets specified by (12-40), Algorithm Global-resynchronize chooses a member of this family that has maximum cardinality, inserts the corresponding delayless resynchronization edge, and removes all synchronization edges that become redundant as a result of inserting this resynchronization edge.

To extend this technique for maximum-throughput resynchronization to the latency-constrained resynchronization problem, we simply replace the subset computation in (12-40) with

$$T \equiv \{\chi(v_1, v_2) | (((v_1, v_2) \notin E) \text{ and } (\rho_G(v_2, v_1) = \infty) \text{ and} \tag{12-41}$$
$$(L'(v_1, v_2) \leq L_{max}))\},$$

where $L'$ is the latency of the synchronization graph $(V, \{E + \{(v_1, v_2)\}\})$ that results from adding the resynchronization edge $(v_1, v_2)$ to $G$.

A pseudocode specification of the extension of Global-resynchronize to the latency-constrained resynchronization problem, called Algorithm **Global-LCR**, is shown in Figure 12.15.

## 12.5.1    Customization to Transparent Synchronization Graphs

In Section 12.2, we mentioned that transparent synchronization graphs are advantageous for performing latency-constrained resynchronization. If the input synchronization graph is transparent, then assuming that $T_{fi(G)}(x, y)$ has been determined for all $x, y \in V$, $L'$ in Algorithm Global-LCR can be computed in $O(1)$ time from

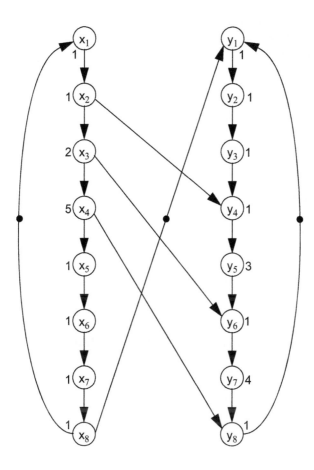

Figure 12.14. The solution derived by Algorithm A when it is applied to the example of Figure 12.9.

$$L'(v_1, v_2) = max(\{(T_{fi(G)}(\upsilon, v_1) + T_{fi(G)}(v_2, o_L)), L_G\}),  \tag{12-42}$$

where $\upsilon$ is the source actor in $fi(G)$, $o_L$ is the latency output, and $L_G$ is the latency of $G$.

Furthermore, $T_{fi(G)}(x, y)$ can be updated in the same manner as $\rho_G$. That

**Function** Global-LCR
**Input**: a reduced synchronization graph $G = (V, E)$
**Output**: an alternative reduced synchronization graph that preserves $G$.

compute $\rho_G(x, y)$ for all actor pairs $x, y \in V$
*complete* = FALSE
**While** not (*complete*)
    *best* = *NULL*, $M = 0$
    **For** $x, y \in V$
        **If** $((\rho_G(y, x) = \infty)$ **and** $((x, y) \notin E))$ **and** $(L'(x, y) \le L_{max})$
            $\chi^* = \chi((x, y))$
            **If** $(|\chi^*| > M)$
                $M = |\chi^*|$
                *best* = $(x, y)$
            **Endif**
        **Endif**
    **Endfor**
    **If** $(best = NULL)$
        *complete* = *TRUE*
    **Else**
        $E = E - \chi(best) + \{d_0(best)\}$
        $G = (V, E)$
        **For** $x, y \in V$             /* update $\rho_G$ */
            $\rho_{new}(x, y) =$
            $min(\{\rho_G(x, y), \rho_G(x, src(best)) + \rho_G(snk(best), y)\})$
        **Endfor**
        $\rho_G = \rho_{new}$
    **Endif**
**Endwhile**
**Return** $G$
**End function**

Figure 12.15. A heuristic for latency-constrained resynchronization.

is, once the resynchronization edge *best* is chosen, we have that for each $(x, y) \in (V \cup \{\upsilon\})$,

$$T_{new}(x, y) = max(\{T_{fi(G)}(x, y), T_{fi(G)}(x, src(best)) + \qquad (12\text{-}43)$$
$$T_{fi(G)}(snk(best), y)\}),$$

where $T_{new}$ denotes the maximum cumulative execution time between actors in the first iteration graph after the insertion of the edge *best* in $G$. The computations in (12-43) can be performed by inserting the simple **for** loop shown in Figure 12.16 at the end of the **else** block in Algorithm Global-LCR. Thus, as with the computation of $\rho_G$, the cubic-time Bellman-Ford algorithm need only be invoked once, at the beginning of the LCR Algorithm, to initialize $T_{fi(G)}(x, y)$. This loop can be inserted immediately before or after the **for** loop that updates $\rho_G$.

## 12.5.2     Complexity

In Section 11.5, it was shown that Algorithm Global-resynchronize has $O(s_{ff}n^4)$ time-complexity, where $n$ is the number of actors in the input synchronization graph, and $s_{ff}$ is the number of feedforward synchronization edges. Since the longest path quantities $T_{fi(G)}(*, *)$ can be computed initially in $O(n^3)$ time and updated in $O(n^2)$ time, it is easily verified that the $O(s_{ff}n^4)$ bound also applies to the customization of Algorithm Global-LCR to transparent synchronization graphs.

In general, whenever the nested **for** loops in Figure 12.15 dominate the computation of the **while** loop, the $O(s_{ff}n^4)$ complexity is maintained as long as $(L'(x, y) \leq L_{max})$ can be evaluated in $O(1)$ time. For general (not necessarily transparent) synchronization graphs, we can use the functional simulation approach described in Section 12.2 to determine $L'(x, y)$ in

---

**For** $x, y \in (V \cup \{\upsilon\})$        /* update $T_{fi(G)}$ */

    $T_{new}(x, y) =$
    $max(\{T_{fi(G)}(x, y), T_{fi(G)}(x, src(best)) + T_{fi(G)}(snk(best), y)\})$
**Endfor**

$T_{fi(G)} = T_{new}$

Figure 12.16. Pseudocode to update $T_{fi(G)}$ for use in the customization of Algorithm Global-LCR to transparent synchronization graphs.

$O(d \times max(\{n, s\}))$ time, where $d = 1 + \rho_{G_0}(x, y)$, and $s$ denotes the number of synchronization edges in $G$. This yields a running time of $O(ds_{ff}n^4 max(\{n, s\}))$ for general synchronization graphs.

The complexity bounds derived above are based on a general upper bound of $n^2$, which is derived in Section 11.5, on the total number of resynchronization steps (**while** loop iterations). However, this $n^2$ bound can be viewed as a very conservative estimate since in practice, constraints on the introduction of cycles severely limit the number of possible resynchronization steps. Thus, on practical graphs, we can expect significantly lower average-case complexity than the worst-case bounds of $O(s_{ff}n^4)$ and $O(ds_{ff}n^4 max(\{n, s\}))$.

### 12.5.3    Example

Figure 12.17 shows the synchronization graph that results from a six-processor schedule of a synthesizer for plucked-string musical instruments in 11 voices based on the Karplus-Strong technique, as shown in Section 11.5. In this example, *exc* and *out* are respectively the latency input and latency output, and the latency is 170. There are ten synchronization edges shown, and none of these is redundant.

Figure 12.18 shows how the number of synchronization edges in the result computed by the heuristic changes as the latency constraint varies. If just over 50 units of latency can be tolerated beyond the original latency of 170, then the heuristic is able to eliminate a single synchronization edge. No further improvement can be obtained unless roughly another 50 units are allowed, at which point the number of synchronization edges drops to 8, and then down to 7 for an additional 8 time units of allowable latency. If the latency constraint is weakened to 382, just over twice the original latency, then the heuristic is able to reduce the number of synchronization edges to 6. No further improvement is achieved over the relatively long range of $(383 - 644)$. When $L_{max} \geq 645$, the minimal cost of 5 synchronization edges for this system is attained, which is half that of the original synchronization graph.

Figure 12.19 and Table 12.1 show how the average iteration period (the reciprocal of the average throughput) varies with different memory access times for various resynchronizations of Figure 12.17. Here, the column of Table 12.1 and the plot of Figure 12.19 labeled $A$ represent the original synchronization graph (before resynchronization); column/plot label $B$ represents the resynchronized result corresponding to the first break-point of Figure 12.18 ($L_{max} = 221$, 9 synchronization edges); label $C$ corresponds to the second break-point of Figure 12.18 ($L_{max} = 268$, 8 synchronization edges); and so on for labels $D$, $E$ and $F$, whose associated synchronization graphs have 7, 6, and 5 synchronization edges, respectively. Thus, as we go from label $A$ to label $F$, the number of synchronization edges in resynchronized solution decreases monotonically.

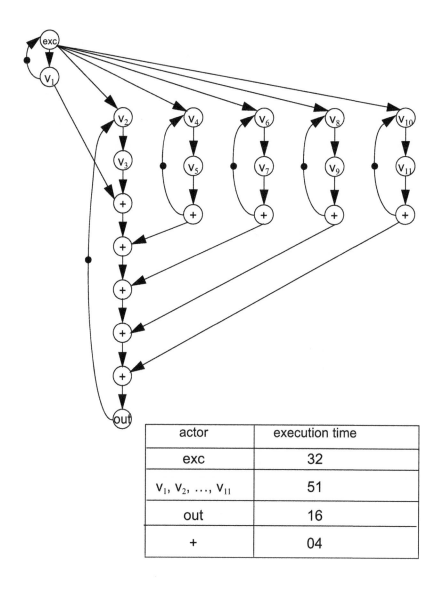

| actor | execution time |
|-------|----------------|
| exc | 32 |
| $v_1, v_2, \ldots, v_{11}$ | 51 |
| out | 16 |
| + | 04 |

Figure 12.17. The synchronization graph that results from a six-processor schedule of a music synthesizer based on the Karplus-Strong technique.

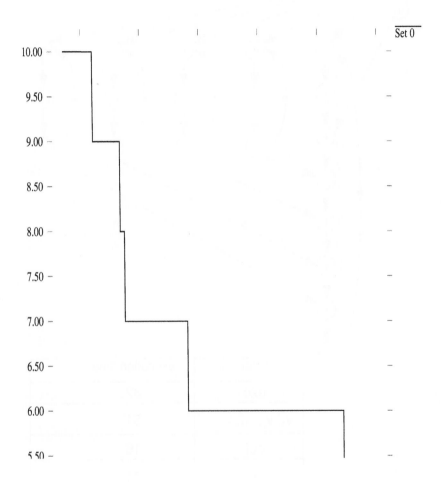

Figure 12.18. Performance of the heuristic on the example of Figure 12.17.

However, as seen in Figure 12.19, the average iteration period need not exactly follow this trend. For example, even though synchronization graph $A$ has one synchronization edge more than graph $B$, the iteration period curve for graph $B$ lies slightly above that of $A$. This is because the simulations shown in the figure model a shared bus, and take bus contention into account. Thus, even though graph $B$ has one less synchronization edge than graph $A$, it entails higher bus contention, and hence results in a higher average iteration period. A similar anomaly is seen between graph $C$ and graph $D$, where graph $D$ has one less synchronization edge than graph $C$, but still has a higher average iteration period. However, we observe such anomalies only within highly localized neighborhoods in which the number of synchronization edges differs by only one. Overall, in a global sense, the figure shows a clear trend of decreasing iteration period with loosening of the latency constraint, and reduction of the number of synchronization edges.

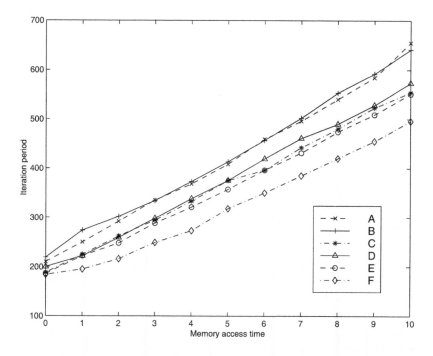

Figure 12.19. Average iteration period (reciprocal of average throughput) vs. memory access time for various latency-constrained resynchronizations of the music synthesis example in Figure 12.17.

It is difficult to model bus contention analytically, and for precise performance data we must resort to a detailed simulation of the shared bus system. Such a simulation can be used as a means of verifying that the resynchronization optimization does not result in a performance degradation due to higher bus contention. Experimental observations suggest that this needs to be done only for cases where the number of synchronization edges removed by resynchronization is small compared to the total number of synchronization edges (i.e., when the resynchronized solution is within a localized neighborhood of the original synchronization graph).

Figure 12.20 shows that the average number of shared memory accesses per graph iteration decreases consistently with loosening of the latency con-

Table 12.1. Performance results for the resynchronization of Figure 12.17. The first column gives the memory access time; "*I*" stands for "average iteration period" (the reciprocal of the average throughput); and "*M*" stands for "memory accesses per graph iteration."

| $T$ | A | | F | | B | | C | | D | | E | |
|---|---|---|---|---|---|---|---|---|---|---|---|---|
| | $I$ | $M$ | $I$ | $M$ | $I$ | $M$ | $I$ | $M$ | $I$ | $M$ | $I$ | $M$ |
| 0 | 210 | 66 | 184 | 47 | 219 | 59 | 188 | 60 | 200 | 50 | 186 | 47 |
| 1 | 250 | 67 | 195 | 43 | 274 | 58 | 225 | 58 | 222 | 50 | 222 | 47 |
| 2 | 292 | 66 | 216 | 43 | 302 | 58 | 262 | 52 | 259 | 50 | 248 | 46 |
| 3 | 335 | 64 | 249 | 43 | 334 | 58 | 294 | 54 | 298 | 50 | 288 | 45 |
| 4 | 368 | 63 | 273 | 40 | 373 | 59 | 333 | 53 | 338 | 48 | 321 | 46 |
| 5 | 408 | 63 | 318 | 43 | 413 | 58 | 375 | 53 | 375 | 49 | 357 | 47 |
| 6 | 459 | 63 | 350 | 43 | 457 | 58 | 396 | 53 | 419 | 50 | 396 | 47 |
| 7 | 496 | 63 | 385 | 43 | 502 | 58 | 442 | 53 | 461 | 51 | 431 | 47 |
| 8 | 540 | 63 | 420 | 43 | 553 | 59 | 480 | 54 | 490 | 50 | 474 | 47 |
| 9 | 584 | 63 | 455 | 43 | 592 | 58 | 523 | 53 | 528 | 50 | 509 | 47 |
| 10 | 655 | 65 | 496 | 43 | 641 | 62 | 554 | 54 | 573 | 51 | 551 | 47 |

straint. As mentioned in Chapter 11, such reduction in shared memory accesses is relevant when power consumption is an important issue, since accesses to shared memory often require significant amounts of energy.

Figure 12.21 illustrates how the placement of synchronization edges changes as the heuristic is able to attain lower synchronization costs.

Note that synchronization graphs computed by the heuristic are not necessarily identical over any of the $L_{max}$ ranges in Figure 12.18 in which the number of synchronization edges is constant. In fact, they can be significantly different. This is because even when there are no resynchronization candidates available that can reduce the net synchronization cost (that is, no resynchronization candidates for which $(|\chi(*)| > 1)$ ), the heuristic attempts to insert resynchronization edges for the purpose of increasing the connectivity; this increases the chance that subsequent resynchronization candidates will be generated for which $|\chi(*)| > 1$ , as discussed in Chapter 11. For example, Figure 12.23 shows the synchronization graph computed when $L_{max}$ is just below the amount needed to permit the minimal solution, which requires only five synchronization edges (solution $F$ ). Comparison with the graph shown in Figure 12.21(d) shows that even though these solutions have the same synchronization cost, the heuristic had

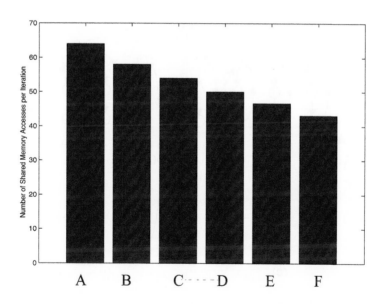

Figure 12.20. Average number of shared memory accesses per iteration for various latency-constrained resynchronizations of the music synthesis example.

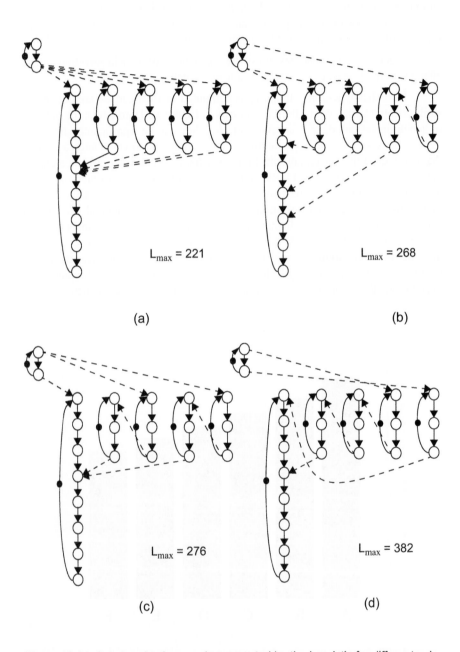

$L_{max} = 221$

(a)

$L_{max} = 268$

(b)

$L_{max} = 276$

(c)

$L_{max} = 382$

(d)

Figure 12.21. Synchronization graphs computed by the heuristic for different values of $L_{max}$.

much more room to pursue further resynchronization opportunities with $L_{max} = 644$, and thus, the graph of Figure 12.23 is more similar to the minimal solution than it is to the solution of Figure 12.21(d).

Earlier, we mentioned that the $O(s_{ff}n^4)$ and $O(ds_{ff}n^4 max(\{n, s\}))$ complexity expressions are conservative since they are based on an $n^2$ bound on the number of iterations of the **while** loop in Figure 12.15, while in practice, the actual number of **while** loop iterations can be expected to be much less than $n^2$. This claim is supported by the music synthesis example, as shown in the graph of Figure 12.22. Here, the $X$-axis corresponds to the latency constraint $L_{max}$, and the $Y$-coordinates give the number of while loop iterations that were executed by the heuristic. We see that between 5 and 13 iterations were required for each execution of the algorithm, which is not only much less than $n^2 = 484$, it is even less than $n$. This suggests that perhaps a significantly tighter bound on the number of while loop iterations can be derived.

## 12.6    Summary

This chapter has discussed the problem of latency-constrained resynchronization for self-timed implementation of iterative dataflow specifications.

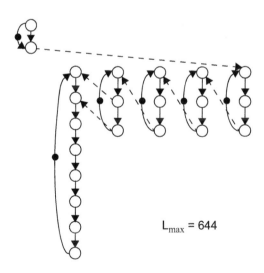

$L_{max} = 644$

Figure 12.23. The synchronization graph computed by the heuristic for $L_{max} = 644$.

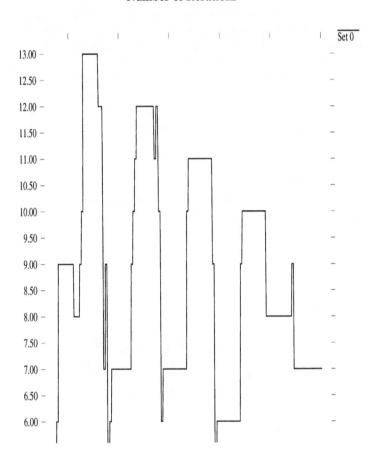

Figure 12.22. Number of resynchronization iterations versus $L_{max}$ for the example of Figure 12.17.

Given an upper bound $L_{max}$ on the allowable latency, the objective of latency-constrained resynchronization is to insert extraneous synchronization operations in such a way that a) the number of original synchronizations that consequently become redundant significantly exceeds the number of new synchronizations, and b) the serialization imposed by the new synchronizations does not increase the latency beyond $L_{max}$. To ensure that the serialization imposed by resynchronization does not degrade the throughput, the new synchronizations are restricted to lie outside of all cycles in the final synchronization graph.

In this chapter, it has been shown that optimal latency-constrained resynchronization is NP-hard even for a very restricted class of synchronization graphs. Furthermore, an efficient, polynomial-time algorithm has been demonstrated that computes optimal latency-constrained resynchronizations for two-processor systems; and the heuristic presented in the Section 11.5 for maximum-throughput resynchronization has been extended to address the problem of latency-constrained resynchronization for general $n$-processor systems. Through an example of a music synthesis system, we have illustrated the ability of this extended heuristic to systematically trade-off between synchronization overhead and latency.

The techniques developed in this chapter and Chapter 11 can be used as a post-processing step to improve the performance of any of the large number of static multiprocessor scheduling techniques for dataflow specifications, such as those described in [BPFC94, CS95, GGD94, Hoa92, LAAG94, PM91, Pri91, SL90, ZKM94].

# 13

---

# INTEGRATED SYNCHRONIZATION
# OPTIMIZATION

---

## 13.1    Computing Buffer Sizes

The previous three chapters have developed several software-based techniques for minimizing synchronization overhead for a self-timed multiprocessor implementation. After all of these optimizations are completed on a given application graph, we have a final synchronization graph $G_s = (V, E_{int} \cup E_s)$ that preserves $G_{ipc}$. Since the synchronization edges in $G_s$ are the ones that are finally implemented, it is advantageous to calculate the self-timed buffer bound $B_{fb}$ as a final step after all the transformations on $G_s$ are completed, instead of using $G_{ipc}$ itself to calculate these bounds. This is because addition of the edges in the *Convert-to-SC-graph* and *Resynchronize* steps may reduce these buffer bounds. It is easily verified that removal of edges cannot change the buffer bounds in (10-1) as long as the synchronizations in $G_{ipc}$ are preserved. Thus, in the interest of obtaining minimum possible shared buffer sizes, we compute the bounds using the optimized synchronization graph. The following theorem tells us how to compute the self-timed buffer bounds from $G_s$.

**Theorem 13.1:** If $G_s$ preserves $G_{ipc}$ and the synchronization edges in $G_s$ are implemented, then for each feedback communication edge $e$ in $G_{ipc}$, the self-timed buffer bound of $e$ ($B_{fb}(e)$) —— an upper bound on the number of data tokens that can be present on $e$ —— is given by:

$$B_{fb}(e) = \rho_{G_s}(snk(e), src(e)) + delay(e), \qquad (13\text{-}1)$$

*Proof:* By Lemma 8.1, if there is a path $p$ from $snk(e)$ to $src(e)$ in $G_s$, then

$$start(src(e), k) \geq end(snk(e), k - Delay(p)). \qquad (13\text{-}2)$$

Taking $p$ to be an arbitrary minimum-delay path from $snk(e)$ to $src(e)$ in $G_s$, we get

$$start(src(e), k) \geq end(snk(e), k - \rho_{G_s}(snk(e), src(e))). \qquad (13\text{-}3)$$

That is, $src(e)$ cannot be more that $\rho_{G_s}(snk(e), src(e))$ iterations "ahead" of $snk(e)$. Thus there can never be more that $\rho_{G_s}(snk(e), src(e))$ tokens more than the initial number of tokens on $e$ —— $delay(e)$. Since the initial number of tokens on $e$ was $delay(e)$, the size of the buffer corresponding to $e$ is bounded above by

$$B_{fb}(e) = \rho_{G_s}(snk(e), src(e)) + delay(e).$$

*QED.*

The quantities $\rho_{G_s}(snk(e), src(e))$ can be computed using Dijkstra's algorithm (Section 3.13.1) to solve the all-pairs shortest path problem on the synchronization graph in time $O(|V|^3)$.

## 13.2    A Framework for Self-Timed Implementation

To present a unified view of multiprocessor implementation issues in a concrete manner that can contribute to the development of future multiprocessor implementation tools, we introduce a flexible framework for combining arbitrary multiprocessor scheduling algorithms for iterative dataflow graphs, including the diverse set that we discussed in Chapter 6, with algorithms for optimizing IPC and synchronization costs of a given schedule, such as those covered in Chapters 10-12.

A pseudocode outline of this framework is depicted in Figure 13.1. In Step 1, an arbitrary multiprocessor scheduling algorithm is applied to construct a parallel schedule for the input dataflow graph. From the resulting parallel schedule, the IPC graph and the initial synchronization graph models are derived in Steps 2 and 3.

Then, in Steps 4-8, a series of transformations is attempted on the synchronization graph. First, Algorithm *RemoveRedundantSynchs* detects and removes all of the synchronization edges in $G_s$ whose associated synchronization functions are guaranteed by other synchronization edges in the graph, as described in Section 10.7. Step 5 then applies resynchronization to the "reduced" graph that emerges from Step 4, and incorporates any applicable latency constraints. Step 6 inserts new synchronization edges to convert the synchronization graph into a strongly connected graph so that the efficient BBS protocol can be used uniformly. Step 7 applies the *Determine Delays* Algorithm discussed in Section 10.9 to determine an efficient placement of delays on the new edges. Finally, Step 8 removes any synchronization edges that have become redundant as a result of the conversion to a strongly connected graph.

After Step 8 is complete, we have a set of IPC buffers (corresponding to the IPC edges of $G_{ipc}$) and a set of synchronization points (the synchronization edges of the transformed version of $G_s$). The main task that remains before mapping the given parallel schedule into an implementation is the determination of the buffer size — the amount of memory that must be allocated — for each IPC

**Function** *ImplementMultiprocessorSchedule*
**Input:** An iterative dataflow graph specification $G$ of a DSP application.
**Output:** An optimized synchronization graph $G_s$, an IPC graph $G_{ipc}$, and IPC buffer sizes $\{B(e) | e$ is an IPC edge in $G_{ipc}\}$.

1. Apply a multiprocessor scheduling algorithm to construct a parallel schedule for $G$ onto the given target multiprocessor architecture. The parallel schedule specifies the assignment of individual tasks to processors, and the order in which tasks execute on each processor.

2. Extract $G_{ipc}$ from $G$ and the parallel schedule constructed in Step 1.

3. Initialize $G_s = G_{ipc}$

4. $G_s = RemoveRedundantSynchs(G_s)$

5. $G_S = Resynchronize(G_s)$

6. $G_s = Convert\text{-}to\text{-}SC\text{-}graph(G_s)$

7. $G_s = DetermineDelays(G_s)$

8. $G_s = RemoveRedundantSynchs(G_s)$

9. Calculate the buffer size $B(e)$ for each IPC edge $e$ in $G_{ipc}$.
   a) Compute $\rho_{G_s}(src(e), snk(e))$, the total delay on a minimum-delay path in $G_s$ directed from $src(e)$ to $snk(e)$
   b) Set $B(e) = 1 + \rho_{G_s}(src(e), snk(e)) + delay(e)$

Figure 13.1. A framework for synthesizing multiprocessor implementations.

edge. From Theorem 13.1, and the definition of the BBS protocol in Section 10.5.1, we can compute these buffer sizes from $G_{ipc}$ and $G_s$ by the procedure outlined in Step 9 of Figure 13.1.

As we have discussed in Chapter 6, optimal derivation of parallel schedules is intractable, and a wide variety of useful heuristic approaches have emerged, with no widely accepted "best choice" among them. In contrast, the technique that we discussed in Section 10.7 for removing redundant synchronizations (Steps 4 and 8) is both optimal and of low computational complexity. However, as discussed in Chapters 11 and 12, optimal resynchronization is intractable, and although some efficient resynchronization heuristics have been developed, the resynchronization problem is very complex, and experimentation with alternative algorithms may be desirable. Similarly, the problems associated with Steps 6 and 7 are also significantly complex to perform in an optimal manner, although no result on the intractability has been derived so far.

Thus, at present, as with the parallel scheduling problem, tool developers are not likely to agree on any single "best" algorithm for each of the implementation subproblems surrounding Steps 5, 6, and 7. For example, some tool designers may wish to experiment with various evolutionary algorithms or other iterative/probabilistic search techniques on one or more of the subproblems [Dre98]. The multiprocessor implementation framework defined in Figure 13.1 addresses the inherent complexity and diversity of the subproblems associated with multiprocessor implementation of dataflow graphs by implementing a natural decomposition of the self-timed synchronization problem into a series of well-defined subproblems, and providing a systematic method for combining arbitrary algorithms that address the subproblems in isolation.

## 13.3    Summary

This section has integrated the software-based synchronization techniques developed in the Chapters 10-12 into a a single framework for the automated derivation of self-timed multiprocessor implementations. The input to this framework is an HSDFG representation of an application. The output is a processor assignment and execution ordering of application subtasks; an IPC graph $G_{ipc} = (V, E_{ipc})$, which represents buffers as communication edges; a strongly connected synchronization graph $G_s = (V, E_{int} \cup E_s)$, which represents synchronization constraints; and a set of shared-memory buffer sizes

$$\{ 1 + B_{fb}(e) | e \text{ is an IPC edge in } G_{ipc} \} . \qquad (13-4)$$

A code generator can accept $G_{ipc}$ and $G_s$ from the output of the *Minimize-SynchCost* framework, allocate a buffer in shared memory for each communication edge $e$ specified by $G_{ipc}$ of size $B_{fb}(e)$, and generate synchronization code for the synchronization edges represented in $G_s$. These synchronizations may be implemented using the *bounded buffer synchronization* (BBS) protocol. The resulting synchronization cost is $2n_s$, where $n_s$ is the number of synchronization edges in the synchronization graph $G_s$ that is obtained after all optimizations are completed.

# 14

## FUTURE RESEARCH DIRECTIONS

This book has explored techniques that minimize interprocessor communication and synchronization costs in statically scheduled multiprocessors for DSP. The main underlying theme is that communication and synchronization in statically scheduled hardware is fairly predictable, and this predictability can be exploited to achieve our aims of low overhead parallel implementation at low hardware cost. The first technique described was the ordered transactions strategy, where the idea is to predict the order of processor accesses to shared resources and enforce this order at run-time. An application of this idea to a shared bus multiprocessor was described, where the sequence of accesses to shared memory is predetermined at compile time and enforced at run-time by a controller implemented in hardware. A prototype of this architecture, called the ordered memory access architecture, demonstrates how low overhead IPC can be achieved at low hardware cost for the class of DSP applications that can be specified as SDF graphs, provided good compile time estimates of execution times exist. We also introduced the IPC graph model for modeling self-timed schedules. This model was used to show that we can determine a particular transaction order such that enforcing this order at run time does not sacrifice performance when actual execution times of tasks are close to their compile time estimates. When actual running times differ from the compile time estimates, the computation performed is still correct, but the performance (throughput) may be affected. We described how to quantify such effects of run time variations in execution times on the throughput of a given schedule.

The ordered transactions approach also extends to graphs that include constructs with data-dependent firing behavior. We discussed how conditional constructs and data-dependent iteration constructs can be mapped to the OMA architecture, when the number of such control constructs is small — a reasonable assumption for most DSP algorithms.

Finally, we described techniques for minimizing synchronization costs in a self-timed implementation that can be achieved by systematically manipulating the synchronization points in a given schedule; the IPC graph construct was used for this purpose. The techniques described include determining when certain synchronization points are redundant, transforming the IPC graph into a strongly connected graph, and then sizing buffers appropriately such that checks for buffer overflow by the sender can be eliminated. We also outlined a technique we call resynchronization, which introduces new synchronization points in the schedule with the objective of minimizing the overall synchronization cost.

The work presented in this book leads to several open problems and directions for further research.

Mapping a general BDF graph onto the OMA architecture to make the best use of our ability to switch between bus access schedules at run time is a topic that requires further study. Techniques for multiprocessor scheduling of BDF graphs could build upon the quasi-static scheduling approach, which restricts itself to certain types of dynamic constructs that need to be identified (for example as conditional constructs or data-dependent iterations) before scheduling can proceed. Assumptions regarding statistics of the Boolean tokens (e.g., the proportion of TRUE values that a control token assumes during the execution of the schedule) would be required for determining multiprocessor schedules for BDF graphs.

The OMA architecture applies the ordered transactions strategy to a shared bus multiprocessor. If the interprocessor communication bandwidth requirements for an application are higher than what a single shared bus can support, a more elaborate interconnect, such as a crossbar or a mesh topology, may be required. If the processors in such a system run a self-timed schedule, the communication pattern is again periodic and we can predict this pattern at compile time. We can then determine the states that the crossbar in such a system cycles through or we can determine the sequence of settings for the switches in the mesh topology. The fact that this information can be determined at compile time should make it possible to simplify the hardware associated with these interconnect mechanisms, since the associated switches need not be configured at run time. Exactly how this compile time information can be made use of for simplifying the hardware in such interconnects is an interesting problem for further study.

In the techniques we proposed in Chapters 10 through 12 for minimizing synchronization costs, no assumptions regarding bounds on execution times of actors in the graph were made. A direction for further work is to incorporate timing guarantees — for example, hard upper and lower execution time bounds, as Dietz, Zaafrani, and O'Keefe use in [DZO92]; and handling of a mix of actors, some of which have guaranteed execution time bounds, and others that have no such guarantees, as Filo, Ku, Coelho Jr., and De Micheli do in [FKCM93]. Such guarantees could be used to detect situations in which data will always be available before it is needed for consumption by another processor.

Also, execution time guarantees can be used to compute tighter buffer size bounds. As a simple example, consider Figure 14.1. Here, the analysis of Section 10.4 yields a buffer size $B_{fb}((A, B)) = 3$, since 3 is the minimum path delay of a cycle that contains $(A, B)$. However, if $t(A)$ and $t(B)$, the execution times of actors $A$ and $B$, are guaranteed to be equal to the same constant, then it is easily verified that a buffer size of 1 will suffice for $(A, B)$. Systematically applying execution time guarantees to derive lower buffer size bounds appears to be a promising direction for further work.

Several useful directions for further work emerge from the concept of self-timed resynchronization described in Chapters 11 and 12. These include investigating whether efficient techniques can be developed that consider resynchronization opportunities within strongly connected components, rather than just across feedforward edges. There may also be considerable room for improvement over the resynchronization heuristics that we have discussed, which are straightforward adaptations of an existing set-covering algorithm. In particular, it would be useful to explore ways to best integrate the heuristics for general synchronization graphs with the optimal chaining method for a restricted class of graphs, and it may be interesting to search for properties of practical synchronization graphs that could be exploited in addition to the correspondence with set covering. The extension of Sarkar's concept of counting semaphores [Sar89] to self-timed, iterative execution, and the incorporation of extended counting semaphores within the framework of self-timed resynchronization, are also interesting directions for further work.

Another interesting problem is applying the synchronization minimization techniques to graphs that contain dynamic constructs, such as graphs expressed using the dynamically-oriented modeling techniques covered in Chapter 4. Sup-

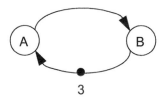

Figure 14.1. An example of how execution time guarantees can be used to reduce buffer size bounds.

pose we schedule a graph that contains dynamic constructs using a quasi-static approach, or a more general approach if one becomes available. Is it still possible to employ the synchronization optimization techniques we discussed in Chapters 10-12? The first step to take would be to obtain an IPC graph equivalent for the quasi-static schedule that has a representation for the control constructs that a processor may execute as a part of the quasi-static schedule. If we can show that the conditions we established for a synchronization operation to be redundant (in Section 10.7) holds for all execution paths in the quasi-static schedule, then we could identify redundant synchronization points in the schedule. It may also be possible to extend the strongly-connect and resynchronization transformations to handle graphs containing conditional constructs; these issues require further investigation.

Also, existing quasi-static scheduling approaches such as Ha's techniques [HL97] do not take the communication overhead of broadcasting control tokens to all the processors in the system into account. Development of quasi-static multiprocessor scheduling approaches that thoroughly take communication overhead into account is an interesting area for future study.

# BIBLIOGRAPHY

[A+87] M. Annaratone et al. The Warp computer: Architecture, implementation, and performance. *IEEE Transactions on Computers*, C-36(12), December 1987.

[A+98] S. Aiello et al. Extending a monoprocessor real-time system in a DSP-based multiprocessing environment. In *Proceedings of the International Conference on Acoustics, Speech, and Signal Processing*, 1998.

[ABU91] Arvind, L. Bic, and T. Ungerer. Evolution of dataflow computers. In *Advanced Topics in Data-flow Computing*. Prentice Hall, 1991.

[ABZ92] A. L. Ambler, M. M. Burnett, and B. A. Zimmerman. Operational versus definitional: A perspective on programming paradigms. *IEEE Computer Magazine*, 25(9), September 1992.

[ACD74] T. L. Adam, K. M. Chandy, and J. R. Dickson. A comparison of list schedules for parallel processing systems. *Communications of the ACM*, 17(12):685–690, December 1974.

[Ack82] W. B. Ackerman. Data flow languages. *IEEE Computer Magazine*, 15(2), February 1982.

[AHU87] A. V. Aho, J. E. Hopcroft, and J. D. Ullman. *Data Structures and Algorithms*. Addison-Wesley, 1987.

[AK87] R. Allen and D. Kennedy. Automatic transformations of FORTRAN programs to vector form. *ACM Transactions on Programming Languages and Systems*, 9(4), October 1987.

[AK98] H. Andrade and S. Kovner. Software synthesis from dataflow models for embedded software design in the G programming language and the LabVIEW development environment. In *Proceedings of the IEEE Asilomar Conference on Signals, Systems, and Computers*, November 1998.

[Amb08] Ambric, Inc. *Ambric Technology Backgrounder*, January 2008. http://www.ambric.com/technology/technology-overview.php.

[AN88] A. Aiken and A. Nicolau. Optimal loop parallelization. In *Proceedings of the ACM Conference on Programming Language Design and Implementation*, 1988.

[AN90] Arvind and R. S. Nikhil. Executing a program on the MIT tagged-token dataflow architecture. *IEEE Transactions on Computers*, 39(3), March 1990.

[Ari91] Ariel Corporation. *User's Manual for the S-56X*, 1991.

[ASI+98] A. Abnous, K. Seno, Y. Ichikawa, M. Wan, and J. Rabaey. Evaluation of a low-power reconfigurable DSP architecture. In *Proceedings of the Reconfigurable Architectures Workshop*, March 1998.

[ATT87] AT&T Bell Laboratories, *Enhanced Modular Signal Processing Arithmetic Processor Programmers Handbook*, 1987.

[B+88] S. Borkar et al. iWarp: An integrated solution to high-speed parallel computing. In *Proceedings of Supercomputing*, 1988.

[BB91] A. Benveniste and G. Berry. The synchronous approach to reactive and real-time systems. *Proceedings of the IEEE*, 79(9):1270–1282, September 1991.

[BB00] B. Bhattacharya and S. S. Bhattacharyya. Quasi-static scheduling of reconfigurable dataflow graphs for DSP systems. In *Proceedings of the International Workshop on Rapid System Prototyping*, pages 84-89, Paris, France, June 2000.

[BB01] B. Bhattacharya and S. S. Bhattacharyya. Parameterized dataflow modeling for DSP systems. *IEEE Transactions on Signal Processing*, 49(10):2408-2421, October 2001.

[BCOQ92] F. Baccelli, G. Cohen, G. J. Olsder, and J. Quadrat. *Synchronization and Linearity*. John Wiley & Sons, Inc., 1992.

[BDTI07] Berkeley Design Technology, Inc. *Massively Parallel Processors Targeting Digital Signal Processing Applications*, August 2007. http://www.bdti.com/articles/0706_arrayprocessor_table.pdf.

[BDTI94] Berkeley Design Technology, Inc. DSP tools get integrated. *Electronic Engineering Times*, February 1994.

[BDW85] J. Beetem, M. Denneau, and D. Weingarten. The GF11 supercomputer. In *International Symposium on Computer Architecture*, June 1985.

[BELP94] G. Bilsen, M. Engels, R. Lauwereins, and J. A. Peperstraete. Static scheduling of multi-rate and cyclo-static DSP-applications. In *Proceedings of the International Workshop on VLSI Signal Processing*, 1994.

[BELP96] G. Bilsen, M. Engels, R. Lauwereins, and J. A. Peperstraete. Cyclo-static dataflow. *IEEE Transactions on Signal Processing*, 44(2):397-408, February 1996.

[BGW03] L. Baumstark, M. Guler, and L. Wills. Extracting an explicitly data-parallel representation of image-processing programs. In *Proceedings of the Working Conference on Reverse Engineering*, 2003.

[BHCF95] S. Banerjee, T. Hamada, P. M. Chau, and R. D. Fellman. Macro pipelining based scheduling on high performance heterogeneous multiprocessor systems. *IEEE Transactions on Signal Processing*, 43(6):1468–1484, June 1995.

[BHLM94] J. T. Buck, S. Ha, E. A. Lee, and D. G. Messerschmitt. Ptolemy: A framework for simulating and prototyping heterogeneous systems. *International Journal of Computer Simulation*, January 1994.

[BHMSVc84] R. K. Brayton, G. D. Hachtel, C. T. McMullen, and A. L. Sangiovanni-Vincentelli. *Logic Minimization Algorithms for VLSI Synthesis*. Kluwer Academic Publishers, 1984.

[BL89] J. Bier and E. A. Lee. Frigg: A simulation environment for multiprocessor DSP system development. In *Proceedings of the International Conference on Computer Design*, pages 280–283, October 1989.

[BL91] B. Barrera and E. A. Lee. Multirate signal processing in Comdisco's SPW. In *Proceedings of the International Conference on Acoustics, Speech, and Signal Processing*, April 1991.

[BL93] S. S. Bhattacharyya and Edward A. Lee. Scheduling synchronous dataflow graphs for efficient looping. *Journal of VLSI Signal Processing*, 1993.

[BL94] S. S. Bhattacharyya and E. A. Lee. Memory management for dataflow programming of multirate signal processing algorithms. *IEEE Transactions on Signal Processing*, 42(5):1190–1201, May 1994.

[Bla87] J. Blazewicz. Selected topics in scheduling theory. In *Surveys in Combinatorial Optimization*. North Holland Mathematica Studies, 1987.

[BML96] S. S. Bhattacharyya, P. K. Murthy, and E. A. Lee. *Software Synthesis from Dataflow Graphs*. Kluwer Academic Publishers, 1996.

[Bok88] S. H. Bokhari. Partitioning problems in parallel, pipelined, and distributed computing. *IEEE Transactions on Computers*, 37(1):48–57, January 1988.

[Bor88] G. Borriello. Combining events and data-flow graphs in behavioral synthesis. In *Proceedings of the International Conference on Computer-Aided Design*, pages 56–59, 1988.

[BPFC94] S. Banerjee, D. Picker, D. Fellman, and P. M. Chau. Improved scheduling of signal flow graphs onto multiprocessor systems through an accurate network modeling technique. In *Proceedings of the International Workshop on VLSI Signal Processing*, 1994.

[Bry86] R. E. Bryant. Graph based algorithms for boolean function manipulation. *IEEE Transactions on Computers*, 35(8):677–691, August 1986.

[BSL90] J. Bier, S. Sriram, and E. A. Lee. A class of multiprocessor architectures for real-time DSP. In *Proceedings of the International Workshop on VLSI Signal Processing*, November 1990.

[BSL96a] S. S. Bhattacharyya, S. Sriram, and E. A. Lee. Latency-constrained resynchronization for multiprocessor DSP implementation. In *Proceedings of the International Conference on Application Specific Systems, Architectures, and Processors*, August 1996. Chicago, Illinois.

[BSL96b] S. S. Bhattacharyya, S. Sriram, and E. A. Lee. Self-timed resynchronization: A post-optimization for static multiprocessor schedules. In *Proceedings of the International Parallel Processing Symposium*, April 1996. Honolulu, Hawaii.

[BSL97] S. S. Bhattacharyya, S. Sriram, and E. A. Lee. Optimizing synchronization in multiprocessor DSP systems. *IEEE Transactions on Signal Processing*, 45(6), June 1997.

[Buc93] J. T. Buck. *Scheduling Dynamic Dataflow Graphs with Bounded Memory using the Token Flow Model*. Ph.D. thesis, Department of Electrical Engineering and Computer Sciences, University of California at Berkeley, September 1993.

[Buck94] J. T. Buck. Static scheduling and code generation from dynamic dataflow graphs with integer-valued control systems. In *Proceedings of the IEEE Asilomar Conference on Signals, Systems, and Computers*, pages 508-513, October 1994.

[BV00] J. Buck and R. Vaidyanathan. Heterogeneous modeling and simulation of embedded systems in El Greco. In *Proceedings of the International Workshop on Hardware/Software Co-Design*, May 2000.

[CDQ85] G. Cohen, D. Dubois, and J. Quadrat. A linear system theoretic view of discrete event processes and its use for performance evaluation in manufacturing. *IEEE Transactions on Automatic Control*, March 1985.

[CG79] R. Cunningham-Green. Minimax algebra. In *Lecture Notes in Economics and Mathematical Systems*. Springer-Verlag, 1979.

[Cha84] M. Chase. A pipelined data flow architecture for digital signal processing: The NEC μPD7281. In *Proceedings of the International Workshop on VLSI Signal Processing*, November 1984.

[Chr85] P. Chretienne. Timed event graphs: A complete study of their controlled executions. In *International Workshop on Timed Petri Nets*, July 1985.

[Chr89] P. Chretienne. Task scheduling over distributed memory machines. In *Proceedings of the International Workshop on Parallel and Distributed Algorithms*, 1989.

[CLC00] N. Cossement, R. Lauwereins, and F. Catthoor. DF*: An extension of synchronous dataflow with data dependency and non-determinism. In *Proceedings of the Forum on Design Languages*, September 2000.

[CL94] M. J. Chen and E. A. Lee. Design and implementation of a multidimensional synchronous dataflow environment. In *Proceedings of the IEEE Asilomar Conference on Signals, Systems, and Computers*, 1994.

[CLR92] T. H. Cormen, C. E. Leiserson, and R. L. Rivest. *Introduction to Algorithms*. MIT Press, 1992.

[Cof76] E. G. Coffman, Jr. *Computer and Job Shop Scheduling Theory*. John Wiley & Sons, Inc., 1976.

[CR92] D. C. Chen and J. M. Rabaey. A reconfigurable multiprocessor IC for rapid prototyping of algorithm-specific high-speed DSP data paths. *IEEE Journal of Solid State Circuits*, 27(12), December 1992.

[CS92] L. F. Chao and E. Sha. Unfolding and retiming data-flow DSP programs for RISC multiprocessor scheduling. In *Proceedings of the International Conference on Acoustics, Speech, and Signal Processing*, April 1992.

[CS93] J. Campos and M. Silva. Structural techniques and performance bounds of stochastic Petri net models. In *Advances in Petri Nets*. Springer-Verlag, 1993.

[CS95] L. Chao and E. H. Sha. Static scheduling for synthesis of DSP algorithms on various models. *Journal of VLSI Signal Processing*, pages 207–223, 1995.

[D+96] R. Davoli et al. Parallel computing in networks of workstations with Paralex. *IEEE Transactions on Parallel and Distributed Systems*, 7(4), April 1996.

[Dal07] B. Dally. Parallel processing simplified -Storm-1 SP16HP: A 112 GMAC, C-programmable media and signal processor. *Microprocessor Forum*, 2007.

[De 94] G. De Micheli. *Synthesis and Optimization of Digital Circuits*. McGraw-Hill, 1994.

[DBC+07] K. Denolf, M. Bekooij, J. Cockx, D. Verkest, and H. Corporaal. Exploiting the expressiveness of cyclo-static dataflow to model multimedia implementations. *EURASIP Journal on Advances in Signal Processing*, 2007:Article ID 84078, 14 pages, 2007.

[Den80] J. B. Dennis. Dataflow supercomputers. *IEEE Computer Magazine*, 13(11), November 1980.

[DG98] A. Dasdan and R. K. Gupta. Faster maximum and minimum mean cycle algorithms for system-performance analysis. *IEEE Transactions on Computer-Aided Design*, 17(10):889–899, October 1998.

[dGH92] S. M. H. de Groot, S. Gerez, and O. Herrmann. Range-chart-guided iterative data-flow graph scheduling. *IEEE Transactions on Circuits and Systems*, pages 351–364, May 1992.

[DIB96] F. Du, A. Izatt, and C. Bandera. MIMD computing platform for a hierarchical foveal machine vision system. In *Proceedings of the IEEE Computer Society Conference on Computer Vision and Pattern Recognition*, pages 720–725, 1996.

[Dij59] E. W. Dijkstra. A note on two problems in connexion with graphs. *Numer. Math.*, pages 269–271, 1959.

[Dre98] R. Drechsler. *Evolutionary Algorithms for VLSI CAD*. Kluwer Academic Publishers, 1998.

[DSBS06] E. F. Deprettere, T. Stefanov, S. S. Bhattacharyya, and M. Sen. Affine nested loop programs and their binary cyclo-static dataflow counterparts. In *Proceedings of the International Conference on Application Specific Systems, Architectures, and Processors*, pages 186-190, Steamboat Springs, Colorado, September 2006.

[DSV+00] E. A. de Kock, W. J. M. Smits, P. van der Wolf, J.-Y. Brunel, W. M. Kruijtzer, P. Lieverse, K. A. Vissers, and G. Essink. YAPI: application modeling for signal processing systems. In *Proceedings of the Design Automation Conference*, 2000.

[Dur91] R. Durett. *Probability: Theory and Examples*. Wadsworth & Brooks/ Cole, 1991.

[DZO92] H. G. Dietz, A. Zaafrani, and M. T. O'Keefe. Static scheduling for barrier MIMD architectures. *Journal of Supercomputing*, 5(4), 1992.

[EJ03a] J. Eker and J. W. Janneck. CAL language report, language version 1.0 — document edition 1. Technical Report UCB/ERL M03/48, Electronics Research Laboratory, University of California at Berkeley, December 2003.

[EJ03b] J. Eker and J. W. Janneck. A structured description of dataflow actors and its application. Technical Report M03/13, Electronics Research Laboratory, University of California at Berkeley, May 2003.

[EK72] J. Edmonds and R. M. Karp. Theoretical improvements in algorithmic efficiency for network flow algorithms. *Journal of the Association for Computing Machinery*, pages 248–264, April 1972.

[EK97] M. Eden and M. Kagan. The Pentium(R) processor with MMX technology. In *Proceedings of the IEEE Computer Society International Conference*, 1997.

[ERL90] H. El-Rewini and T. G. Lewis. Scheduling parallel program tasks onto arbitrary target machines. *Journal of Parallel and Distributed Computing*, pages 138–153, 1990.

[F+97] R. Fromm et al. The energy efficiency of IRAM architectures. In *International Symposium on Computer Architecture*, June 1997.

[FKCM93] D. Filo, D. C. Ku, C. N. Coelho Jr., and G. De Micheli. Interface optimization for concurrent systems under timing constraints. *IEEE Transactions on Very Large Scale Integration (VLSI) Systems*, 1(3), September 1993.

[FKM92] D. Filo, D. C. Ku, and G. De Micheli. Optimizing the control-unit through the resynchronization of operations. *INTEGRATION, the VLSI Journal*, pages 231–258, 1992.

[Fly66] M. J. Flynn. Very high-speed computing systems. *Proceedings of the IEEE*, December 1966.

[FWM07] S. Fischaber, R. Woods, and J. McAllister. SoC memory hierarchy derivation from dataflow graphs. In *Proceedings of the IEEE Workshop on Signal Processing Systems*, October 2007.

[G+91] M. Gokhale et al. Building and using a highly programmable logic array. *IEEE Computer Magazine*, 24(1):81–89, January 1991.

[G+92] A. Gunzinger et al. Architecture and realization of a multi signal processor system. In *Proceedings of the International Conference on Application Specific Array Processors*, pages 327–340, 1992.

[G+07] J. Glossner, D. Iancu, M. Moudgill, G. Nacer, S. Jinturkar, S. Stanley, and M. Schulte. The Sandbridge SB3011 platform. *EURASIP Journal on Embedded Systems*, 2007.

[GB91] J. Gaudiot and L. Bic, editors. *Advanced Topics in Data-flow Computing*. Prentice Hall, 1991.

[GB04] M. Geilen and T. Basten. Reactive process networks. In *Proceedings of the International Workshop on Embedded Software*, pages 137-146, September 2004.

[Ger95] S. Van Gerven. Multiple beam broadband beamforming: Filter design and real-time implementation. In *Proceedings of the IEEE ASSP Workshop on Applications of Signal Processing to Audio and Acoustics*, 1995.

[GGA92] K. Guttag, R. J. Grove, and J. R. Van Aken. A single-chip multiprocessor for multimedia: the MVP. *IEEE Computer Graphics and Applications*, 12(6),

November 1992.

[GGD94] R. Govindarajan, G. R. Gao, and P. Desai. Minimizing memory requirements in rate-optimal schedules. In *Proceedings of the International Conference on Application Specific Array Processors*, August 1994.

[GJ79] M. R. Garey and D. S. Johnson. *Computers and Intractability: A Guide to the Theory of NP-Completeness*. W. H. Freeman and Company, 1979.

[GLL99] A. Girault, B. Lee, and E. A. Lee. Hierarchical finite state machines with multiple concurrency models. *IEEE Transactions on Computer-Aided Design of Integrated Circuits and Systems*, 18(6):742-760, June 1999.

[GMN96] B. Gunther, G. Milne, and L. Narasimhan. Assessing document relevance with run-time reconfigurable machines. In *Proceedings of the IEEE Symposium on FPGAs for Custom Computing Machines*, pages 10–17, April 1996.

[Gra69] R. L. Graham. Bounds on multiprocessing timing anomalies. *SIAM Journal of Applied Math*, 17(2):416–429, March 1969.

[Gri84] J.D. Grimm, J.A. Eggert, and G.W. Karcher, Distributed signal processing using dataflow techniques, in *Proc. 17th Hawaii Intl. Conf. Syst. Sci.*, Jan. 1984, pp. 2938.

[Gri88] C. M. Grinstead. Cycle lengths in $A^k b^*$. *SIAM Journal on Matrix Analysis*, October 1988.

[GS92] F. Gasperoni and U. Schweigelshohn. Scheduling loops on parallel processors: A simple algorithm with close to optimum performance. In *Proceedings of the International Conference on Vector & Parallel Processors*, September 1992.

[GVNG94] D. J. Gajski, F. Vahid, S. Narayan, and J. Gong. *Specification and Design of Embedded Systems*. Prentice Hall, 1994.

[GW92] B. Greer and J. Webb. Real-time supercomputing on iWarp. In *Proceedings of the SPIE*, 1992.

[GY92] A. Gerasoulis and T. Yang. A comparison of clustering heuristics for scheduling directed graphs on multiprocessors. *Journal of Parallel and Distributed Computing*, 16:276–291, 1992.

[H+93] J. A. Huisken et al. Synthesis of synchronous communication hardware in a multiprocessor architecture. *Journal of VLSI Signal Processing*, 6:289–299, 1993.

[Ha92] S. Ha. *Compile Time Scheduling of Dataflow Program Graphs with Dynamic Constructs*. Ph.D. thesis, Department of Electrical Engineering and Computer Sciences, University of California at Berkeley, April 1992.

[Hal93] N. Halbwachs. *Synchronous Programming of Reactive Systems*. Kluwer Academic Publishers, 1993.

[Hav91] B. R. Haverkort. Approximate performability analysis using generalized stochastic Petri nets. In *Proceedings of the International Workshop on Petri Nets and Performance Models*, pages 176–185, 1991.

[HCA89] J. J. Hwang, Y. C. Chow, and F. D. Anger. Scheduling precedence graphs in systems with inter-processor communication times. *SIAM Journal of Computing*, 18(2):244–257, April 1989.

[HCK+07] C Hsu, I. Corretjer, M. Ko., W. Plishker, and S. S. Bhattacharyya. Dataflow interchange format: Language reference for DIF language version 1.0, user's guide for DIF package version 1.0. Technical Report UMIACS-TR-2007-32, Institute for Advanced Computer Studies, University of Maryland at College Park, June 2007.

[HCRP91] N. Halbwachs, P. Caspi, P. Raymond, and D. Pilaud. The synchronous data flow programming language LUSTRE. *Proceedings of the IEEE*, September 1991.

[HFHK97] S. Hauck, T. W. Fry, M. M. Hosler, and J. P. Kao. The Chimaera reconfigurable functional unit. In *Proceedings of the IEEE Symposium on FPGAs for Custom Computing Machines*, pages 105–116, April 1997.

[HFK+07] C. Haubelt, J. Falk, J. Keinert, T. Schlichter, M. Streubühr, A. Deyhle, A. Hadert, and J. Teich. A SystemC-based design methodology for digital signal processing systems. *EURASIP Journal on Embedded Systems*, 2007:Article ID 47580, 22 pages, 2007.

[HGHP96] K. Herrmann, K. Gaedke, J. Hilgenstock, and P. Pirsch. Design of a development system for multimedia applications based on a single chip multiprocessor array. In *Proceedings of the International Conference on Electronics, Circuits, and Systems*, pages 1151–1154, 1996.

[HKB05] C. Hsu, M. Ko, and S. S. Bhattacharyya. Software synthesis from the dataflow interchange format. In *Proceedings of the International Workshop on Software and Compilers for Embedded Systems*, pages 37-49, Dallas, Texas, September 2005.

[HL97] S. Ha and E. A. Lee. Compile-time scheduling of dynamic constructs in dataflow program graphs. *IEEE Transactions on Computers*, 46(7), July 1997.

[HO93] M. Hartmann and J. B. Orlin. Finding minimum cost to time ratio cycles with small integral transit times. *Networks*, September 1993.

[HO98] L. Hammond and K. Olukotun. Considerations in the design of hydra: A multiprocessor-on-a-chip microarchitecture. Technical Report CSL-TR-98-749,

Stanford University Computer Systems Lab, February 1998.

[Hoa92] P. Hoang. *Compiling Real Time Digital Signal Processing Applications onto Multiprocessor Systems*. Ph.D. thesis, Department of Electrical Engineering and Computer Sciences, University of California at Berkeley, June 1992.

[HR92] P. Hoang and J. Rabaey. Hierarchical scheduling of DSP programs onto multiprocessors for maximum throughput. In *Proceedings of the International Conference on Application Specific Array Processors*, August 1992.

[HR96] M. Hunt and J. A. Rowson. Blocking in a system on a chip. *IEEE Spectrum Magazine*, pages 35–41, November 1996.

[Hu61] T. C. Hu. Parallel sequencing and assembly line problems. *Operations Research*, 9, 1961.

[HW97] J. R. Hauser and J. Wawrzynek. Garp: A MIPS processor with a reconfigurable coprocessor. In *Proceedings of the IEEE Symposium on FPGAs for Custom Computing Machines*, pages 24–33, April 1997.

[Joh74] D. S. Johnson. Approximation algorithms for combinatorial problems. *Journal of Computer and System Sciences*, pages 256–278, 1974.

[KA96] Y. Kwok and I. Ahmad. Dynamic critical path scheduling: An effective technique for allocating task graphs to multiprocessors. *IEEE Transactions on Parallel and Distributed Systems*, 7(5):506–521, May 1996.

[Kah74] G. Kahn. The semantics of a simple language for parallel programming. In *Proceedings of the IFIP Congress*, 1974.

[Kar78] R. Karp. A note on the characterization of the minimum cycle mean in a digraph. *Discrete Mathematics*, 23, 1978.

[KB88] S. J. Kim and J. C. Browne. A general approach to mapping of parallel computation upon multiprocessor architectures. In *Proceedings of the International Conference on Parallel Processing*, pages 1–8, 1988.

[KBB06] M. Khandelia, N. K. Bambha, and S. S. Bhattacharyya. Contention-conscious transaction ordering in multiprocessor DSP systems. *IEEE Transactions on Signal Processing*, 54(2):556-569, February 2006.

[KD01] B. Kienhuis and E. F. Deprettere. Modeling stream-based applications using the SBF model of computation. In *Proceedings of the IEEE Workshop on Signal Processing Systems*, pages 385-394, September 2001.

[KFHT07] J. Keinert, J. Falk, C. Haubelt, and J. Teich. Actor-oriented modeling and simulation of sliding window image processing algorithms. In *Proceedings of the IEEE Workshop on Embedded Systems for Real-Time Multimedia*, October 2007.

[KHT06] J. Keinert, C. Haubelt, and J. Teich. Modeling and analysis of windowed synchronous algorithms. In *Proceedings of the International Conference on Acoustics, Speech, and Signal Processing,* May 2006.

[KHT07] J. Keinert, C. Haubelt, and J. Teich. Simulative buffer analysis of local image processing algorithms described by windowed synchronous data flow. In *Proceedings of the International Conference on Embedded Computer Systems: Architectures, Modeling, and Simulation,* July 2007.

[KHT08] J. Keinert, C. Haubelt, and J. Teich. Synthesis of multi-dimensional high-speed FIFOs for out-of-order communication. In *Proceedings of the International Conference on Architecture of Computing Systems,* February 2008.

[Kim88] S. J. Kim. *A General Approach to Multiprocessor Scheduling.* Ph.D. thesis, Department of Computer Science, University of Texas at Austin, 1988.

[KL93] A. Kalavade and E. A. Lee. A hardware/software codesign methodology for DSP applications. *IEEE Design and Test of Computers Magazine,* 10(3):16–28, September 1993.

[KLL87] S. Y. Kung, P. S. Lewis, and S. C. Lo. Performance analysis and optimization of VLSI dataflow arrays. *Journal of Parallel and Distributed Computing,* pages 592–618, 1987.

[KM66] R. M. Karp and R. E. Miller. Properties of a model for parallel computations: Determinacy, termination, queueing. *SIAM Journal of Applied Math,* 14(6), November 1966.

[KM92] D.C. Ku and G. De Micheli. Relative scheduling under timing constraints: Algorithms for high-level synthesis of digital circuits. *IEEE Transactions on Computer-Aided Design,* 11(6):696–718, June 1992.

[Koh75] W. H. Kohler. A preliminary evaluation of critical path method for scheduling tasks on multiprocessor systems. *IEEE Transactions on Computers,* pages 1235–1238, December 1975.

[Koh90] W. Koh. *A Reconfigurable Multiprocessor System for DSP Behavioral Simulation.* Ph.D. thesis, Department of Electrical Engineering and Computer Sciences, University of California at Berkeley, June 1990.

[Kru87] B. Kruatrachue. *Static Task Scheduling and Grain Packing in Parallel Processing Systems.* Ph.D. thesis, Department of Computer Science, Oregon State University, 1987.

[KS83] K. Karplus and A. Strong. Digital synthesis of plucked-string and drum timbres. *Computer Music Journal,* 7(2):56–69, 1983.

[KSB08] M. Ko, C. Shen, and S. S. Bhattacharyya. Memory-constrained block processing for DSP software optimization. *Journal of Signal Processing Systems*, 50(2):163-177, February 2008.

[Kun88] S. Y. Kung. *VLSI Arrays Processors*. Prentice Hall, Englewood Cliffs, N.J., 1988.

[KZP+07] M. Ko, C. Zissulescu, S. Puthenpurayil, S. S. Bhattacharyya, B. Kienhuis, and E. Deprettere. Parameterized looped schedules for compact representation of execution sequences in DSP hardware and software implementation. *IEEE Transactions on Signal Processing*, 55(6):3126-3138, June 2007.

[LAAG94] G. Liao, E. R. Altman, V. K. Agarwal, and G. R. Gao. A comparative study of DSP multiprocessor list scheduling heuristics. In *Proceedings of the Hawaii International Conference on System Sciences*, 1994.

[Lam86] L. Lamport. The mutual exclusion problem: Part I and II. *Journal of the Association for Computing Machinery*, 33(2):313–348, April 1986.

[Lam88] M. Lam. Software pipelining: An effective scheduling technique for VLIW machines. In *Proceedings of the ACM Conference on Programming Language Design and Implementation*, pages 318–328, June 1988.

[Lam89] M. Lam. *A Systolic Array Optimizing Compiler*. Kluwer Academic Publishers, 1989.

[Lap91] P. D. Lapsley. Host interface and debugging of dataflow DSP systems. Master's thesis, Department of Electrical Engineering and Computer Sciences, University of California at Berkeley, December 1991.

[Law76] E. L. Lawler. *Combinatorial Optimization: Networks and Matroids*. Holt, Rinehart and Winston, 1976.

[LB90] E. A. Lee and J. C. Bier. Architectures for statically scheduled dataflow. *Journal of Parallel and Distributed Computing*, 10:333–348, December 1990.

[LBSL94] P. Lapsley, J. Bier, A. Shoham, and E. A. Lee. *DSP Processor Fundamentals*. Berkeley Design Technology, Inc., 1994.

[LDK98] S. Y. Liao, S. Devadas, and K. Keutzer. Code density optimization for embedded DSP processors using data compression techniques. *IEEE Transactions on Computer-Aided Design*, 17(7):601–608, July 1998.

[LEAP94] R. Lauwereins, M. Engels, M. Ade, and J. A. Peperstraete. Grape-ii: Graphical rapid prototyping environment for digital signal processing systems. In *Proceedings of the International Conference on Signal Processing Applications and Technology*, 1994.

[Lee86] E. A. Lee. *A Coupled Hardware and Software Architecture for Programmable DSPs*. Ph.D. thesis, Department of Electrical Engineering and Computer Sciences, University of California at Berkeley, May 1986.

[Lee88a] E. A. Lee. Programmable DSP architectures — Part I. *IEEE ASSP Magazine*, 5(4), October 1988.

[Lee88b] E. A. Lee. Recurrences, iteration, and conditionals in statically scheduled block diagram languages. In *Proceedings of the International Workshop on VLSI Signal Processing*, 1988.

[Lee91] E. A. Lee. Consistency in dataflow graphs. *IEEE Transactions on Parallel and Distributed Systems*, 2(2), April 1991.

[Lee93] E. A. Lee. Representing and exploiting data parallelism using multidimensional dataflow diagrams. In *Proceedings of the International Conference on Acoustics, Speech, and Signal Processing*, pages 453–456, April 1993.

[Lee96] R. B. Lee. Subword parallelism with MAX2. *IEEE Micro*, 16(4), August 1996.

[Lei92] F. T. Leighton. *Introduction to Parallel Algorithms and Architectures: Arrays, Trees, Hypercubes*. Morgan Kaufmann Publishers Inc., 1992.

[LEP90] R. Lauwereins, M. Engels, J.A. Peperstraete, E. Steegmans, and J. Van Ginderdeuren. GRAPE: A CASE tool for digital signal parallel processing. *IEEE ASSP Magazine*, 7(3), April 1990.

[Leu02] R. Leupers. Compiler design issues for embedded processors. *IEEE Design and Test of Computers Magazine*, August 2002.

[LH89] E. A. Lee and S. Ha. Scheduling strategies for multiprocessor real time DSP. In *Global Telecommunications Conference*, November 1989.

[LLG+92] D. Lenoski, J. Laudon, K. Gharachorloo, W. D. Weber, and J. Hennessey. The Stanford DASH multiprocessor. *IEEE Computer Magazine*, March 1992.

[LM87] E. A. Lee and D. G. Messerschmitt. Static scheduling of synchronous dataflow programs for digital signal processing. *IEEE Transactions on Computers*, February 1987.

[LM95] Y. S. Li and S. Malik. Performance analysis of embedded software using implicit path enumeration. In *Proceedings of the Design Automation Conference*, 1995.

[LM96] E. Lemoine and D. Merceron. Run-time reconfiguration of FPGA for scanning genomic databases. In *Proceedings of the IEEE Symposium on FPGAs for Custom Computing Machines*, pages 90–98, April 1996.

[LMKJ07] C. Lucarz, M. Mattavelli, J. Thomas-Kerr, and J. Janneck. Reconfigurable media coding: A new specification model for multimedia coders. In *Proceedings of the IEEE Workshop on Signal Processing Systems*, October 2007.

[Lou93] J. Lou. Application development on the Intel iWarp system. In *Proceedings of the SPIE*, 1993.

[Lov75] L. Lovasz. On the ratio of optimal integral and fractional covers. *Discrete Mathematics*, pages 383–390, 1975.

[LP81] H. R. Lewis and C. H. Papadimitriou. *Elements of the Theory of Computation*. Prentice Hall, 1981.

[LP82] H. R. Lewis and C. H. Papadimitriou. *Elements of the Theory of Computation*. Prentice Hall, 1982.

[LP95] E. A. Lee and T. M. Parks. Dataflow process networks. *Proceedings of the IEEE*, pages 773–799, May 1995.

[LP98] W. Liu and V. K. Prasanna. Utilizing the power of high-performance computing. *IEEE Signal Processing Magazine*, 15(5):85–100, September 1998.

[LPP00] K. N. Lalgudi, M. C. Papaefthymiou, and M. Potkonjak. Optimizing computations for effective block-processing. *ACM Transactions on Design Automation of Electronic Systems*, 5(3):604-630, July 2000.

[LS91] C. E. Leiserson and J. B. Saxe. Retiming synchronous circuitry. *Algorithmica*, pages 5–35, 1991.

[M+92] N. Morgan et al. The ring array processor: A multiprocessing peripheral for connectionist applications. *Journal of Parallel and Distributed Computing*, 14(3):248–259, March 1992.

[Mad95] V. Madisetti. *VLSI Digital Signal Processors*. IEEE Press, 1995.

[Math07] MathStar, Inc. *MathStar FPOA Overview,* April 2007. http://www.mathstar.com/Documentation/ProductBriefs/FPOAOverview_ProductBrief_REL_V1.5.pdf.

[MCR01] P. K. Murthy, E. G. Cohen, and S. Rowland. System Canvas: A new design environment for embedded DSP and telecommunication systems. In *Proceedings of the International Workshop on Hardware/Software Co-Design*, April 2001.

[MCVL01] S. Mostert, N. Cossement, J. van Meerbergen, and R. Lauwereins. DF*: Modeling dynamic process creation and events for interactive multimedia applications. In *Proceedings of the International Workshop on Rapid System Prototyping*, 2001.

[MD97] E. Mirsky and A. DeHon. MATRIX: A reconfigurable computing device with configurable instruction distribution and deployable resources. In *Proceedings of the Hot Chips Symposium*, August 1997.

[Mes88] D. G. Messerschmitt. Breaking the recursive bottleneck. In J. K. Skwirzynski, editor, *Performance Limits in Communication Theory and Practice*. Kluwer Academic Publishers, 1988.

[MKTM94] C. L. McCreary, A. A. Khan, J. J. Thompson, and M. E. McArdle. A comparison of heuristics for scheduling DAGS on multiprocessors. In *Proceedings of the International Parallel Processing Symposium*, pages 446–451, 1994.

[ML02] P. K. Murthy and E. A. Lee. Multidimensional synchronous dataflow. *IEEE Transactions on Signal Processing*, (8):2064-2079, August 2002.

[MM92] F. Moussavi and D. G. Messerschmitt. Statistical memory management for digital signal processing. In *Proceedings of the International Symposium on Circuits and Systems*, pages 1011–1014, May 1992.

[Mol82] M. K. Molloy. Performance analysis using stochastic Petri nets. *IEEE Transactions on Computers*, September 1982.

[Mot89] Motorola Inc. *DSP96002 IEEE Floating-Point Dual-Port Processor User's Manual*, 1989.

[Mot90] Motorola Inc. *DSP96000ADS Application Development System Reference Manual*, 1990.

[Mou96] G. Mouney. Parallel solution of linear ODE's: Implementation on transputer networks. *Concurrent Systems Engineering Series*, 1996.

[Mur89] T. Murata. Petri nets: Properties, analysis, and applications. *Proceedings of the IEEE*, pages 39–58, January 1989.

[ND08] H. Nikolov and E. F. Deprettere. Parameterized stream-based functions dataflow model of computation. In *Proceedings of the Workshop on Optimizations for DSP and Embedded Systems*, 2008.

[Nic89] D. M. Nicol. Optimal partitioning of random programs across two processors. *IEEE Transactions on Computers*, 15(2):134–141, 1989 1989.

[NL04] S. Neuendorffer and E. Lee. Hierarchical reconfiguration of dataflow models. In *Proceedings of the International Conference on Formal Methods and Models for Codesign,* June 2004.

[O+96] K. Olukotun et al. The case for a single-chip multiprocessor. *SIGPLAN Notices*, 31(9):2–11, September 1996.

[O+07] J. D. Owens, D. Luebke, N. Govindaraju, M. Harris, J. Kruger, A. E. Lefohn, and T. Purcell. A survey of general-purpose computation on graphics hardware. *Computer Graphics Forum*, 26(1): 80-113, 2007.

[Ols89] G. J. Olsder. Performance analysis of data-driven networks. In J. McCanny, J. McWhiter, and E. Swartzlander Jr., editors, *Systolic Array Processors; Contributions by Speakers at the International Conference on Systolic Arrays*. Prentice Hall, 1989.

[ORVK90] G. J. Olsder, J. A. C. Resing, R. E. De Vries, and M. S. Keane. Discrete event systems with stochastic processing times. *IEEE Transactions on Automatic Control*, 35(3):299–302, March 1990.

[Ous94] J. K. Ousterhout. *An Introduction to Tcl and Tk*. Addison-Wesley, 1994.

[P+97] D. Patterson et al. A case for intelligent RAM: IRAM. *IEEE Micro*, April 1997.

[Pap90] G. M. Papadopoulos. Monsoon: A dataflow computing architecture suitable for intelligent control. In *Proceedings of the 5th IEEE International Symposium on Intelligent Control*, 1990.

[Pap91] A. Papoulis. *Probability, Random Variables and Stochastic Processes*. McGraw-Hill, 1991.

[PBL95] J. L. Pino, S. S. Bhattacharyya, and E. A. Lee. A hierarchical multiprocessor scheduling system for DSP applications. In *Proceedings of the IEEE Asilomar Conference on Signals, Systems, and Computers*, November 1995.

[Pet81] J. L. Peterson. *Petri Net Theory and the Modeling of Systems*. Prentice Hall, 1981.

[PH96] D. A. Patterson and J. L. Hennessey. *Computer Architecture: A Quantitative Approach*. Morgan Kaufmann Publishers Inc., second edition, 1996.

[PHLB95] J. Pino, S. Ha, E. A. Lee, and J. T. Buck. Software synthesis for DSP using Ptolemy. *Journal of VLSI Signal Processing*, 9(1), January 1995.

[Pico08] picoChip Designs, Ltd. *picoChip Product Briefs*, January 2008. http://www.picochip.com/info/datasheets.

[PLN92] D. B. Powell, E. A. Lee, and W. C. Newman. Direct synthesis of optimized DSP assembly code from signal flow block diagrams. In *Proceedings of the International Conference on Acoustics, Speech, and Signal Processing*, March 1992.

[PM91] K. K. Parhi and D. G. Messerschmitt. Static rate-optimal scheduling of iterative data-flow programs via optimum unfolding. *IEEE Transactions on Computers*, 40(2):178–194, February 1991.

[PPL95] T. M. Parks, J. L. Pino, and E. A. Lee. A comparison of synchronous and cyclo-static dataflow. In *Proceedings of the IEEE Asilomar Conference on Signals, Systems, and Computers*, November 1995.

[Pra87] M. Prastein. Precedence-constrained scheduling with minimum time and communication. Master's thesis, University of Illinois at Urbana-Champaign, 1987.

[Pri91] H. Printz. *Automatic Mapping of Large Signal Processing Systems to a Parallel Machine*. Ph.D. thesis, School of Computer Science, Carnegie Mellon University, May 1991.

[Pri92] H. Printz. Compilation of narrowband spectral detection systems for linear MIMD machines. In *Proceedings of the International Conference on Application Specific Array Processors*, August 1992.

[PS94] M. Potkonjac and M. B. Srivastava. Behavioral synthesis of high performance, and low power application specific processors for linear computations. In *Proceedings of the International Conference on Application Specific Array Processors*, pages 45–56, 1994.

[PSK+08] W. Plishker, N. Sane, M. Kiemb, K. Anand, and S. S. Bhattacharyya. Functional DIF for rapid prototyping. In *Proceedings of the International Workshop on Rapid System Prototyping*, Monterey, California, June 2008.

[Pto98] Department of Electrical Engineering and Computer Sciences, University of California at Berkeley. *The Almagest: A Manual for Ptolemy*, 1998.

[Pur97] S. Purcell. Mpact 2 media processor, balanced 2X performance. In *Proceedings of SPIE*, 1997.

[PY90] C. Papadimitriou and M. Yannakakis. Toward an architecture-independent analysis of parallel algorithms. *SIAM Journal of Computing*, pages 322–328, 1990.

[Rao85] S. Rao. *Regular Iterative Algorithms and their Implementation on Processor Arrays*. Ph.D. thesis, Stanford University, October 1985.

[RCG72] C. V. Ramamoorthy, K. M. Chandy, and M. J. Gonzalez. Optimal scheduling strategies in multiprocessor systems. *IEEE Transactions on Computers*, February 1972.

[RCHP91] J. M. Rabaey, C. Chu, P. Hoang, and M. Potkonjak. Fast prototyping of datapath intensive architectures. *IEEE Design and Test of Computers Magazine*, 8(2):40–51, June 1991.

[Reg94] D. Regenold. A single-chip multiprocessor DSP solution for communications applications. In *Proceedings of the IEEE International ASIC Conference and Exhibit*, pages 437–440, 1994.

[Rei68] R. Reiter. Scheduling parallel computations. *Journal of the Association for Computing Machinery*, October 1968.

[RH80] C. V. Ramamoorthy and G. S. Ho. Performance evaluation of asynchronous concurrent systems using Petri nets. *IEEE Transactions on Software Engineering*, SE-6(5):440–449, September 1980.

[RN81] M. Renfors and Y. Neuvo. The maximum sampling rate of digital filters under hardware speed constraints. *IEEE Transactions on Circuits and Systems*, March 1981.

[RPM92] S. Ritz, M. Pankert, and H. Meyr. High level software synthesis for signal processing systems. In *Proceedings of the International Conference on Application Specific Array Processors*, August 1992.

[RPM93] S. Ritz, M. Pankert, and H. Meyr. Optimum vectorization of scalable synchronous dataflow graphs. In *Proceedings of the International Conference on Application Specific Array Processors*, October 1993.

[RS94] S. Rajsbaum and M. Sidi. On the performance of synchronized programs in distributed networks with random processing times and transmission delays. *IEEE Transactions on Parallel and Distributed Systems*, 5(9), September 1994.

[RS98] S. Rathnam and G. Slavenburg. Processing the new world of interactive media. *IEEE Signal Processing Magazine*, 15(2), March 1998.

[RSB97] S. Ramaswamy, S. Sapatnekar, and P. Banerjee. A framework for exploiting task and data parallelism on distributed memory muticomputers. *IEEE Transactions on Parallel and Distributed Systems*, 8(11), November 1997.

[RSR06] I. Radojevic, Z. Salcic, and P. Roop. Design of heterogeneous embedded systems using DFCharts model of computation. In *Proceedings of the International Conference on VLSI Design*, January 2006.

[RWM95] S. Ritz, M. Willems, and H. Meyr. Scheduling for optimum data memory compaction in block diagram oriented software synthesis. In *Proceedings of the International Conference on Acoustics, Speech, and Signal Processing*, pages 2651-2654, May 1995.

[Sar88] V. Sarkar. Synchronization using counting semaphores. In *Proceedings of the International Symposium on Supercomputing*, 1988.

[Sar89] V. Sarkar. *Partitioning and Scheduling Parallel Programs for Multiprocessors*. MIT Press, 1989.

[SBGC07] S. Stuijk, T. Basten, M. C. W. Geilen, and H. Corporaal. Multiprocessor resource allocation for throughput-constrained synchronous dataflow graphs.

In *Proceedings of the Design Automation Conference*, 2007.

[SBSV92] N. Shenoy, R. K. Brayton, and A. L. Sangiovanni-Vincentelli. Graph algorithms for clock schedule optimization. In *Proceedings of the International Conference on Computer-Aided Design*, pages 132–136, 1992.

[Sch88] H. Schwetman. Using CSIM to model complex systems. In *Proceedings of the 1988 Winter Simulation Conference*, pages 246–253, 1988.

[Sha89] P. L. Shaffer. Minimization of interprocessor synchronization in multi-processors with shared and private memory. In *Proceedings of the International Conference on Parallel Processing*, 1989.

[Sha98] M. El Sharkawy. Multiprocessor 3d sound system. In *Proceedings of the Midwest Symposium on Circuits and Systems*, 1998.

[SHL+97] D. Shoemaker, F. Honore, P. LoPresti, C. Metcalf, and S. Ward. A unified system for scheduled communication. In *Proceedings of the International Conference on Parallel and Distributed Processing Techniques and Applications*, July 1997.

[SI85] D. A. Schwartz and T. P. Barnwell III. Cyclo-static solutions: Optimal multiprocessor realizations of recursive algorithms. In *Proceedings of the International Workshop on VLSI Signal Processing*, pages 117–128, June 1985.

[Sih91] G. C. Sih. *Multiprocessor Scheduling to account for Interprocessor Communication*. Ph.D. thesis, Department of Electrical Engineering and Computer Sciences, University of California at Berkeley, April 1991.

[SL90] G. C. Sih and E. A. Lee. Scheduling to account for interprocessor communication within interconnection-constrained processor networks. In *Proceedings of the International Conference on Parallel Processing*, 1990.

[SL93a] G. C. Sih and E. A. Lee. A compile-time scheduling heuristic for inter-connection-constrained heterogeneous processor architectures. *IEEE Transactions on Parallel and Distributed Systems*, 4(2):75–87, February 1993.

[SL93b] G. C. Sih and E. A. Lee. Declustering: A new multiprocessor scheduling technique. *IEEE Transactions on Parallel and Distributed Systems*, 4(6), June 1993.

[SL94] S. Sriram and E. A. Lee. Statically scheduling communication resources in multiprocessor DSP architectures. In *Proceedings of the IEEE Asilomar Conference on Signals, Systems, and Computers*, November 1994.

[SOIH97] W. Sung, M. Oh, C. Im, and S. Ha. Demonstration of hardware software codesign workflow in PeaCE. In *Proceedings of the International Conference on VLSI and CAD*, October 1997.

[SPB06] S. Saha, S. Puthenpurayil, and S. S. Bhattacharyya. Dataflow transformations in high-level DSP system design. In *Proceedings of the International Symposium on System-on-Chip*, pages 131-136, Tampere, Finland, November 2006.

[SPRK04] N. Shah, W. Plishker, K. Ravindran, and K. Keutzer. NP-Click: a productive software development approach for network processors. *IEEE Micro*, 24(5), 2004.

[Sri92] M. B. Srivastava. *Rapid-Prototyping of Hardware and Software in a Unified Framework*. Ph.D. thesis, Department of Electrical Engineering and Computer Sciences, University of California at Berkeley, June 1992.

[Sri95] S. Sriram. *Minimizing Communication and Synchronization Overhead in Multiprocessors for Digital Signal Processing*. Ph.D. thesis, Department of Electrical Engineering and Computer Sciences, University of California at Berkeley, 1995.

[Ste97] R. S. Stevens. The processing graph method tool (PGMT). In *Proceedings of the International Conference on Application Specific Systems, Architectures, and Processors*, pages 263-271, July 1997.

[Sto77] H. S. Stone. Multiprocessor scheduling with the aid of network flow algorithms. *IEEE Transactions on Software Engineering*, 3(1):85–93, January 1977.

[Sto91] A. Stolzle. *A Real Time Large Vocabulary Connected Speech Recognition System*. Ph.D. thesis, Department of Electrical Engineering and Computer Sciences, University of California at Berkeley, December 1991.

[Str08] Stretch, Inc. *Software-Configurable Processors: Solving The Embedded System Design Dilemma*, 2008. http://www.stretchinc.com/technology/.

[STZ+99] K. Strehl, L. Thiele, D. Ziegenbein, R. Ernst, and J. Teich. Scheduling hardware/software systems using symbolic techniques. In *Proceedings of the International Workshop on Hardware/Software Co-Design*, 1999.

[SW92] R. R. Shively and L. J. Wu. Application and packaging of the AT&T DSP3 parallel signal processor. In *Proceedings of the International Conference on Application Specific Array Processors*, pages 316–326, 1992.

[SZT+04] T. Stefanov, C. Zissulescu, A. Turjan, B. Kienhuis, and E. Deprettere. System design using Kahn process networks: the Compaan/Laura approach. In *Proceedings of the Design, Automation and Test in Europe Conference and Exhibition*, February 2004.

[T+95] A. Trihandoyo et al. Real-time speech recognition architecture for a multi-channel interactive voice response system. In *Proceedings of the Interna-*

*tional Conference on Acoustics, Speech, and Signal Processing*, 1995.

[Tex98] Texas Instruments, Inc. *TMS320C62X/C67X CPU and Instruction Set Reference Guide*, March 1998.

[Tex07] Texas Instruments, Inc. *Product Bulletin: TMS320TCI6488 DSP Platform*, 2007. http://focus.ti.com/lit/ml/sprt415/sprt415.pdf.

[Tex08] Texas Instruments, Inc. *TNETV3020 Carrier Infrastructure Platform*, 2008. http://focus.ti.com/lit/ml/spat174a/spat174a.pdf.

[TGB+06] B. D. Theelen, M. C. W. Geilen, T. Basten, J. P. M. Voeten, S. V. Gheorghita, and S. Stuijk. A scenario-aware data flow model for combined long-run average and worst-case performance analysis. In *Proceedings of the International Conference on Formal Methods and Models for Codesign*, July 2006.

[Tho86] VLSI CAD Group, Stanford University. *Thor Tutorial*, 1986.

[Til07] Tilera Corp. *Technology Brief: Tile Processor Architecture*, 2007. http://www.tilera.com/pdf/ArchBrief_Arch_V1_Web.pdf.

[TJM+07] J. Thomas-Kerr, J. Janneck, M. Mattavelli, I. Burnett, and C. Ritz. Reconfigurable media coding: Self-describing multimedia bitstreams. In *Proceedings of the IEEE Workshop on Signal Processing Systems*, October 2007.

[TKA02] W. Thies, M. Karczmarek, and S. Amarasinghe. StreamIt: A language for streaming applications. In *Proceedings of the International Conference on Compiler Construction*, 2002.

[TKD04] A. Turjan, B. Kienhuis, and E. Deprettere. Approach to classify interprocess communication in process networks at compile time. In *Proceedings of the International Workshop on Software and Compilers for Embedded Systems*, September 2004.

[TONL96] M. Tremblay, J. M. O'Connor, V. Narayanan, and H. Liang. VIS speeds new media processing. *IEEE Micro*, 16(4), August 1996.

[TSZ+99] L. Thiele, K. Strehl, D. Ziegenbein, R. Ernst, and J. Teich. FunState — An internal representation for codesign. In *Proceedings of the International Conference on Computer-Aided Design*, November 1999.

[TTL95] J. Teich, L. Thiele, and E. A. Lee. Modeling and simulation of heterogeneous real-time systems based on a deterministic discrete event model. In *Proceedings of the International Symposium on Systems Synthesis*, pages 156–161, 1995.

[V+96] J. E. Vuillemin et al. Programmable active memories: Reconfigurable systems come of age. *IEEE Transactions on Very Large Scale Integration (VLSI) Systems*, 4(1), March 1996.

[Vai93] P. P. Vaidyanathan. *Multirate Systems and Filter Banks.* Prentice Hall, 1993.

[VLS86] VLSI CAD Group, Stanford University. *Thor Tutorial,* 1986.

[VNS07] S. Verdoolaege, H. Nikolov, and T. Stefanov. pn: A tool for improved derivation of process networks. *EURASIP Journal on Embedded Systems,* 2007, 2007. Article ID 75947, 13 pages.

[VPS90] M. Veiga, J. Parera, and J. Santos. Programming DSP systems on multi-processor architectures. In *Proceedings of the International Conference on Acoustics, Speech, and Signal Processing,* April 1990.

[WLR98] A.Y. Wu, K.J.R. Liu, and A. Raghupathy. System architecture of an adaptive reconfigurable DSP computing engine. *IEEE Transactions on Circuits and Systems for Video Technology,* February 1998.

[W+97] E. Waingold et al. Baring it all to software: Raw machines. *IEEE Computer Magazine,* pages 86–93, September 1997.

[YG94] T. Yang and A. Gerasoulis. DSC: Scheduling parallel tasks on an unbounded number of processors. *IEEE Transactions on Parallel and Distributed Systems,* 5(9):951–967, September 1994.

[YM96] J. S. Yu and P. C. Mueller. On-line cartesian space obstacle avoidance scheme for robot arms. *Mathematics and Computers in Simulation,* August 1996.

[Yu84] W. Yu. *LU Decomposition on a Multiprocessing System with Communication Delay.* Ph.D. thesis, University of California at Berkeley, 1984.

[YW93] L. Yao and C. M. Woodside. Iterative decomposition and aggregation of stochastic marked graph Petri nets. In G. Rosenberg, editor, *Advances in Petri Nets 1993.* Springer-Verlag, 1993.

[ZKD05] C. Zissulescu, B. Kienhuis, and E. Deprettere. Expression synthesis in process networks generated by laura. In *Proceedings of the International Conference on Application Specific Systems, Architectures, and Processors,* July 2005.

[ZKM94] V. Zivojnovic, H. Koerner, and H. Meyr. Multiprocessor scheduling with a-priori node assignment. In *Proceedings of the International Workshop on VLSI Signal Processing,* 1994.

[ZRM94] V. Zivojnovic, S. Ritz, and H. Meyr. Retiming of DSP programs for optimum vectorization. In *Proceedings of the International Conference on Acoustics, Speech, and Signal Processing,* April 1994.

[ZS89] A. Zaky and P. Sadayappan. Optimal static scheduling of sequential loops on multiprocessors. In *Proceedings of the International Conference on Parallel Processing,* pages 130–137, 1989.

[ZVSM95] V. Zivojnovic, J. M. Velarde, C. Schlager, and H. Meyer. DSP-STONE: A DSP-oriented benchmarking methodology. In *Proceedings of the International Conference on Signal Processing Applications and Technology*, 1995.

# INDEX

**Numerics**
1024 162
1-subsumes 295
2LCR 288
2-subsume 295

**A**
actor ordering step 86
actors 6
Acyclic Precedence Expansion Graph (APEG) 43
admissible schedule 98
Algorithm A 296
Algorithm B for general 2LCR 302
Algorithm Global-resynchronize 255
application graph 44
Asymptotic complexity ("O") notation 49
average iteration period 85

**B**
back-off time 215
backward constraint 270
BBS 214
blocked schedule 89
blocking factor 91
Boolean dataflow 7, 78
bounded buffer synchronization 214
branch actors 121

**C**
CAL 75
category A resynchronization 296
category B resynchronization 296
chainable 264

## W

# ABOUT THE AUTHORS

### Sundararajan Sriram

Sundararajan Sriram received a Bachelor of Technology degree in Electrical Engineering from the Indian Institute of Technology, Kanpur, in 1989, and a Ph.D. in Electrical Engineering and Computer Sciences from the University of California at Berkeley in 1995. Between 1995 and 2005 he was a Senior Member of Technical Staff with Texas Instruments, Dallas, Texas, where he worked on system design for cellular communications, including single chip multiprocessors for wireless base stations. At present he is with Ozmo Devices Inc. where he is leading the DSP group, designing low power wireless communications chips. His research interests are in design of algorithms and VLSI architectures for applications in digital signal processing and communications, an area in which he has over 15 publications and over 12 patents.

Dr. Sriram is a member of the IEEE Communications and Signal Processing societies, and has served as an Associate Editor for the *IEEE Transactions on Circuits and Systems — II.*

### Shuvra S. Bhattacharyya

Shuvra S. Bhattacharyya is a Professor in the Department of Electrical and Computer Engineering, University of Maryland at College Park. He holds a joint appointment in the University of Maryland Institute for Advanced Computer Studies (UMIACS), and an affiliate appointment in the Department of Computer Science. Dr. Bhattacharyya is coauthor or coeditor of five books and the author or co-author of more than 100 refereed technical articles. His research interests include VLSI signal processing; biomedical circuits and systems; embedded software; and hardware/software co-design. He received the B.S. degree from the University of Wisconsin at Madison, and the Ph.D. degree from the University of California at Berkeley. Dr. Bhattacharyya has held industrial positions as a Researcher at the Hitachi America Semiconductor Research Laboratory (San Jose, California), and Compiler Developer at Kuck & Associates (Champaign, Illinois).